Y0-CDD-960

MYCOBACTERIA OF CLINICAL INTEREST

MYCOBACTERIA OF CLINICAL INTEREST

Proceedings of the International Symposium
on Mycobacteria of Clinical Interest,
Cordoba, Spain, 27–28 September 1985

Editor:

MANUEL CASAL

1986

EXCERPTA MEDICA, Amsterdam – New York – Oxford

International Congress Series No. 697
ISBN 0 444 80808 6

Published by:
Elsevier Science Publishers B.V. (Biomedical Division)
P.O. Box 211
1000 AE Amsterdam
The Netherlands

Sole distributors for the USA and Canada:
Elsevier Science Publishing Company Inc.
52 Vanderbilt Avenue
New York, NY 10017
USA

Library of Congress Cataloging in Publication Data

International Symposium on Mycobacteria of Clinical
Interest (1985 : Córdoba, Spain)
Mycobacteria of clinical interest.

(International congress series ; no. 697)
Includes bibliographies and index.
1. Mycobacterial diseases--Congresses. 2. Mycobacteria--Congresses. 3. Tuberculosis--Congresses. 4. Leprosy--Congresses. I. Casal, Manuel. II. Title. III. Series. [DNLM: 1. Mycobacterium--congresses. 2. Mycocacterium Infections--congresses.
W3 EX89 no. 697 / WC 302 I61m 1985]
RC116.M8I5 1985 616.9′2 86-11589
ISBN 0-444-80808-6 (U.S.)

Printed in The Netherlands

PREFACE

The extraordinary progress in our knowledge of mycobacteria of clinical interest in recent years led us to organise an international Symposium which was held in Córdoba, Spain, on the 27th and 28th of September, 1985.

This Symposium was organised by the Department of Microbiology of the Faculty of Medicine in Córdoba, which was co-sponsored by the World Health Organization (WHO); International Union Against Tuberculosis (IUAT); European Society of Mycobacteriologists (ESM); Sociedad Española de Enfermedades Infecciosas y Microbiología Clínica (SEIMC); Sociedad Española de Patología Respiratoria (SEPAR); Sociedad Española de Microbiología (SEM); Sociedad Española de Higiene y Medicina Preventiva Hospitalaria (SEHMPH) ...etc., and had the collaboration of Ministerio Educación y Ciencia (CAICYT); Minsterio Sanidad (FISS); Instituto de Cooperación Iberoamericano (ICI); Consejeria Educación y Ciencia; Junta de Andalucia; Consejeria Salud y Consumo; Junta de Andalucia, Universidad de Córdoba ...etc.

More than 300 participants from 35 countries and all five continents were present.

The scientific programme included invited lectures as well as posters and round table discussions which covered the following themes: immunopathology and diagnosis of tuberculosis; treatment of tuberculosis; experimental chemotherapy in tuberculosis and leprosy; automated methods of laboratory techniques; antimicrobial agents; epidemiology; taxonomy; pathogenicity; mycobacteriosis; clinical aspects of tuberculosis, and leprosy.

The multidisciplinary participation of, among others, microbiologists, clinicians, immunologists, epidemiologists etc., was fundamental to the success of the Symposium.

Novel topics of great interest to this Symposium were the new antimicrobial agents active against mycobacteria and new therapeutic possibilities; a system of rapid diagnosis of tuberculosis and mycobacteriosis; mycobacteriosis in AIDS; progress in immunopathology of tuberculosis and leprosy; progress in bacteriology and vaccination in leprosy; progress in immunological diagnosis and new epidemiological biovars of *M. tuberculosis*.

As organisers, we believed it was important to publish the proceedings of the Symposium in order to give all those who where present, those who could not attend, and other interested people, the opportunity to share in the interesting information presented in the invited papers and free communications contained in this volume.

ACKNOWLEDGEMENTS

I would like to thank every participant and all present at the Symposium, especially the invited speakers and presidents of sessions and, last but not least, the authors of the papers in this book for their collaboration in the Symposium and the preparation of the manuscripts.

I should also like to thank all the national and international organisations, either scientific or governmental, institutional or private companies, which collaborated with us to make this Symposium possible. In a special way, I would like to thank Merrell-Dow España, S.A. for their financial help in publishing this book and the Organising Committee for their hard work.

Manuel Casal

CONTENTS

IMMUNOPATHOLOGY AND DIAGNOSIS

TAXONOMY

AUTOMATED METHODS OF LABORATORY DIAGNOSIS

EPIDEMIOLOGY

TREATMENT

MYCOBACTERIOSIS

IMMUNOPATHOLOGY AND DIAGNOSIS

Mycobacteria of Clinical Interest. M. Casal, editor

THE IMMUNOPATHOLOGY OF TUBERCULOSIS

G.A.W. Rook.

Dept. Microbiology, Middlesex Hospital Medical School, London W1, England.

Introduction

The major unsolved problems of tuberculosis are:-

1) The failure of BCG to protect in some environments (1,2)

2) "Persister" organisms which although sensitive, are not readily killed by drugs. This results in a need for prolonged therapy, and problems with patient compliance and relapse.

3) The necrotising response to the antigens of M. tuberculosis, which causes tissue damage, but which is evidently unable to kill persisters, even after the load of bacilli has been drastically reduced by chemotherapy.

This paper is limited to those stages of the pathways of cell-mediated immunity in which recent developments have led to some insight into these problems.

Protective antigens.

Progress with molecular biology and monoclonal antibodies has not yet identified these. It is often assumed that the protective anitgens will be species specific ones. There is no logic behind this belief and no evidence for it, whereas there is clear evidence for the view that the common antigens can be protective. Thus in animal models the mycobacterial species used for vaccination is not important (3), and the same appears to be true in man since BCG can protect against leprosy (4), and environmental slow growers can protect against tuberculosis (5). These facts suggest that the property which makes an antigen protective is not species specificity. What therefore is the important property? A strong possibility is that the most important antigens are those which are released by living organisms. Therefore we have used cloning of T lymphocytes by limiting dilution to study the frequency of T cells able to respond to sonicated M. tuberculosis, in the peripheral blood of normal donors and of ex-tuberculosis cases. We have then analysed the ability of the sonicate responsive clones obtained to recognise live M. tuberculosis, and totally unrelated mycobacteria. The results show that T cells which recognise sonicate do not necessarily recognise live bacilli, and that those which do recognise live bacilli may show a degree of specificity for M. tuberculosis, but are often responsive to common mycobacterial antigens (6).

We therefore propose the hypothesis (Fig. 1) that T cells which recognise antigens released by live bacilli are potentially protective, and that those which recognise only antigens which are released from dead or disrupted organisms may lead to immunopathology. If antigens recognised by these T cells are poorly metabolisable particulate cell wall fragments, the result may be chronic granulomatous inflammation such as is seen in Lupus vulgaris or Tuberculoid leprosy. Meanwhile if the T cells from the same individual are failing to recognise the antigens released by live organisms these could be proliferating unrecognised elsewhere in the lesions. This mechanism could also provide one explanation for the survival of the "persisters" which cause such problems in the treatment of both leprosy and tuberculosis. Two other possible mechanisms will be discussed later.

Fig 1. HYPOTHESIS SUGGESTED BY THE FAILURE OF ABOUT 50% OF SONICATE-RESPONSIVE T CELL CLONES TO RECOGNISE LIVE M. TUBERCULOSIS.*

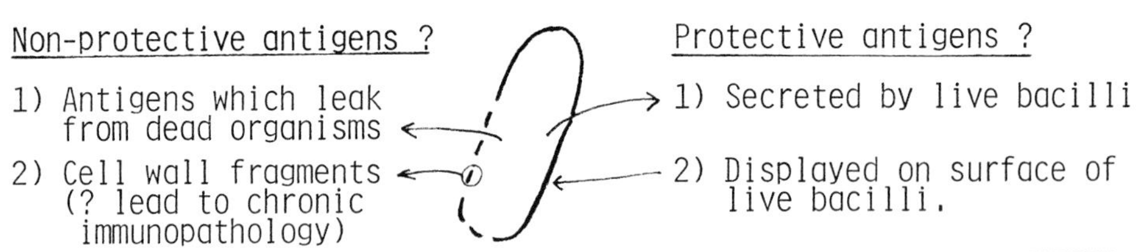

*N.B. Data on which this suggestion is based in Ref. 6.

Antigen presentation.

Whatever the identity of the protective antigens of M. tuberculosis, it is certain that in order to prime a response, they must be "presented" to T cells by antigen-presenting cells (APC) in association with products of the major histocompatibility complex (MHC). It appears likely that any cells which express the relevant MHC products can present antigen, though at present most attention is focussed on macrophages and dendritic cells, which may or may not be of similar lineage. There is evidence that high or low responsiveness of inbred mouse strains to BCG can be related to inherent characteristics of the antigen-presenting fuc-tion of their macrophages (7). Moreover it is clear that presentation of antigen by different types of antigen-presenting cell, for instance dendritic cells rather than macrophages, can lead to quite different types of response (8). The type of antigen presentation which occurs in a mycobacterial infection will be affected by the route of challenge.

This point may have considerable implications. Thus the relative susceptibility to mycobacterioses of C57Bl and CBA mice is reversed if challange is subcutaneous rather than intravenous (9,10) and C3H mice which normally give little granulomatous response after intravenous challenge, will do so if first primed subcutaneously (11). However further studies are required to determine whether the mode of initial antigen presentation is in fact the explanation of these effects.

As far as man is concerned, we have argued previously (12) that the very heavy oral intake of mycobacteria to which some human populations are exposed must modulate subsequent responses to mycobacterial antigens encountered by other routes. In a mouse model, there is firm evidence that oral intake can alter the subsequent response to BCG (13).

Before outlining our first steps towards elucidating the relevance of antigen-presenting cell subpopulations to these effects in vivo, I must first explain the recent revelations of Askenase and his colleagues (14). These workers have established that there is an early 2-hour component to the delayed hypersensitivity (DTH) response in the mouse, which is a prerequisite for the occurrence of the subsequent, and more widely recognised 24-hour component. The mechanism is understood in some detail (Fig. 2). A T cell population which is distinct from that which mediates the 24-hour component of DTH, releases an antigen-specific factor which sensitises tissue mast cells, in a manner which is broadly analogous to the effects of IgE. When antigen is introduced, it interacts with this T cell derived factor to cause release of 5-hydroxytryptamine. This in turn causes vascular changes which lead to the 2-hour peak of swelling, and facilitate the arrival at the site of challenge of the MHC-restricted T cells which mediate the late components of DTH. If the 2-hour component is blocked pharmacologically, the 24-hour component does not occur (14).

Fig 2. THE TWO HOUR COMPONENT OF DELAYED HYPERSENSITIVITY (Ref 14)

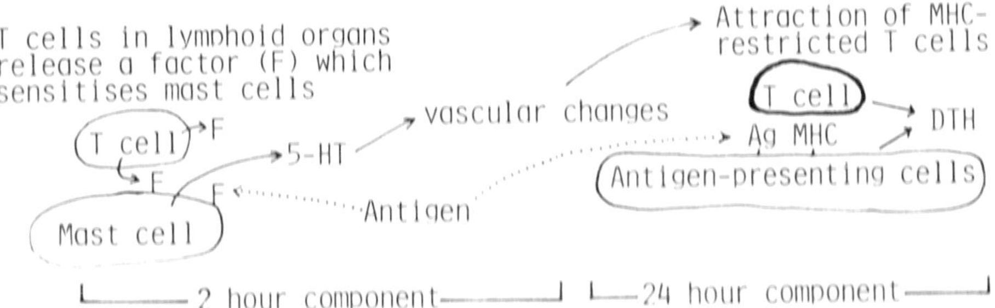

The equivalent of the 2-hour response in man could provide the "missing link" between the presence of antigen responsive T cells in the peripheral blood, and the presence of DTH, or the rapid localisation of these cells in sites of infection, which presumably involve similar mechanisms. For example we have been puzzled by the fact that T cell cloning reveals that all normal donors have circulating T cells which recognise the common mycobacterial antigens, but only 15-20% manifest positive skin-test responses to these antigens. Lack of the tissue-sensitising T cell factor is an attractive theoretical explanation. Therefore it is clearly of importance to establish the existence of the 2-hour component in responses to mycobacterial antigens, and to discover how it is regulated.

To achieve this, in collaboration with S. Mukherjee and D.R. Katz (Dept. Histopathology, Middlesex Hospital Medical School), we have studied very early events following priming of delayed hypersensitivity to PPD in mice. We have primed the mice using single subcutaneous injections of PPD-pulsed antigen-presenting cell populations purified from the spleen. This protocol was adopted to allow analysis of the different roles of different types of antigen-presenting cell in this system.

Details of ths protocols and results will be published elsewhere. The most relevant point for the present discussion is shown in the table below.

Table 1. DIFFERING ROLE OF TWO TYPES OF ANTIGEN-PRESENTING CELL IN THE PRIMING OF 2 HOUR AND 24 HOUR COMPONENTS OF DTH.

Cell type used for priming [a]	Type of cell primed in recipients [b]
Macrophages	2 hour and 24 hour
T cells	none
Dendritic cells	24 hour only

[a]Fractionated spleen cell populations, pulsed with PPD (150ug/ml) in vitro, washed, and injected s.c. into recipients

[b]See Fig. 2 for explanation of these T cell functions.

The 2-hour component of the DTH response is readily demonstrated, and is seen in animals tested as early as 1 day after suitable priming. PPD-pulsed macrophages will prime both the 2-hour and the 24-hour components. However PPD-pulsed dendritic cells will prime only the cells which mediate the MHC-restricted 24-hour component. Thus animals primed with pulsed dendritic cells alone do not show positive DTH responses unless this is revealed by evoking the 2-hour component by another means. I believe that these results may lead to an understanding of how it is possible to have primed circulating antigen-responsive cells, which fail to localise in lesions or skin-test sites.

Obviously these semi-in vitro experiments underestimate the complexity of the regulation of antigen presentation during true infections, where induction of increased class 2 MHC expression by gamma interferon, and depression of MHC expression in infected macrophages may be important.

The protective T cells.

Not only are the specificity and antigen-presenting cell interactions of the protective T cells unclear, but so is their phenotype. There is a widespread belief that T cells showing class 2 MHC restriction (T4+ cells in man) are all effector or helper cells, and that T cells showing class 1 MHC-restriction (T8+ in man) are cytotoxic or suppressor cells. It has however become clear that both of these major T cell subsets can have both effector and regulatory effects, and we cannot automatically assume that T8+ T cells in lesions of leprosy or tuberculosis are suppressor cells. Recent work in mice infected with Listeria monocytogenes has suggested that some protection can be transferred with Class 1 restricted (Lyt2+) T cells, and that these are particularly involved in granuloma formation (15). Thus both major T cell subsets appear to be involved in this infection. There is also one report of a protective role for transferred Class 1 restricted (Lyt2+) cells in mice challenged with M. tuberculosis (16), and in collaboration with B.R. Champion (Dept. Immunology, Middlesex Hospital Medical School) we have been able to show that a Class 2 MHC-restricted PPD-responsive T cell line can also cause significant protection. Moreover it is clear that there are functionally different subsets within the Class 2 MHC-restricted population.(20) Thus since the factor producing T cells discussed in the previous section seem to be functionally distinct from both of the effector populations discussed here, it seems likely that in tuberculosis as in listeriosis, synergistic effects of several types of T cell are required.

T cell/macrophage interactions.

Some murine PPD-responsive T cell lines cultured in vitro for several months by B.R. Champion will recognise and activate macrophages infected with M. tuberculosis (17). (The protocol used for this work is outlined in Fig. 3.) Appropriate recognition can occur through the common mytobacterial antigens. The phenomenon is Class 2 MHC restricted (I-A) and can result in a degree of inhibition of the bacilli compatible with 100% stasis (17,18). We did not see total destruction of the organisms.

It was important to confirm whether the effect seen in these assays was indeed stasis, or whether it was kill of some bacilli balanced by continuing proliferation of others. We have evidence that using M. avium strains sensitive to amoxycillin the effect is indeed stasis (18). Thus lymphokines added to infected macrophages oppose the effect of amoxycillin which is a bactericidal antibiotic, which kills only replicating organisms. If lymphokines caused kill of some bacilli, while allowing proliferation of others, the effect of the antibiotic would be additive. I feel that there is now some doubt as to whether kill of pathogenic mycobacteria by mycrophages has ever been demonstrated in vitro.

Most of the published experiments cannot distinguish between stasis and kill, though there is one report of lymphokine-induced kill of M. microti by murine macrophages (19).

Fig 3. ASSAY FOR ACTIVATION BY LYMPHOKINE OR T CELL LINES, OF ANTI-TUBERCULOSIS EFFECTS IN MACROPHAGES (Refs 17,18)

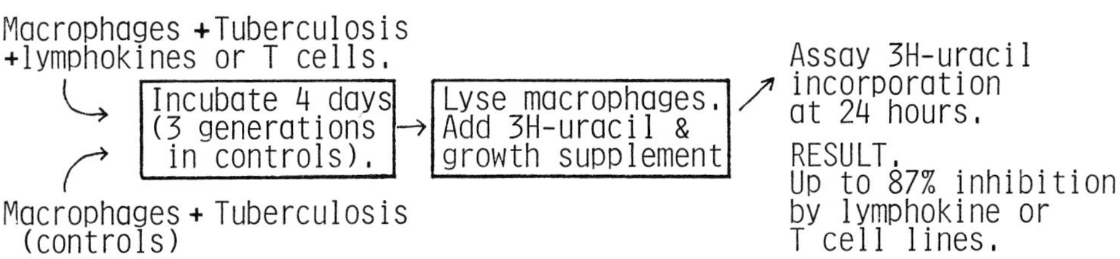

If the effect which such in vitro assays measure is indeed stasis, one must be cautious about regarding the assays as correlates of a mechanism which would be protective in vivo. A mechanism leading to stasis of M. tuberculosis might merely be another cause of persisters. The existence of such T cell-induced stasis could explain the reactivation of persisters by steroids in man, and also those rare cases of indolent tuberculosis which only respond to chemotherapy if steroids are also given.

An alternative to the direct addition of T cells to macrophage monolayers is the addition of crude T cell supernatants, purified lymphokines, or defined lymphokines produced by recombinant DNA technology. It has been reported that murine macrophages will control proliferation of M. microti when activated by recombinant Gamma interferon in vitro, and we have similar data using the 3H-uracil uptake assay and M. tuberculosis. This does not mean that other lymphokines or macrophage modulating agents are not also involved.

It seems to be much more difficult to demonstrate lymphokine-induced activation of anti-tuberculosis effects in human macrophages. Table 2 below summarises data obtained with human alveolar macrophages and with fresh, or pre-cultured human monocytes. Conditions were as used in the experiments with murine peritoneal macrophages described earlier. A complete dose/response curve with lymphokine or recombinant gamma interferon (a gift from Genentech) was performed with the cells from each donor. The data are calculated from the optimum inhibition seen in each case, relative to the growth in the same macrophages cultured without lymphokine. The figures can be shown to indicate that complete stasis, such as is seen with murine peritoneal macrophages, was never achieved. Moreover in fresh monocytes from some donors, M. tuberculosis grew faster in the presence of gamma interferon than in its absence.

Table 2. TUBERCULOSIS INHIBITORY EFFECTS OF HUMAN MACROPHAGES.[a]

Cell type	Donors	Gamma interferon[b]	Crude lymphokine
Monocyte	6	-9.8±34(-58 to +25)	-18.5±19 (-42 to+6)
Macrophage[c]	8	-43±15 (-68 to -25)	-40±20 (-62 to -19)
Alveolar	8	-18±17 (-38 to - 1)	-8.7±10 (-25 to +5)

[a] Protocol as in Fig. 3.   [c] Monocytes precultured for 5-8 days
[b] Mean±SD (ranges) of optimum % inhibition from dose/response curves. Calculated relative to growth in non-stimulated cells.

There are several possible interpretations of these results.

1) Human macrophages may be inherently less able to control proliferation of M. tuberculosis. This could explain the greater susceptibility of man.

2) The experimental conditions may be irrelevant to the in vivo situation.

3) Human monocytes may not be at an appropriate stage of maturation. They may therefore fail to be activated appropriately by the lymphokines.

Induction of macrophage maturation by Vitamin D3 metabolites.

Vitamin D is best known for its role in calcium metabolism. However vitamin D3 (cholecalciferol) metabolites are now known to be important cell regulatory hormones, with widespread effects. Some cause phenotypic differentiation of monocytes, or monocytic and myeloid cell lines into more mature cells (21). The possibility that human monocytes are not sufficiently mature to respond appropriately to lymphokines therefore prompted us to investigate the effects of cholecalciferol metabolites on the ability of human monocytes to control proliferation of M. tuberculosis in the presence or absence of gamma interferon. The results (22) indicate that incubation of monocytes with three cholecaliciferol metabolites induced anti-tubercolosis activity to an extent which correlated with their binding affinities to the intracellular receptor protein. This effect was additive rather than synergistic with the effect of gamma interferon (IFN-Y). In collaberative experiments with Dr. L. Fraher (Dept. Medicine, Middlesex Hospital Medical School) we also observed that monocytes incubated in gamma interferon develop enhanced 25-(OH) D3 1-hydroxylase activity. Thus such interferon-activated monocytes can convert the inactive circulating form of vitamin D3 into the most active form in vivo. This would result in local amplification of the anti-tuberculosis effect. The interactions suggested by these results are indicated in Fig. 5.

Fig 4. INTERACTION OF GAMMA INTERFERON AND VITAMIN D3 IN MACROPHAGE ACTIVATION.

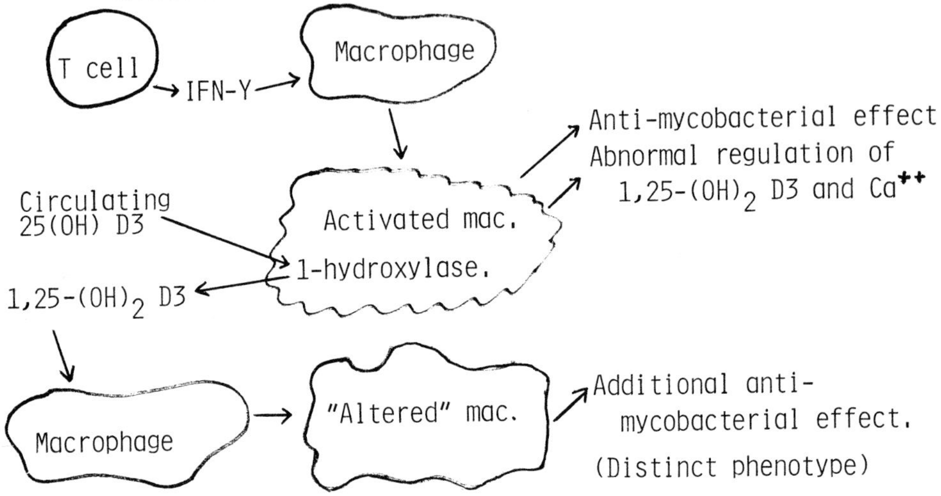

T cell → IFN-Y → Macrophage

Activated mac. 1-hydroxylase.

Circulating 25(OH) D3

$1,25-(OH)_2$ D3

Anti-mycobacterial effect

Abnormal regulation of $1,25-(OH)_2$ D3 and Ca^{++}

Macrophage → "Altered" mac.

Additional anti-mycobacterial effect.

(Distinct phenotype)

These findings help to explain claims for the efficacy of vitamin D in the treatment of some forms of tuberculosis (23), and the occasional finding of raised serum calcium, and disturbed vitamin D metabolism in these patients (24). Of particular interest is the unquestionable efficacy of vitamin D in the treatment of Lupus vulgaris (23). Published descriptions suggest that high doses of vitamin D caused in these patients a raised ESR, fever, malaise, and necrosis of the lesion, followed by scarring and cure. In contrast to the beneficial effects seen in Lupus vulgaris, vitamin D therapy used without concomitant chemotherapy in cases of pulmonary tuberculosis made the disease worse (25). Was the vitamin D converting the indolent granulomatous responses of Lupus into the necrotising Koch phenomenon?

The koch phenomenon.

The Koch phenomenon is the immunologically mediated necrosis which occurs 24-48 hours after intradermal injection of particulate or soluble antigen into not only tuberculous guinea-pigs (the species in which it was first described) but also the majority of past or present human tuberculosis cases. This necrosis is not an inevitable correlate of cell-mediated immunity, because it does not occur in normal individuals protected by BCG vaccination, or in Tuberculoid leprosy cases or healthy contacts. The necrotic component of the lesions of tuberculosis, for which the Koch phenomenon must be at least partly responsible, provides yet another potential site for persisters, and when necrotic tissue liquefies it becomes a medium in which rapid growth of the organisms can occur. It would clearly be of help to know the mechanism of the necrosis, and how it is regulated. We have hypothesised elsewhere (12) that it could be due to inappropriate triggering of secretory functions of activated macrophages, perhaps due to the activity of an inappropriate T cell subset, to disturbance of a regulatory mechanism, or to a pharmacological activity of the organism itself. There are not yet any answers to these questions, but I feel that vitamin D may be providing some clues.

A further relevant piece of information has emerged recently. It has been known for some years (discussed in 12) that the immune response to mycobacteria results in priming of macrophages for release of Tumour Necrosis Factor (the gene for which has now been cloned). It has now been revealed that this molecule is identical to Cachectic Factor, which causes weight loss and decreased activity of Lipoprotein Lipase (26). Thus the weight loss seen in tuberculosis could be a sign of chronic triggering of activated macrophages.

Conclusions.

The problems highlighted in the introduction were, 1) Failed BCG trials; 2) Persisters- and 3) The necrotising response.

1) The failed trials remain mysterious. However the fact that common antigens can act as protective antigens in both in vitro and in vivo models is compatible with the suggestion that heavy exposure by unusual routes (such as oral) to highly immunogenic mycobacteria containing these antigens, must be a relevant factor. Similarly it is noteworthy that T cell clones which respond to live bacilli can do so through common antigens.

2) I have suggested that factors contributing to persisters could be:-

a) Failure to recognise antigens released by live bacilli.

b) T cell-induced bacteriostasis by macrophages.

c) Organisms trapped necrotic tissue.

The first of these is a soluble problem. The technology exists for determining which antigens are actively released. Preliminary data suggests that there are very few of these, and it will soon be possible to test the hypothesis (Fig. 1). The other two (b.& c) are more difficult. It seems remarkable that we still do not know the mechanism of stasis or kill of mycobacteria by mactophages or the mechanism of the tissue necrosis.

3) The mechanism of the Koch phenomenon remains unknown but investigation of the macrophage-maturing effects of cholecalciferol metabolites seems a hopeful approach.

References.

1. Tuberculosis prevention trial in Madras (1979) Indian J. Med. Res. 70:349
2. Bechelli LM, Kyaw Lwin, Gallego GP, Mg Mg Gyi, Uemura K, Sundaresan R, Tamondong C, Matejka M, Sansarricq H, Walter J (1974) Bull. WHO 51:93
3. Edwards ML, Goodrich JM, Muller D, Pollack A, Ziegler JE, Smith DW (1982) J. Inf. Dis. 145:733
4. Brown JAK, Stone MD, Sutherland I (1968) Brit. Med. J. 1:24
5. Comstock GW, Palmer CE (1966) Am. Rev. Resp. Dis. 93:171
6. Rook GAW, Steele J, Barnass S, Mace, J, Stanford JL (1985) Clin. Exp. Immunol. (In press)
7. Nakamura RM, Tanaka H, Tokunaga T (1982) Immunology 47:729
8. Britz JS, Askenase PW, Ptak W, Steinman RM, Gershon RK (1982) J. Exp. Med. 155:1344
9. Brown IN, Glynn AA, Plant J (1982) Immunology 47:149
10. Alexander J, Curtis J (1979) Immunology 36:563
11. Kakinuma M, Onoe K, Yasumizu R, Yamamoto K (1983) Immunology 50:423
12. Rook GAW (1983) Bull. I.U.A.T. 58:60
13. Brown CA, Brown I, Swinburne S (1985) Tubercle (In press)
14. Van Loveren H, Askenase PW (1984) J. Immunol. 133:2397
15. Naher H, Sperling U, Hahn H (1985) J. Immunol. 134:569
16. Orme IM, Collins FM (1984) Cell. Immunol. 84:113
17. Rook GAW, Champion BR, Steele J. Varey AM, Stanford JL (1985) Clin. Exp. Immunol. 59:414
18. Altes C, Steele J, Stanford J, Rook GAW (1985) Tubercle (In press)
19. Walker L, Lowrie DB (1981) Nature 293:69
20. Rees ADM, Knott G, Lamb JR (1985) Behring Institut Mit. 77:(In press)
21. Mangelsdorf DJ, Koeffler HP, Donaldson CA, Pike JW, Haussler MR (1984 J. Cell. Biol. 98:391
22. Rook GAW, Steele J, Fraher L, Barker S, Karmali R, O'Riordan J, Stanford JL (1985) Immunology (In press)
23. Macrae DE, (1947) Br. J. Dermatol. 59:333
24. Epstein S, Stern PH, Bell NH, Dowdeswell I, Turner RT (1984) Calcif, Tissue. Int. 36:541
25. Fanielle G (1951) Bruxelles Med. 31:475
26. Beutler B, Greenwald D, Hulmes JD, Chang M, Pan YCE, Mathison J, Ulevitch R, Cerami A (1985) Nature 316:552

Acknowledgements.

The work discussed was supported in part by grants from the Medical Research Council, and the World Health Organisation.

Mycobacteria of Clinical Interest. M. Casal, editor

STIMULATION OF SUPEROXIDE OUTPUT AND CHEMILUMINESCENCE BY MYCOLIC AND OTHER CARBOXYLIC ACID ESTERS IN HUMAN NEUTROPHILS

IVAN TARNOK, ZSUZSA TARNOK and EVA ROHRSCHEIDT-ANDRZEJEWSKI

Division of Bacterial Physiology, Dept of Immunochemistry and Biochemical Microbiology, Research Institute Borstel, Institute for Experimental Biology and Medicine, 2061 Borstel (F.R.G.)

INTRODUCTION

Neutrophils (polymorphonuclear leukocytes) interact with different stimuli: during the reaction, superoxide (SUP) output and chemiluminescence (CHL) occur (1-3). As far as we know, comprehensive investigations have not been performed into the stimulating activity of different carboxylic acid esters. In order to obtain data in this regard, a micromethod using light reflection spectrophotometry has been developed (4) which allows (a) measurement in a low number of neutrophils (5000 - 50000 cells in 50 μl) and (b) investigation of the interaction of neutrophils with water-insoluble compounds.

On the other hand, conventional CHL determination was slightly modified for our purpose.

MATERIAL AND METHODS

Neutrophils. These were isolated according to (5); the cells were kept at 30°C before use.

SUP determination. Cytochrome c reduction by SUP was measured according to (4) and (5). 50 μl neutrophil suspension containing 5000-80000 cells, 10 μg ferricytochrome c, 33 μg D-glucose and 33 μg bovine serum albumin in Dulbecco-buffer were pipetted into a specially developed teflon sample holder of 50-70 μl capacity and the change in light reflection of the suspension after stimulation was recorded (in transmission units; at 417 nm). During cytochrome c reduction, light transmission decreases at this wavelength.

Stimulation of SUP output by N-formylmethionyl-leucyl-phenylalanine (FMLP). 2 μl of a 10 μM FMLP solution (in water) were added to 50 μl neutrophil suspension and measurement was carried out as described (5,6).

Stimulation of SUP output by esters. Table I summarizes the compounds tested. Except for mycolic acids, esters were used in the same molar amounts. Mycolic acid methyl esters were isolated from mycobacteria according to (7). If not otherwise noted, free carboxylic acids were purchased from Fluka A.G. and Sigma Chemie (F.R.G.) and the esters prepared in our laboratory. O-myristyl derivatives of 2- and 3-hydroxymyristic acids were synthesized by Wollenweber (8). To investigate the stimulating activity of the esters, these were dissolved in n-hexane or $CHCl_3$, pipetted into the teflon sample holder, and $CHCl_3$ was evaporated under a slow stream of nitrogen.

In this way, the sample holder became coated with a thin layer of the ester in question. Subsequently, the neutrophil suspension was added and the change in light reflection recorded.

Controls were run using $CHCl_3$ without esters; before the addition of neutrophils, the solvent was evaporated in the sample holder as described.

Stimulation of CHL by esters. The ester solution (in $CHCl_3$) under investigation was added to the bottom of a 12 x 60 mm glass reagent tube and the solvent evaporated. The amount of ester remaining in the reagent tube was 1-100 µg for mycolic acid methyl esters and 10 ng - 200 µg for other esters which were used in the same molar amounts. A 50 µl neutrophil suspension (10000 - 70000 cells; without cytochrome c) and 10 µl (= 1 µg) of the luminescence amplifier lucigenin (Sigma Chemie, F.R.G.) were added and CHL was recorded as c.p.m. in the Berthold Biolumat LB 9500.

Influence of esters on SUP output or CHL stimulation by FMLP. The glass reagent tube or teflon sample holder was coated with the compound in question and neutrophil suspension and FMLP were added. The differences in c.p.m. from the results obtained without the compound (inhibiting or stimulating effect) were calculated.

Superoxide dismutase (SOD) sensitivity of SUP output and CHL. Prior to filling the teflon sample holder or reagent tube (both prepared with the ester in question) with the neutrophil suspension, 0.006 units of SOD (in 5 µl Dulbecco buffer) were added and measurements carried out as described.

RESULTS

Stimulation of SUP output by carboxylic acid esters. Table I summarizes the relevant observations. Neutrophils brought into reaction with a mycolic acid methyl ester layer in the teflon sample holder produced considerable amounts of SUP as measured by the cytochrome c reduction technique. In comparison with FMLP, the velocity of the output was lower.

Methyl, ethyl and propyl esters of C_{10}-C_{18}-fatty acids or their 2- or 3-hydroxylated derivatives were found to be inactive as tested by SUP output stimulation. In contrast, the butyl esters of non-substituted C_{12}-C_{18}-acids were found to be active, an effect was measurable using about 3 µg ester/50 µl neutrophil suspension (0.1 - 0.02 µmol ester).

The activity of both hydroxybutyl esters and 2- or 3(O-myristyl)-myristic acid methyl and ethyl esters could be easily measured at a concentration of 2-5 µg/50 µl neutrophil suspension (3-OH myristic acid butyl ester).

Stimulation of CHL. Mycolic acid methyl esters were strongly active; about 5 µg in the reagent tube stimulated CHL. Similarly to results obtained for SUP output

TABLE I

STIMULATION OF SUPEROXIDE OUTPUT (SUP) AND CHEMILUMINESCENCE (CHL) BY ESTERS

carboxylic acid	methyl	ethyl	propyl	butyl	remarks
mycolic	CHL/SUP	nt	nt	nt	M.smegmatis
C1-C6	inact	inact	nt	SUP	
C7-C13	nt	nt	nt	CHL/SUP	
C14	inact	inact	inact	CHL/SUP	
C15-C18	nt	nt	nt	CHL/SUP	
C10-(3-OH)	inact	nt	nt	nt	
C12-(2-OH)	inact	nt	nt	nt	
C14-(2-OH)	inact	nt	nt	CHL/SUP	
C14-(3-OH)	inact	nt	nt	CHL/SUP	
C14-(3-O-myristyl)	CHL/SUP	CHL/SUP	nt	nt	component of endotoxins

nt: not tested, inact: no activity; CHL,SUP: stimulation of chemiluminescence and/or superoxide output

Remark: inactivity: used at the same molar concentration as myristic acid butyl ester the compound induces the same or less activity as a chemotactic peptide (FMLP)

stimulation, methyl, ethyl and propyl esters of carboxylic acids have proved considerably less active or even inactive if used in the same molar amounts.

On the other hand, µg amounts (0.1 - 0.02 µmol) of butyl esters even of non-substituted acids were generally active.

O-Acylated 3-hydroxymyristic acid methyl and ethyl esters stimulated both SUP output and CHL.

SUP output or CHL inhibition by inactive compounds. None of the inactive compounds was able to inhibit stimulation by FMLP. However, active compounds markedly inhibited stimulation by FMLP.

Inhibition of SUP output and CHL by superoxide dismutase (SOD). Both tests became negative when SOD was present in the reaction mixture.

DISCUSSION

When the light reflection technique and the teflon sample holder were used to measure the interaction of neutrophils with water-insoluble compounds, biologically important mycolic acid methyl esters (related to the esters in the cell-wall of mycobacteria) were found to be highly active as regards SUP output and CHL stimulation.

The methyl, ethyl and propyl esters of the straight-chain carboxylic acids which were tested act as inhibitors in neutrophils stimulated with FMLP.

In contrast, butyl esters of non-substituted and OH-substituted acids stimulated SUP output (already the formic acid butyl ester); CHL stimulation by butyl esters started if there were 7 or more C-atoms in the acid molecule. 3-OH-myristic acid esters and their O-acyl derivatives (active in both tests) are related to amide- or ester-bound components of lipopolysaccharides of Gram-negative bacteria (9) which themselves induce SUP output and CHL. As far as we know, the SUP output and CHL stimulating activity of these esters has not yet been reported. It would be interesting to investigate whether these parts of the lipopolysaccharide molecule might be mainly responsible for their biological activities.

None of the inactive compounds inhibited SUP output or CHL stimulation by FMLP; thus, they do not influence the respiratory burst in this respect. As regards the active compounds, the effect of SOD showed that both cytochrome c reduction and CHL activation were caused by SUP anions generated in neutrophils; however, the reaction mechanism is at present unknown.

ACKNOWLEDGEMENTS

We are indebted to Drs U. Zähringer, H. Brade and H.-W. Wollenweber for supplying the different compounds used in the experiments.

REFERENCES

1. Allen RC, Stjernholm LR, Steele RH (1972) Biochem Biophys Res Comm 47:679-684
2. Babior BM, Kipnes RS, Curnutte J (1973) J Clin Invest 52:741-744
3. Johnston RB jr, Keele BB jr, Misra HP, Lehmeyer JE, Webb LS, Baener RL, Rajagopalan KV (1975) J Clin Invest 55:1357-1372
4. Tarnok I, Tarnok Zs, Krallmann-Wenzel U, Röhrscheidt-Andrzejewski E (1985) In: Frontiers in Microbiology (in press)
5. Moo-Yeen West, Sinclair D, Southwell-Kelly P (1981) Biochem Biophys Res Comm 100:212-218
6. Schiffmann E, Corcoran BA, Wahl SM (1975) Proc Natl Acad Sci (USA) 72:1059-1062
7. Minnikin DE, Alshamaony L, Goodfellow M (1975) J Gen Microbiol 88:200-204
8. Wollenweber H-W (1982) Thesis. Albert-Ludwigs University, Freiburg, FRG
9. Rietschel E, Lüderitz O (1980) Forum Mikrobiologie 3:12-20

ISOZYMES OF PEROXIDASE AND SUPEROXIDE DISMUTASE FROM MYCOBACTERIUM PHLEI AND MYCOBACTERIUM SMEGMATIS

Z. GONZALEZ-LAMA, P. LUPIOLA, O.E. SANTANA and R.H. LOPEZ-ORGE

Departamento de Microbiologia, Colegio Universitario de Medicina, Apartado de Correos 550, Las Palmas de Gran Canaria (Spain)

INTRODUCTION

Superoxide dismutase (SOD), catalase and peroxidase (POX) play an important role in H_2O_2 metabolism. Superoxide dismutase has been observed in all mycobacteria so far studied (1-5); this enzyme appears to be a prerequisite for aerobiosis (6). Catalase and peroxidase are present in most mycobacteria (5,7). Mycobacterial catalases have been used in taxonomic studies (8-10).

In the present paper we report on a study of the isozymes, by isoelectric-focusing in polyacrylamide gel, of the peroxidase and the superoxide dismutase of one *Mycobacterium phlei* strain in comparison with one *Mycobacterium smegmatis* strain.

MATERIAL AND METHODS

The strains were grown in Brain Heart Infusion (BHI, Difco). Cells grown for 48 h at 37ºC with vigorous aeration on a gyratory incubator shaker were harvested by centrifugation. The crude enzyme preparation was obtained by sonication and centrifugation to 48.000 xg 30 min in a refrigerated centrifuge. Supernatant was used as crude extract. Peroxidase and superoxide dismutase isozymes were separated using a flat bed of polyacrylamide gel electrofocusing, pH 3.5-10. The pH was determined by use of a flat surface electrode at 0.5 cm intervals. Superoxide dismutase activity in the gel was located by the photochemical method (11) and peroxidase activity was detected by covering the gel with a mixture containing H_2O_2 and benzidine hydrochloride (12). The prosthetic metal ion of superoxide dismutase was determined as described elsewhere (13).

RESULTS AND DISCUSSION

Figure 1 is a diagram of isoelectricfocusing of peroxidase and superoxide dismutase activities in crude extracts of *M. phlei* and *M. smegmatis*.

Two catalase activities have been described in *M. phlei* (14) and a single peroxidase (15) by acrylamide gel electrophoresis. We separated two similar isozymes of peroxidase in *M. phlei* and *M. smegmatis* with a pI of 4.0 & 4.2. The major component, in both, is an isozyme with pI of 4.0. In *M. phlei* we found two isozymes of superoxide dismutase with pI of 4.0 & 4.2, the major component is an isozyme with pI of 4.2, but in *M. smegmatis* we saw only one isozyme of superoxide

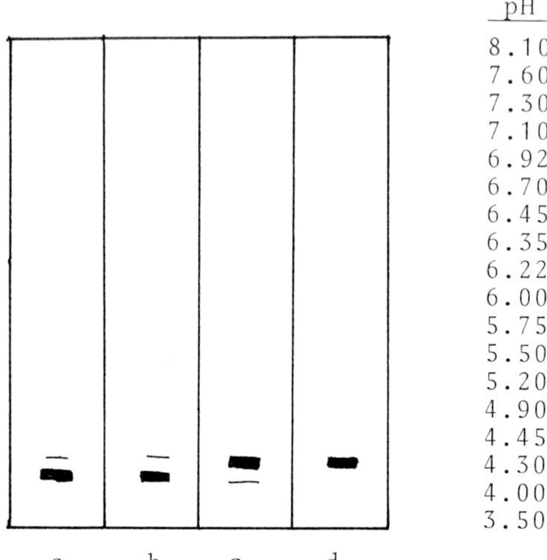

Figure 1. Isoelectricfocusing of peroxidase and superoxide dismutase from crude extracts of M. *phlei* and M. *smegmatis*.

a: Peroxidase isozymes of M. *phlei*
b: Peroxidase isozymes of M. *smegmatis*
c: Superoxide dismutase isozymes of M. *phlei*
d: Superoxide dismutase isozyme of M. *smegmatis*

dismutase with a pI of 4.2.

It has been demonstrated that treatment with cyanide and H_2O_2 results in loss of Cu/Zu-SOD and Fe-SOD activity but not in Mn-SOD activity (16,17). *M. leprae* (4), *M. lepraemurium* (18) and *M. smegmatis* (1) have Mn-SOD, whereas *M. tuberculosis* (2) has Fe-SOD. Superoxide dismutases of *M. phlei* and *M. smegmatis* were unaffected after incubation with C^N and H_2O_2. From this experiment, *M. phlei* and *M. smegmatis* appear to be an Mn-SOD.

REFERENCES

1. Kusonose E, Noda Y, Ichihara K, Kusunose M (1976) Arch Microbiol 108:65-73
2. Kusonose E, Ichihara K, Noda Y, Kusonose M (1976) J Biochem 80:1343-1352
3. Kusonose E, Kusonose M, Ichihara K, Izumi S (1980) J Gen Appl Microbiol 26:369-372
4. Kusunose E, Kusunose M, Ichihara K, Izumi S (1981) FEMS Microbiol Letters 10:49-52
5. Wheeler PR, Gregory D (1980) J Gen Microbiol 121:457-464
6. Fridovich I (1975) Ann Rev Biochem 44:147-159
7. Tirunarayanan MO, Vischer WA (1957) Ann Rev Tuberc 75:62-70
8. Wayne LG, Diaz GA (1976) Int J Syst Bacteriol 26:38-44
9. Wayne LG, Diaz GA (1979) Int J Syst Bacteriol 29:19-24
10. Wayne LG, Diaz GA (1982) Int J Syst Bacteriol 32:296-304
11. Beauchamp CO, Fridovich I (1971) Anal Biochem 44:276-287
12. Gonzalez-Lama Z, Feinstein RN (1977) Proc Soc Exp Biol Med 154:322-324
13. Gonzalez-Lama Z, Santana OE, Betancor P (1981) In : Allen A (ed) Electrophoresis'81 Walter de Gruyter & Co Berlin-New York, pp. 699-702
14. Stavri H, Stavri D (1975) Arch Roum Path Exp Microbiol 34:85-89
15. Davis WB, Phillips DM (1977) Antimicrob Agents Chemother 12: 529-533
16. Beauchamp CO, Fridovich I (1973) Biochim Biophys Acta 317:50-64
17. Asada K, Yoshikawa K, Takahashi M, Maeda Y, Enmanji K (1975) J Biol Chem 250:2801-2807
18. Ichihara K, Kusonose E, Kusonose M, Mori T (1977) J Biochem 81:1427-1433

Mycobacteria of Clinical Interest. M. Casal, editor

A STUDY OF CELL-MEDIATED IMMUNITY IN TUBERCULOSIS USING THE LYMPHOCYTE MIGRATION INHIBITION TEST

P. CARRION, D. HARRIS, A. CASTILLO, P. BARRANCO and M.C. MAROTO
Department of Microbiology and Parasitology, School of Medicine, Granada (Spain)

INTRODUCTION

As the World Health Organisation has stated, due to the technical difficulties involved, cell-mediated immunity in tuberculosis is still unclear. This has been the main reason for undertaking this study.

MATERIAL AND METHODS

A total of 92 individuals were studied. They were divided into three groups:

- 55 cases of tuberculosis: 42 acute patients
  13 chronic patients
- 28 healthy individuals used as a control group
- 9 persons who had had possible contact with tuberculous patients

They were all subjected to various studies:

- lymphocyte migration inhibition test (LMIT); phytohaemagglutin was used as a nonspecific mitogen (PHA); purified protein derivate (PPD) from the State Serum Institute, Copenhagen, was used as a specific antigen.
- bacilloscopy using the Ziehl-Nielsen stain
- Mendel-Mantoux intradermic reaction (2 U.T.).

RESULTS

The results obtained with the LMIT are shown in Table I. Eighty-two percent showed a positive reaction to phytohaemagglutinin whereas the specific reaction with PPD was positive only in 18.4%.

The results obtained with bacilloscopy are shown in Table II. Twenty-six percent of the acute patients and 7% of the chronic patients were positive, representing 21.8% of all the tuberculous patients. The control group were all negative.

The results of the Mendel-Mantoux intradermic reaction are shown in Table III. Fifty-seven percent of all the individuals tested were positive (n=65).

In Tables III and IV the results obtained by applying the LMIT and the intradermic reaction, and the LMIT and bacilloscopy respectively, are shown. In the former, 49.23% coincided with one another, being either positive or negative. In the latter, 75% of the results coincided.

We found only 49.23% of the results of all individuals tested using the Mantoux and the LMIT to coincide, being either positive or negative. Similarly,

TABLE I

SPECIFIC AND NONSPECIFIC IMMUNE RESPONSE IN THE DIFFERENT GROUPS STUDIED

	Total	SCI positive	%	ICI positive	%
Acute patients	42	10	23.8	35	83.5
Chronic patients	13	3	23	11	84.6
Control group	28	2	7.1	23	82.1
Contacts with TB patients	9	2	22.2	7	77.7
Total	92	17	18.4	76	82.6

SCI = Specific cell-mediated immunity to tuberculin (Serum Institute). Dose 100 μg/ml
ICI = Nonspecific cell-mediated immunity to phytohaemagglutinin.

TABLE II

RESULTS OF BACILLOSCOPY ACCORDING TO CLASSIFICATION OF THE PATIENTS (n=92)

	Total	Bacilloscopy (+)	%	Bacilloscopy (-)	%
Acute patients	42	11	26	31	73.8
Chronic patients	13	1	7.6	12	92.3
Control group	28	0	0	28	100
Contacts with TB patients	9	0	0	9	100
Total	92	12	13	80	86.9

TABLE III

RELATIONSHIP BETWEEN RESULTS OF THE INTRADERMIC REACTION AND THE SPECIFIC IMMUNE RESPONSE (n=65)

	Total	Mantoux positive	%	Mantoux negative	%
SCI (+)	10	7	70	3	30
SCI (-)	55	30	54.5	25	45.5
Total	65	37	57	28	43

TABLE IV
RELATIONSHIP BETWEEN THE RESULTS OF BACILLOSCOPY AND THE SPECIFIC CELLULAR RESPONSE

	Total	Bacilloscopy positive	%	Bacilloscopy negative	%
SCI (+)	17	3	17.6	14	82.3
SCI (-)	75	9	12	66	88
Total	92	12	13	80	86.9

the results of bacilloscopy coincided in 75% of those tested.

CONCLUSIONS

Only 70% of the 57% of patients with positive Mantoux results showed a positive cell-mediated reaction.

The specific cell-mediated reaction was positive more frequently (22-23%) in tuberculous patients, whatever their clinical presentation, whereas it was positive in only 7% of the controls.

On the basis of these results, we would question the validity of a specific cell-mediated immunological test using the LMIT in tuberculosis.

STUDY OF SPECIFIC CELL-MEDIATED IMMUNITY IN TUBERCULOSIS USING THE LYMPHOBLAST TRANSFORMATION TEST

I. GALAN, D. HARRIS, F. GARRIDO, G. PIEDROLA and M.C. MAROTO
Department of Microbiology and Parasitology, School of Medicine, Granada (Spain)

INTRODUCTION

Due to the doubtful nature of cell-mediated immunity in tuberculosis, and the wide variety of techniques employed to study it, we have tried to increase our knowledge with a test described as useful in the literature, namely lymphoblast transformation test (LTT).

MATERIALS AND METHODS

A total of 33 individuals were studied, 20 of whom had tuberculosis; 17 of these were acute patients and 3 chronic. Thirteen healthy subjects constituted the control group.

All were subjected to the LTT using two different types of mitogen:

- a nonspecific mitogen, phytohaemagglutinin (PHA), with which the classical technique was used (Table I).
- a specific antigen, purified protein derivate (PPD), RT 23 strain, 100 UT (Llorente Lab.) and 50,000 UT (State Serum Institute, Copenhagen). In this case, due to the technical difficulties involved, some variations were introduced:

(1) The concentration of cells per well was increased from 5×10^4 to 5×10^5.
(2) Variations in the tuberculin units per well from 1 to 100 UT were made.
(3) The incubation time in the CO_2 chamber was increased from 48 to 112 hours.
(4) The concentration of timidine trithiade was increased from 0.25 to 0.50 µCi.

The 43 individuals studied were also subjected to the Mendel-Mantoux intradermic reaction, using 2 UT. This test was considered positive when the diameter of the induration was 10 mm or more, on reading at 72 hours.

RESULTS

Table II shows that PHA stimulation was positive in 100% of cases at an optimum dose of 1/4.

The LTT with PPD was negative in all individuals with different variations in the test, except in two healthy subjects who were Mantoux positive at an optimum dose of 2 and 8 UT. In these cases, the concentration of cells per well was 5×10^5 and the incubation time in the CO_2 chamber was 112 hours.

In these individuals, although the test was considered positive, the lymphocyte stimulation curve was flatter than that with PHA and the CPM was not so high.

TABLE I

THE LYMPHOBLAST TRANSFORMATION TEST

Extraction of 20 ml of peripheral blood + 0.1 ml heparin

↓

Dilute the medium with Hanks or sterile PBS

↓

Separate lymphocytes with Ficoll-Hypaque medium

↓

Adjust with Neubauer chamber at 2.5×10^5 cells in RPMI 1640 medium and adding glutamine, gentamycin and ampicillin

↓

Add 200 μl/well in the microtiter dishes

↓

Add 5 μl of the different dilutions of PHA (1/1 - 1/64), - three times

↓

Incubate in the carbon dioxide chamber for 48 hours

↓

Add timidine trithiade 0.25 μCi/well

↓

Incubate in carbon dioxide chamber for 24 hours

↓

Wash the dish three times using the Harvested

↓

The incorporation of TH_3 into cellular DNA is quantified using toluol, and the results expressed as the average of the three determinations, in counts per minute (CPM)

↓

Lymphocyte proliferation is considered positive when the CPM becomes progressively greater until an optimum dose of mitogen is reached. This is the dose the CPM of which is 2.5 times greater than that of cells not stimulated.

N.B. The whole process must be carried out under strictly sterile conditions.

TABLE II

	TTL		
	PHA +	PPD +	Mantoux +
Acute tuberculosis (n=17)	17	-	10
Chronic tuberculosis (n=3)	3	-	3
Control group (n=13)	13	2	10
Total	33	2	23

CONCLUSIONS

(1) Although the LTT with PHA gives optimum results, when specific antigens are used, namely PPD, a revision and adjustment of the technique is needed in the majority of cases.

(2) The value of TTL as a diagnostic test is arguable, and further investigations are needed in this field.

Mycobacteria of Clinical Interest. M. Casal, editor

ADJUVANT AND SUPPRESSIVE EFFECTS OF MYCOBACTERIA IN EAE INDUCTION MAY BE RELATED TO DIFFERENT DETERMINANTS

MARIJA MOSTARICA STOJKOVIC, MILICA PETROVIC, STANISLAV VUKMANOVIC and MIODRAG L. LUKIC

Institute of Microbiology and Immunology, School of Medicine and Institute of Physiology, School of Pharmacy, University of Belgrade, Belgrade (Yugoslavia)

INTRODUCTION

Complete Freund's adjuvant (CFA) containing tubercle bacilli in mineral oil enhances the production of experimental allergic encephalomyelitis (EAE) when emulsified with an aqueous suspension of central nervous system (CNS) tissue or with myelin basic protein (MBP) extracted from CNS tissue (1,2). However, there is also ample evidence that CFA prevents the induction of EAE in guinea pigs (3) and rats (4). This suppressive effect of adjuvant was long lasting, was related to the inductive phase of the autoimmune response and could be abrogated by pretreatment of the rats with cyclophosphamide (4). This latter finding indicated that some component of tubercle bacilli may induce non-antigen-specific suppressor cells which regulate the induction of autoimmunity. We have observed that 6,6'-trechalose dimycolate (TDM) represents an adjuvant active structure of Mycobacteria. In this paper we provide evidence that despite its excellent adjuvanticity, TDM does not exhibit the capacity to prevent EAE, thus suggesting that the suppressive and adjuvant effects of Mycobacteria may be related to different factors in the microorganism.

MATERIAL AND METHODS

Animals

Inbred DA rats, 12-16 weeks old, obtained from the animal colony maintained at the Institute for Biological Research, Belgrade, were used in all experiments.

Encephalitogens

Myelin basic protein (MBP) isolated from rat spinal cord or bovine spinal cord (BSC) were used in the induction of EAE.

Adjuvants

Complete Freund's adjuvant (CFA) containing 10 mg *M.phlei*/ml or incomplete Freund's adjuvant (IFA) supplemented with 1 mg/ml 6,6'-trechalose dimycolate (TDM, kindly donated by prof. E.Lederer, Institut de Biochemie, Orsay, France) were used in the procedure of induction or prevention of EAE.

EAE induction, prevention and evaluation

EAE was induced and clinical signs of neurological dysfunction scored as described elsewhere(5). Rats were pretreated with CFA or TDM-IFA by i.d.

injection of 0.1 ml emulsion prepared by mixing adjuvants and saline ana partes. A rechallenge with BSC-CFA was performed 35 days after first immunization or pretreatment, by injecting 0.1 ml of encephalitogenic emulsion in the opposite hindfoot pad.

RESULTS

TDM can replace Mycobacteria in adjuvant active moiety

DA rats exhibited clinically and histologically verified EAE when immunized with appropriate antigen (BSC, MBP) emulsified in CFA. Adjuvant activity of TDM incorporated in IFA was indistinguishable from that seen with CFA as evaluated by the day of onset and grade of clinical signs of EAE (Table I).

TABLE I

TDM CAN REPLACE MYCOBACTERIA IN CFA IN THE INDUCTION OF EAE

Adjuvant	Encephalitogen	Clinical EAE		
		Incidence[a]	Onset[b]	Grade[c]
CFA	BSC	5/5	9	3.4
TDM-IFA	BSC	5/5	8.2	3.4
TDM	BSC	5/5	8.8	3
CFA	MBP	15/15	9.2	2.9
TDM-IFA	MBP	14/14	9.2	2.6
TDM	MBP	0/5	-	0

[a] Number of rats with signs / total number of rats

[b] Average day after immunization on which signs were first observed

[c] Average of grades 0 to 4

Furthermore, TDM alone (without IFA) was sufficient for EAE induction when BSC was used as encephalitogen (Table I).

TDM can support antigen-specific resistance to reinduction of EAE

As previously shown by others in various strains, rats that recovered from EAE are resistant to reinduction of the disease (6). DA rats exhibit complete resistance to reinduction of the disease when TDM in IFA was used as adjuvant irrespective of whether heterologous nervous tissue (BSC) or purified encephalitogen was used for first immunization (data not shown).

TDM does not possess determinants responsible for the suppression of EAE

As shown in Table II, pretreatment of DA rats with CFA abolished their susceptibility to EAE induction. However, when whole Mycobacteria were re-

TABLE II

PRETREATMENT WITH TDM DOES NOT PREVENT INDUCTION OF EAE

Pretreatment (day -35)	Clinical EAE Incidence[a]	Onset[b]	Grade[c]
CFA	3/5	9	1.2
TDM-IFA	5/5	8.8	4
-	5/5	7	3.6

[a,b,c] See legend, Table I

placed by TDM in CFA, resistance to the induction of EAE with appropriate immunization was not observed.

DISCUSSION

We were able to demonstrate that TDM incorporated in IFA is sufficient to induce EAE when used in emulsion with appropriate antigen. However, when applied alone prior to EAE induction, TDM, in contrast to whole Mycobacteria, did not exhibit any preventive effect. Thus, our study suggests that two types of mechanism, one antigen-dependent and the other adjuvant-dependent, are operative in resistance to reinduction of EAE.

Our results indicate that the adjuvant and suppressive capacities of Mycobacteria may be related to different factors in the microorganism. These results are in agreement with a recent report showing that a unique epitope on *M. leprae* is responsible for activation of T suppressor cells in lepromatous patients (7), as well as with our own findings (8) that autoantigens possess suppressor factors different from those seen in helper T lymphocytes.

REFERENCES

1. Morgan I (1947) J Exp Med 85:131-140

2. Kies MW, Alvord EC (1959) In: Kies MW, Alvord EC (eds) Allergic Encephalomyelitis. Charles C Thomas Co, Springfield, pp 293-299

3. Kies MW, Alvord EC (1958) Nature 182:1106

4. Hempel K, Freitag A, Freitag B, Endres B, May B, Leibaldt G (1985) Int Archs Allergy appl Immun 76:193-199

5. Mostarica Stojković M, Petrović M, Lukić ML (1982) Clin Exp Immunol 50:311-317

6. Willenborg DO (1979) J Immunol 123:1145-1150

7. Mehra V, Brennan PJ, Rada E, Convit J, Bloom BR (1984) Nature 308:194-196

8. Lukić ML, Mitchison NA (1984) Eur J Immunol 14:766-768

THE USE OF MONOCLONAL ANTIBODIES FOR THE IDENTIFICATION OF MYCOBACTERIA AND THE DIAGNOSIS OF MYCOBACTERIAL DISEASES; LEPROSY AND TUBERCULOSIS.

AREND H.J. KOLK, WIM VAN SCHOOTEN, RAYMOND EVERS, JELLE E.R. THOLE*,SJOUKJE KUIJPER, MADELEINE Y.L. DE WIT, TEUNIS A. EGGELTE AND PAUL R. KLATSER.

Laboratorium of Tropical Hygiene, Royal Tropical Institute, Meibergdreef 39, 1105 AZ Amsterdam (The Netherlands).
*National Institute for Public Health and Environmental Hygiene, Bilthoven, (The Netherlands).

INTRODUCTION

A battery of monoclonal antibodies (Moabs) specific for different mycobacteria would provide a useful tool for the rapid diagnosis of mycobacterial diseases by: a. Direct identification of the causative agent; b. Identification of antigens in body fluids released by the mycobacteria; c. Detection of specific antibodies in patients' sera; d. The identification of antigens involved in cellular immunity which could lead to the development of a specific skin test and a new generation of vaccines.

MATERIALS AND METHODS

Production and characterization of monoclonal antibodies.
Moabs have been made and characterized as previously been described by Kolk and coworkers (1-3). For the Moabs against M.avium the BALB/c mice were immunized with a mixture of 7 strains of M.avium with different serotypes.

Isolation of mycobacterial antigens.

Lipids . Lipid antigens were isolated by extraction of the mycobacteria with chloroform/methanol (C:M = 2:1). The extracts were tested with the Moabs in ELISA .

Proteins. Protein antigens recognized by F67-13 and F47-9 were isolated from a 100,000 g supernatant of sonicated M.leprae by ion exchange chromatography (Klatser et al, unpublished) and a 64 kD protein from M.tuberculosis. A 64 kD protein from a M.bovis BCG recombinant clone (4) was isolated from E.coli pressate by affinity chromatography with Moab F67-13.

Table I

CHARACTERIZATION OF MONOCLONAL ANTIBODIES BY ELISA, IMMUNOFLUORESCENCE, MOLECULAR WEIGHT OF THE ANTIGEN AND IG CLASS

Species/strains[1]	Immunogen: M.tuberculosis Clone[2] F23-49	F29-29	F67-9	F67-13	F67-19
M.tuberculosis	+/4+[3]	±/3+	-/-	+/±	-/-
M.leprae	-/-	-/-	-/-	+/-	-/-
M.bovis BCG	+/3+	±/-	-/-	+/-	-/-
M.W.antigen	15kD	45kD	35kD	64kD	40kD
Ig class	IgG2a	IgG2a	IgG3	IgG1	IgG1

1. 20 species/strains were tested(1,3).
2. Selection out of 30 clones.
3. The figure before the line indicates the results in the ELISA, the figure after the line the results in the immunofluorescence.

Table II

CHARACTERIZATION OF MONOCLONAL ANTIBODIES BY ELISA, IMMUNOFLUORESCENCE, MOLECULAR WEIGHT OF THE ANTIGEN AND IG CLASS

Species/strains[1]	Immunogen: M.avium Clone[2] F75-3[4]	F75-5[4]	F75-13[4]	Imm.: M. leprae F47-9	F47-21
M.avium 1	5/4+[3]	5/4+	5/4+	-/-	-/-
M.avium 7	4/4+	-/-	-/-	-/-	-/-
M.leprae	3/2+	-/-	-/-	8/-	-/3+
M.tuberculosis	3/3+	-/-	-/-	-/-	-/-
M.W.antigen	-	-	-	36kD	lipid
Ig class	IgM	IgM	IgM	IgG1	IgG1

1. 20 species/strains were tested (1,3).
2. Selection out of 20 clones.
3. The same as for table I.
4. Tested on lipid extract except for M.leprae for which sonicate was used.

T-cell lines and T-cell clones. T-cell lines and T-cell clones from Mantoux positive individuals and leprosy patients were obtained from peripheral blood lymphocytes (PBLs) stimulated in vitro with M.tuberculosis or M.leprae. After culturing for 10-14 days the T-cells were cloned by limiting dilution. Clones were screened for antigen specificity after 14 days in a proliferation assay measuring the H-thymidine incorporation. Autologous PBLs or autologous Epstein Barr virus transformed B-cells were used as antigen presenting cells. Details of this procedure will be published elsewhere.

Table III
RECOGNITION OF PURIFIED MYCOBACTERIAL ANTIGENS BY T-CELLS

T-cells	H-thymidine incorporation (cpm x 10^{-3}) medium	64kd M.tub	64kd[4] r-BCG	64kd M.lep	36kd M.lep	PPD	M. leprae sonicate
Kk[1] T-cell line from mantoux pos.	0.3	2.4[3]	1.5[3]	0.8	0.8	3.5[3]	0.4
Kk 1-1[1] T-cell clone mantoux pos.	0.4	6.8[3]	7.3[3]	nt	nt	3.3[3]	0.4
Br 2-1[2] T-cell clone Leprosy patient	1.5	nt	nt	nt	8.8[3]	5.6[3]	4.2[3]
Mm 2[2] T-cell clone leprosy patient	4.3	4.7	9.9[3]	nt	7.1[3]	nt	8.2[3]

1. These cells were induced to proliferate with autologous Epstein Barr virus transformed B cells as antigen presenting cells.
2. These cells were induced to proliferate with autologous PBLs as antigen presenting cells.
3. Significant higher than background.
4. Purified by affinity chromatography from material obtained by recombinant DNA technology (4).

RESULTS AND DISCUSSION

Table I represents the characterization of Moabs against M.tuberculosis; Moab F23-49, F29-29 and F67-9 were reacting with M.tuberculosis and M.bovis BCG only . F67-19 reacted only in the SGIP assay with M.tuberculosis and only very weakly with M.bovis BCG. F67-13 reacted with a common epitope on a 64 kD protein present in many mycobacteria and also with a 64 kD protein from a M.bovis BCG recombinant clone (4).

The characterization of Moabs against M.avium and M.leprae is presented in table II. Moab F75-3 reacted with all mycobacteria whereas F75-5 and F75-13 were strain specific. The Moabs against M.leprae, F47-9 and F47-21, were both M.leprae specific. Moab F47-9 recognized a 36 kD protein and was only reactive in SGIP and ELISA, whereas F47-21 reacted in immunofluorescence and on a lipid extract of M.leprae. This Moab F47-21 proved to be directed against the M.leprae specific phenolic gycolipid (Fig.1) and can be used for the identification of M.leprae in skin smears (3). Moab F75-13 can be used for the identification of M.avium (strain 1) specific lipids (Fig 1) with a sensitivity of 5.10^5 bacteria.

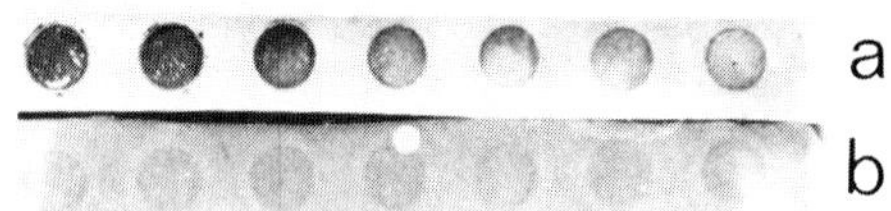

Fig. 1. Reaction of Moab F47-21 (specific for M.leprae)
a) on C:M extract of M.leprae
b) on C:M extract of M.tuberculosis.

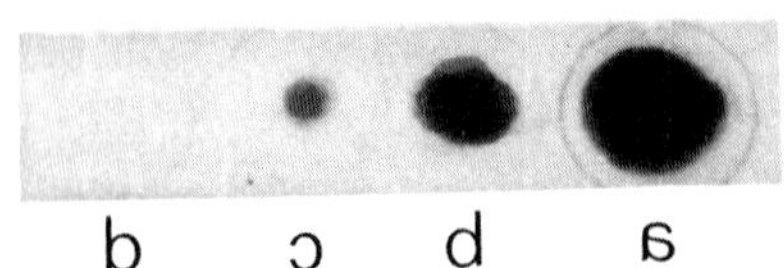

Fig. 2. Reaction of Moab F75-13 (specific for M.avium) with a C:M extract of M.avium
a) 10^8 b) 10^7 c) 10^6 bacteria
d) M.tuberculosis extract 10^8 bacteria.

Table III shows that the M.tuberculosis specific T-cell clone Kk 1-1 could be stimulated by a purified 64 kD protein from M.tuberculosis and a recombinant DNA obtained 64 kD BCG protein. This indicates that on the 64 kD protein from M.tuberculosis a specific determinant is recognized by T-cell clone Kk 1-1 (table III) and a common determinant by the Moab F67-13 (table I). The crossreacting clones Br 2-1 and Mn 2 could be stimulated by a common mycobacterial determinant present on the r-DNA BCG 64 kD protein and the 36 kD protein from M.leprae.From these results we conclude that on the 36 kD protein from M.leprae a specific determinant is recognized by Moab F47-9 (table II) and a common determinant by the T-cell clones Br 2-1 and Mn 2 (table III).

The results with the Moabs are promissing for the detection of mycobacterial antigens in body fluids and for the development of a vaccine and specific skin tests for mycobacterial diseases.

ACKNOWLEDGEMENTS

This work was supported by the Netherlands Leprosy Relief Association,the Associazione Italiana "Amici di R. Follereau", the Q.M.Gastmann Wichers Foundation, the Immunology of Leprosy (IMMLEP) component of the UNDP/World Bank/WHO Special Programme for Research and Training in Tropical Diseases and by the Commission of the European Communities Directorate-General for Science Research and Development,TSD.043.

REFERENCES

1. Kolk AHJ, Ho ML,Klatser PR, Eggelte TA, Kuijper S, de Jonge S, van Leeuwen J (1984) Clin. exp. Immunol. 58, 511-521
2. Klatser PR, van Rens MM, Eggelte TA (1984) Clin. exp. Immunol. 56, 537-544
3. Kolk AHJ, Ho ML, Klatser PR, Eggelte TA, Portaels F (1985) Ann. Microbiol. (Inst. Pasteur) In press
4. Thole JER, Dauwerse HG, Das PK, Groothuis DG, Schouls LM, van Embden JDA (1985) Infect. Immun.Submitted for publication.

Mycobacteria of Clinical Interest. M. Casal, editor

ANTIGENE CROSSREACTIVITY AMONG PATHOGENIC AND OPPORTUNISTIC MYCOBACTERIA

MILOS JOVANOVIC,OLGA BERGER, LJUBISA MARKOVIC and VERA LEPOSAVIC,
Institute of Microbiology and Immunology, Medical Faculty, University of Belgrade, 1100 Belgrade,P.O.Box 497 (Yugoslavia)

INTRODUCTION

While assessing the degree of cell-mediated immunity (CMI) during an infection caused by various mycobacteria (My), several problems drew our attention: (a) the immunogenic capacity of opportunistic My in relation to *M.tuberculosis*; (b) the impact of some tuberculostatics on the state of CMI. It was noted some time ago that there was a difference between skin reactivity to PPD and sensitivity of the My in question (1), that intraparenteral infection leads to increased resistance to TB (2) and that tuberculostatics change skin reactivity to tuberculin, i.e. PPD (3).

MATERIAL AND METHODS

For our study we used standard strains of My obtained from Prague: *M.tuberculosis* (My Tbc 1-47), *M.bovis* (Tbc_b3/50), *M.avium* (My 66/72), *M.kansasii* (My 235/80), *M.scrofulaceum* (My 189/75),*M.smegmatis* (My 2/44), *M.phlei* (My 204/75), *M. fortuitum* (My 22/63), *M.terrae* (My 238/80), BCG strain (Torkak, Belgrade) and a fresh virulent strain of *M.tuberculosis*. Outbred guinea-pigs maintained in our own animal facilities were used. Skin tests were carried out by i.c. inoculation with 0.1 ml of 1:10 *Tuberculinum vetus Kochi*, Torlak, Belgrade. The results were read after 4 and 24 hours respectively. Infiltrates of 5 mm diameter or larger indicated a positive tuberculin reaction. The LIF test was carried out by Federlin's micromethod (4) which was adjusted in our laboratory (5). Guinea-pigs were treated with isoniazid and Rifadin in dosages of 10 mg/kg of body weight. Myambutol was given in dosages of 50 mg/kg body weight.

The immunogenic capacity of individual "atypical" My was examined in guinea-pigs (8 animals in each group) by inoculating each animal with 0.2 mg of pure culture of various My in the inguinal area. When CMI was achieved, all animals including a newly introduced group were inoculated with the virulent strain of *M.tuberculosis* (See Table I, Control 1). The same procedure was repeated 180 days and 330 days respectively (See Table I, Control 2 and 3) after the start of the experiment when new control groups had also been introduced. The immunogenic capacity of particular My was assessed from the general condition of an animal and the time of survival of guinea-pigs previously inoculated by some opportunistic My and subsequently by a virulent strain of *M. tuberculosis*.

RESULTS

TABLE I

TIME OF SURVIVAL OF GUINEA-PIGS PREVIOUSLY INOCULATED BY SOME OPPORTUNISTIC MYCOBACTERIA

Mycobacterial species and controls	The time of survival X n=8
Control 1	36
Control 2	41
Control 3	39
M. scrofulaceum	225
M. smegmatis	275
M. phlei	275,5
M. avium	393
M. fortuitum	419
M. kansasii	465

Table I shows that previous inoculation of guinea-pigs by some 'atypical' My leads to a very statistically significant period of survival after inoculation by a virulent strain of *M. tuberculosis*.

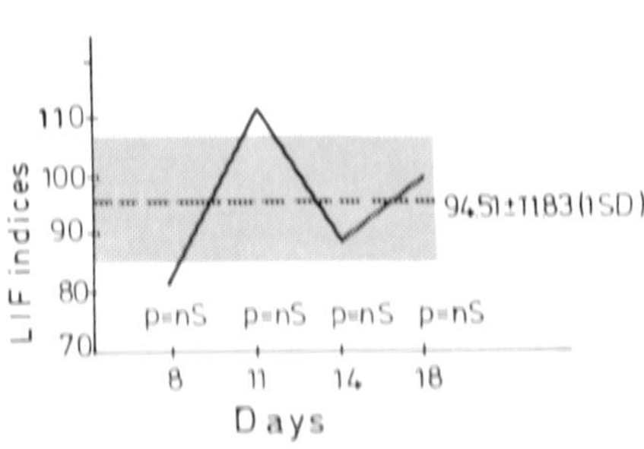

Fig.1. LIF indices of peripheral blood of healthy guinea-pigs

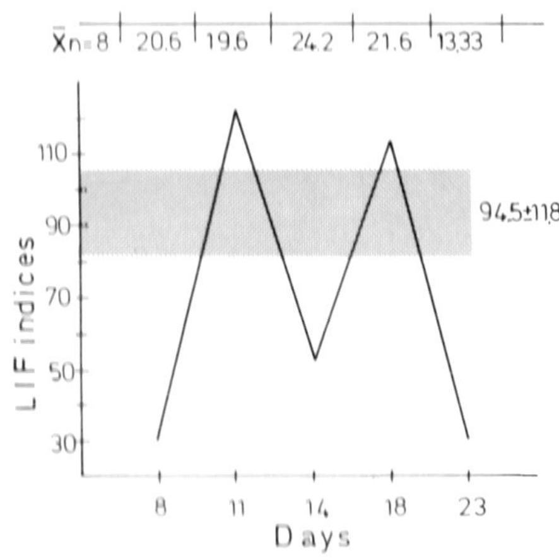

Fig.2. Untreated tuberculosis infection: LIF indices and skin tests

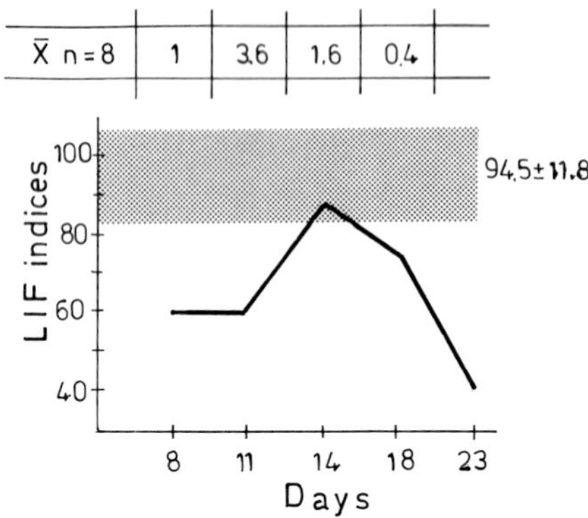

Fig.3. Tuberculosis infection treated with Isoniazid

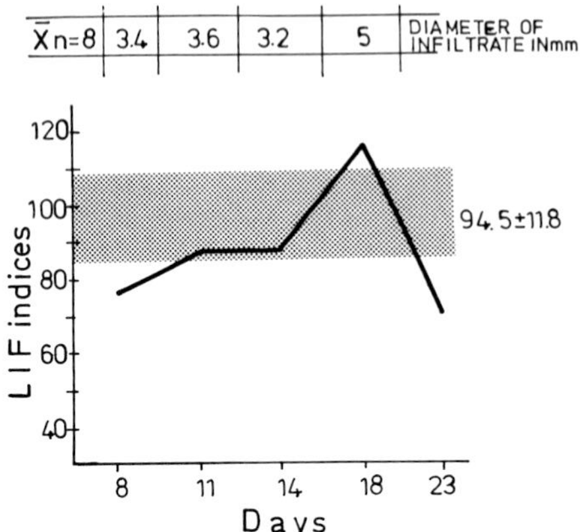

Fig.4. Tuberculosis infection treated with Mymbutol

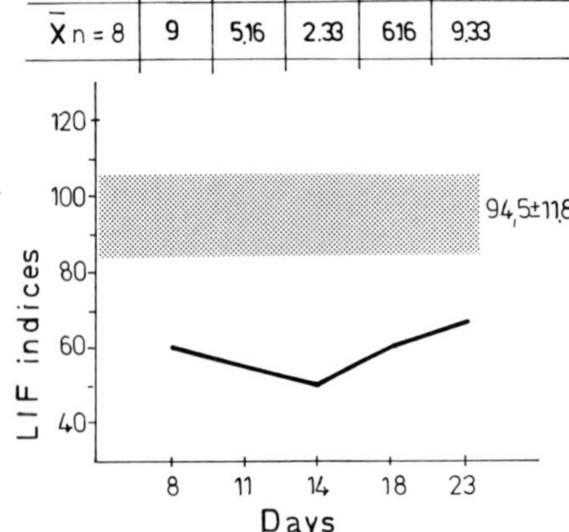

Fig.5. Tuberculosis infection treated with Rifadin

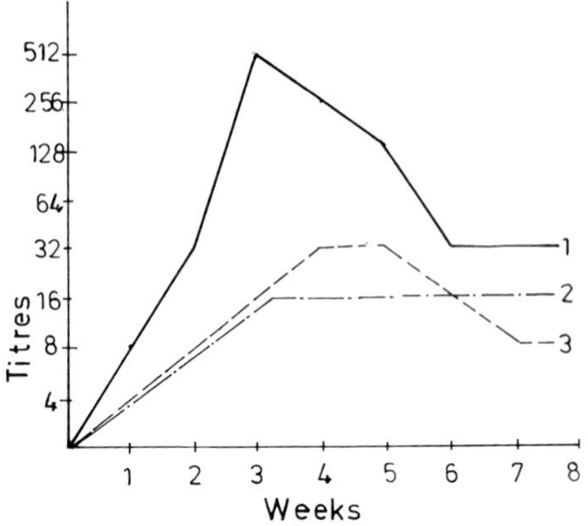

Fig.6. Titres of antibodies during the infection caused by M.tuberculosis (1), M. kansasii (2) and M.fortuitum

As shown in Figures 1-5, the LIF indices change under different conditions and these changes depend on the tuberculostatics used. There is a correlation between skin reactivity to tuberculin and LIF indices of untreated animals (see Fig.2).

Figure 6 shows the dynamics of antibody production during infections caused by different My. There are obvious differences in the dynamics of antibody formation depending on the virulence of the strain in question.

Figure 7 shows the degree of skin reactivity to tuberculin and values of LIF indices for *M.kansasii* during long-term observation.

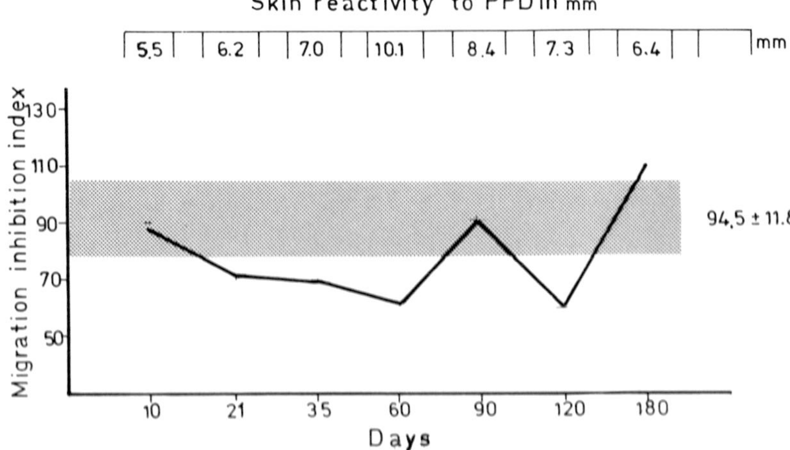

Fig.7. LIF indices and skin reactivity to tuberculine for M.kansasii infection during a long-term observation

CONCLUSIONS

Opportunistic mycobacteria are able to increase resistance to infection caused by a virulent strain of *M. tuberculosis*.

The tuberculostatics which were used change the immune response during treatment of experimental tuberculosis infection.

Tuberculostatics have an influence on skin reactivity to tuberculin which either decreases or vanishes, and when assessing the state of cell-mediated immunity during tuberculosis treatment the activity of some lymphokines should be assessed.

REFERENCES

1. Magnusson M. (1969) Z.Immun.Forsch., 137,177
2. Grigg E.R.N. (1958) Am. Rev.Resp.Dis.,113,78
3. Jovanović M. and Simić M. (1980) Glas CCCXV de l' Academie Serbe des Science et des Arts,Classe des Sciences medicales,32,229
4. Federlin K.R.N. (1971) J.Clin.Pathol.,24,533
5. Jovanović M. (1977), Doktorska disertacija,Beograd

Mycobacteria of Clinical Interest. M. Casal, editor

CELLULAR IMMUNE RESPONSE TO COMPLEX AND FRACTIONATED MYCOBACTERIAL ANTIGENS IN RABBITS IMMUNIZED BY MYCOBACTERIA

MILAN KUBIN, EVA WISINGEROVA, JAN PEKAREK, BOHUMIR PROCHAZKA
Institute of Hygiene and Epidemiology and Institute for Sera and Vaccines, Prague (Czechoslovakia)

INTRODUCTION

Migration inhibition of spleen macrophages from rabbits immunized with various mycobacteria has shown a specific response when tested with homologous cellular complex antigens. However, some cross-reactivity was still recorded in heterologous antigens, especially in those prepared from taxonomically related organisms (1,2). It seemed reasonable therefore to detect in complex cellular antigens a fraction responsible for inhibition of the migration of hypersensitive macrophages and to use purified cellular antigens in testing cellular hypersensitivity in mycobacterial infections.

MATERIAL AND METHODS

Mycobacteria. *Mycobacterium tuberculosis*, strain 1005 and *M. kansasii* ATCC 12478 were propagated in the surface pellicle on Sauton medium for 8 weeks, the bacterial mass was separated by centrifugation, washed three times with 10mM tris (hydroxymethyl)-aminomethane buffer (pH 7,8) and stored at -20°C.

Complex antigens. Wet bacteria, 10 g in 40 ml of tris buffer, were disintegrated in an MSE-Mk-2 ultrasound generator at 150 W for 15 min. Bacteria were spun down at 15,000 x g for 45 min., the supernatant centrifuged at 25,000 g for 45 min., concentrated by evaporation to 10 ml, and sterilized by filtration. The protein content was determined by the Kjeldahl method and adjusted to 10 mg/ml, and the products thus obtained were freeze-dried.

Fractionated antigens. Twenty mg of the freeze-dried complex antigen were dissolved in 2 ml of 0.1 m Tris-0.4 M NaCl buffer (pH 8.0) and applied to column packed with Sephadex G 150 (size 3.5 x 40 cm, void volume 70 ml, flow rate 26 ml/hour, fraction volume 2.2 ml) and eluted with the same buffer. The fractions were collected in an ultrarack fraction collector in combination with Uvicord (registration in the UV region at 254 nm) and pooled according to the formed peaks, dialyzed against distilled water and freeze-dried. Gel chromatography on Sephadex G 150 separated the complex antigens into three fractions according to their absorbance at 254 nm. A rough determination of their molecular weights was made on a Sephadex G 150 column (Fig 1.).

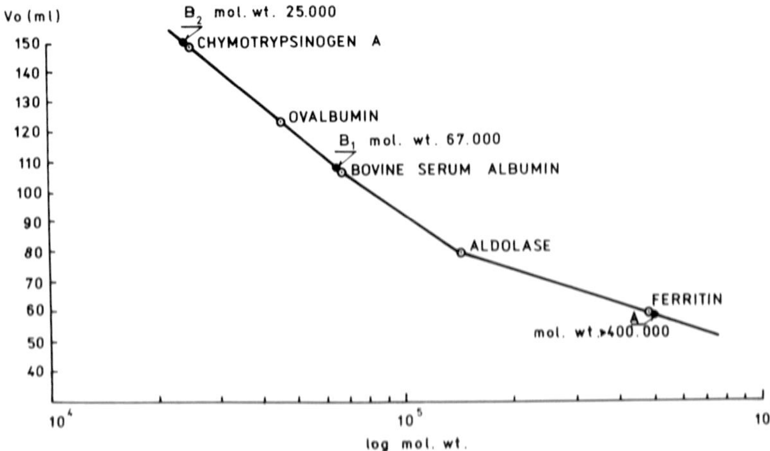

Fig.1. Elution volumes (V_0) plotted against log of the weight estimation of commercial reference preparations and of cellular antigen fractions A, B_1 and B_2 of *M. tuberculosis* and *M. kanasii* on Sephadex G 150 columns (3.5 x 40 cm, pH 8.0)

The position of fraction A in the elution curve profile, as well as its molecular weight (m.w.) suggested that it is composed of substances of high m.w., probably more than 400,000, of a very complex nature (proteins, polysaccharides, nucleic acids); the ratio 260/280 - 1.8 suggested a high proportion of nucleoproteins. The second lower peak of fraction B with a 260/280 ratio of about 1.0 indicated its probable protein character with a m.w. of about 32,000; this fraction separated into two subfractions: B_1 - V_0 108 ml, m.w. 67,000, and B_2 - V_0 150 ml, m.w. 25,000. Fraction C has a low molecular weight, approx. 10,000.

Immunization of rabbits (groups of 15 animals) was performed by intratarsal innoculation of *M. tuberculosis* and *M. kansasii* as described in a previous paper (1).

Direct and indirect migration inhibition test using the spleen fragment method was performed as described in the same paper (1). The results were expressed in migration indices (MI), i.e. the ratio of the migration distance in the control to the migration distance in the presence of the antigen.

RESULTS

In the direct migration inhibition test, the migration activity of spleen macrophages of immunized rabbits was significantly reduced by homologous complex antigens (Table 1). The migration indices (MI) in the group of animals immunized with *M. tuberculosis* group were 0.40 and 0.44 in the *M. kansasii* group; the responses against heterologous complex antigens were significantly higher in both groups but lower than in the control animals. Inhibitory activity was located in the material eluted with the B2 fraction. The respective MI were 0.50 and 0.55, being slightly weaker than those obtained with the complex antigens; heterologous responses were significantly higher but lower than those in control animals. The migration activity of spleen cells of immunized animals was not significantly affected by fractions A, B_1 or C.

TABLE 1

Average values and 95% confidence intervals of migration indices in migration inhibition tests in rabbits immunized with *M. tuberculosis* and *M. kansasii* and tested with homologous and heterologous complex and fractionated antigens.

Antigen employed	Direct test		Indirect test		Control animals
	Animals immunized by:				
	M. tubercul.	*M. kansasii*	*M. tubercul.*	*M. kansasii*	
M. tuberculosis					
Complex	0.40 0.37-0.42	0.46 0.44-0.47	0.58 0.55-0.61	0.73 0.72-0.75	0.95 0.89-1.00
Fraction A	0.91 0.88-0.94	0.85 0.82-0.88	0.99 0.96-1.02	1.05 1.03-1.08	0.98 0.90-1.06
Fraction B1	0.89 0.86-0.93	0.84 0.81-0.86	0.90 0.86-0.93	1.06 1.04-1.08	0.95 0.88-1.02
Fraction B2	0.50 0.47-0.52	0.58 0.55-0.60	0.77 0.72-0.81	0.75 0.72-0.78	0.98 0.93-1.03
Fraction C	0.94 0.92-0.97	0.88 0.86-0.91	n.p.	0.89 0.87-0.90	1.01 1.00-1.03
M. kansasii					
Complex	0.51 0.49-0.53	0.44 0.41-0.46	0.78 0.74-0.82	0.67 0.65-0.69	0.92 0.84-0.99
Fraction A	0.90 0.87-0.93	0.85 0.82-0.88	1.00 0.96-1.04	0.96 0.94-0.98	0.94 0.88-1.01
Fraction B1	0.87 0.84-0.91	0.79 0.77-0.82	0.86 0.84-0.88	0.92 0.90-0.93	0.91 0.85-0.97
Fraction B2	0.63 0.61-0.66	0.55 0.53-0.58	0.78 0.74-0.82	0.67 0.65-0.70	0.93 0.84-1.02
Fraction C	0.99 0.96-1.02	0.88 0.86-0.90	n.p.	0.88 0.86-0.91	0.92 0.74-1.11

In the indirect test, the overall responses were analogous to those in the direct test (Table 1). The differences between homologous and heterologous complex antigens were here even more pronounced: MI with homologous antigens were 0.58 and 0.67 resp. and with heterologous complexes 0.73 and 0.78 resp. As in the direct test, inhibition activity was detected in the B2 fraction:

The MI in homologous systems were 0.77 and 0.67 resp., in the heterologous 0.75 and 0.78 resp.; the inhibiting capacity was also slightly weaker than in complex antigens, as in the direct test. There was no significant inhibition of the migratory activity of spleen cells when fractions A, B_1 and C were employed.

DISCUSSION

The use of gel chromatography on Sephadex G 150 revealed inhibitory activity in a purely protein fraction (B_2) of molecular weight about 25,000. The slightly weaker immunogenic capacity of this fraction, when compared with the complex antigen, might be related to its possible overlapping with both adjoining fractions B_1 and C. All remaining fractions did not show activity in the inhibition migration tests. However, the desired increase in specificity in migration inhibition tests was not achieved when fraction B_2 was employed. Its activity, although significantly weaker in heterologous than in homologous systems, was still relatively high in heterologous responses and differed significantly from those obtained in control animals. These crossreactive activities can be explained by the close relatedness of *M.tuberculosis* and *M.kansasii* used for immunization and, further, by the presence of common epitopes in the B_2 fractions of both these mycobacterial species.

ACKNOWLEDGEMENT

This study was supported by the technical Service Agreement No.V25/181/13 with the World Health Organization, Geneva, Switzerland.

REFERENCES

1. Kubín M, Pekárek J, Švejcar, Procházka B /1981/ In vitro cellular immune response to homologous and heterologous antigens in rabbits sensitized by five species of *Mycobacterium* and *Nocardia asteroides*. Infection and Immunity 33:725-727
2. Thorel MF, Pekárek J, Švejcar J, Kubín M, Procházka B /1982/ Utilisation du test d´inhibition de la migration des macrophages dans l´étude des relations entre des antigènes mycobactériens. Ann Immunol /Inst Pasteur/ 133 C:325-332

Mycobacteria of Clinical Interest. M. Casal, editor

DIAGNOSIS OF MYCOBACTERIAL INFECTIONS USING GAS CHROMATOGRAPHY AND GAS CHROMATOGRAPHY/MASS SPECTROMETRY

LENNART LARSSON[1] and GÖRAN ODHAM[2]

[1]Department of Medical Microbiology, University of Lund, S-223 62 LUND, Sweden and [2]Laboratory of Ecological Chemistry, University of Lund, S-223 62 LUND, Sweden

INTRODUCTION

Chromatographic analysis of cellular lipids and lipid constituents plays an important role in studies on mycobacterial taxonomy. Both thin-layer chromatography (TLC) and gas chromatography (GC) have proved useful in routine species identification of mycobacteria e.g. in hospital laboratories. Moreover, the potential of GC in combination with mass spectrometry (MS) to directly demonstrate presence of mycobacteria in clinical specimens by analysis of specific marker molecules has been reported. In this communication, methodological aspects as well as the benefit of GC and GC/MS analysis in the diagnosis of mycobacterial infections (including species identification) will be discussed.

DEMONSTRATION OF MYCOBACTERIA IN CLINICAL SPECIMENS

Mycobacteria synthesize a variety of lipid components characteristic for the genus and sometimes also for species. Determination of such specific compounds e.g. in cultures, body fluids, tissues and other complex environments would be valuable for quantitatively establishing presence of these bacteria and bacterial fragments. For this purpose we selected two branched-chain fatty acids to serve as chemical markers: Tuberculostearic (10-methyloctadecanoic) acid which is considered typical for microorganisms belonging to *Actinomycetales* and C_{32} mycocerosic (2,4,6,8-tetramethyloctacosanoic) acid which has been found exclusively in a limited number of species of *Mycobacterium*, namely *M. africanum*, *M. bovis*, *M. kansasii*, *M. marinum* and *M. tuberculosis* (1a). Demonstration of one or both of these fatty acids directly in clinical specimens or short-term incubated cultures would be diagnostically very helpful for these infections considering the time-consuming procedures often necessary for diagnosis using traditional methods.

Detection using GC/MS

GC analysis is an excellent technique for separation of molecules of moderate molecular weight. For optimal sensitivity and separation efficiency, unsplit injection on to a capillary column is recommended (1b). The detection system usually employed e.g. in fatty acid profiling of mycobacteria for species characterisation (see below), the flame ionisation detector (FID), provides a sensitivity down to about 1 ng per component and responds to practically all organic molecules. This limited sensitivity and lack of selectivity makes the FID unsuitable for demonstrating trace amounts of cellular chemical markers in complex environments. However, by connecting the gas chromatograph on-line to a mass spectrometer and focusing the GC/MS at only one or a limited number of pre-selected ions, characteristic for the individual component(s) studied, the sensitivity of the detection can be increased by three to four orders of magnitude depending upon the nature of the analyte, method of ionization etc. This technique is known as selected ion monitoring (SIM). The ultimate response is achieved when ions of a single mass number of ions are monitored. When using SIM analysis of compounds in the low picogram range, the chromatographic retention time must be accurately measured as such small amounts can not be positively identified by recording spectra over the complete mass range of ions. For tuberculostearic acid, for example, several other acids with 19 carbon atoms occur in nature, all producing molecular ions of m/z 312 upon electron impact (EI) ionization of the methyl ester. Only high resolution capillary columns will ascertain identity, and their use is under all circumstances strongly recommended in ultrasensitive quantitative MS analysis (see ref 2 for details on quantitative MS applications in microbiology).

A result from our early studies using SIM analysis of tuberculostearic and C_{32} mycocerosic acids (analyzed as methyl esters) in a five day incubated culture of a sputum specimen from a patient with pulmonary tuberculosis is shown (Fig. 1). The ionization modes utilized were EI monitoring at the positive molecular ion of tuberculostearate and chemical ionization (CI) monitoring at the protonated molecular ion of C_{32} mycocerosate. Strong responses for the two acids were observed indicating presence of mycobacteria in the sputum with a limited choice as to species (3).

We also reported detection of tuberculostearic acid in a cerebrospinal fluid specimen from a patient with tuberculous meningitis (4). The sample was subjected to acid methanolysis, purified by TLC and analyzed by capillary column GC and CI-SIM (Fig. 2). The direct demonstration of specific mycobacterial fatty acids in CSF fluid specimens implies a possibility of rapidly diagnosing this life-threatening disease.

In order to further increase the sensitivity and selectivity of the SIM technique in analyses of microbial fatty acids, alternative derivatives and MS detection modes have recently been tested. Analysis of negative ions (NI) of compounds possessing high electron affinity has proved particularly useful for this purpose leading to formation of abundant NI in the ion source. NIMS

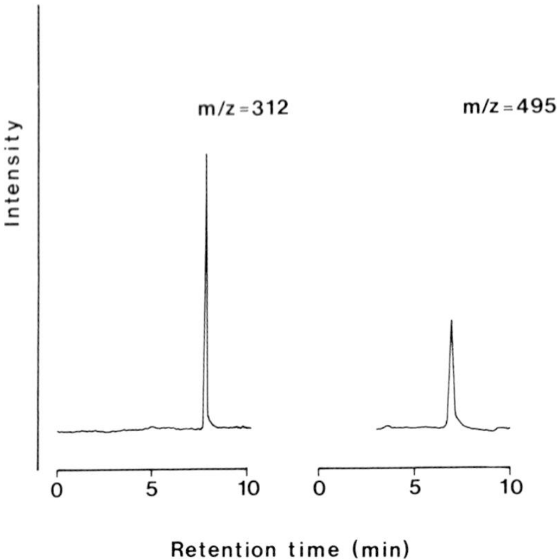

Fig. 1. Mass fragmentograms of a five day incubated culture of a tuberculous sputum specimen demonstrating methyl esters of tuberculostearic acid, EI *m/z* 312 (left) and C_{32} mycocerosic acid, CI isobutane *m/z* 495 (right). From ref. 3 with permission.

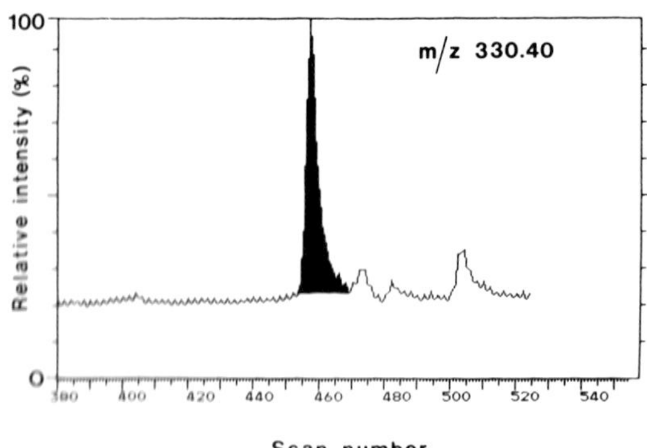

Fig. 2. Detection of methyl ester of tuberculostearic acid using SIM monitoring at *m/z* 330 (CI, ammonia) in a cerebrospinal fluid specimen from a patient with tuberculous meningitis. From ref. 4 with permission.

analysis of pentafluorobenzyl (PFB) esters probably represents the ultimate sensitivity and selectivity available today providing possibilities to determine lipid markers in the femtogram range (5). In case of tuberculostearic acid this sensitivity limit would correspond to approximately one hundred mycobacterial cells. Fig. 3 illustrates an example where tuberculostearic acid was demonstrated in a lymphatic gland of a patient with acquired immune deficiency syndrome (AIDS) and disseminated mycobacterial infection (6). The analyses were performed monitoring at *m/z* 297 representing the negative carboxylate ions (formed by CI) of the PFB ester. The observed peak, appearing at a retention time identical with that of authentic tuberculostearate, indicates substantial amounts of this acid in the sample. This technique can be very valuable in cases when alternative methods for diagnosis of mycobacterial infections in connection with AIDS are unsuitable due to the health risks when handling specimens suspected to be infected with HTLV III virus in the laboratory.

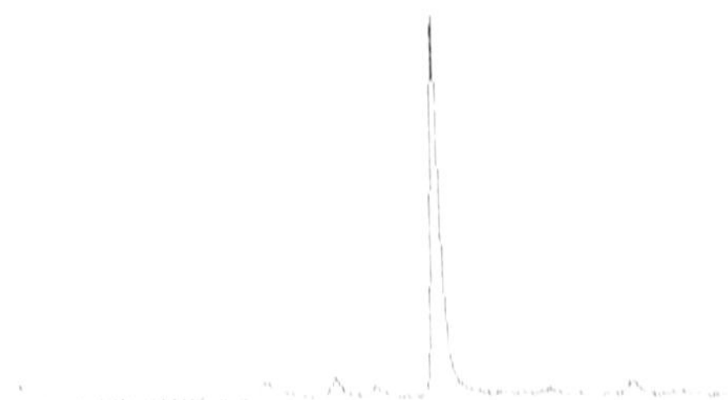

Fig. 3. Detection of tuberculostearic acid as PFB derivative using NIMS-SIM (monitoring at *m/z* 297) in a lymphatic gland of a patient with AIDS and mycobacterial infection. From ref. 6 with permission.

Data from the investigations of clinical specimens are continuously being accumulated (2 - 10). NI-SIM appears to be a particularly promising approach and in the near future, comprehensive evaluations of this technique´s diagnostic value will be presented.

SPECIES IDENTIFICATION OF MYCOBACTERIA

Use of GC analysis for mycobacterial characterisation was reported as early as in the 1960s. Thus, chromatograms representing pyrolysis fragments of whole mycobacterial cells were found characteristic according to the various strains and species analyzed (11 - 13). In subsequent studies using improved methodology (high resolution columns, computerized evaluation of the chromatograms etc), pyrolytic GC methods were further applied for microbial differentiation although difficulties with interlaboratory and time-to-time reproducibility have been reported. On-line connection of the pyrolysis unit to a MS has proved useful for mycobacterial differentiation, but as this technique does not involve chromatographic separation it will not be further discussed here. Combined analysis of cellular fatty acids and carbohydrates yields chromatograms of high differential diagnostic capacity. This technique therefore should be useful in mycobacterial taxonomic studies although care must be taken to avoid hydrolysis of the moisture-sensitive carbohydrate derivatives for achieving reproducible results (14 - 16).

Analysis of mycobacterial fatty acids

Cellular fatty acid analysis is a widely employed GC method for characterization of mycobacteria (17 - 34) and appears, as will be shown, to be suitable for use in routine differential diagnosis. The resulting chromatograms are, however, influenced by several parameters as discussed below.

Cultivation conditions. The fatty acid profiles are dependent upon the composition of the culture medium. Löwenstein-Jensen medium, for example, contains large amounts of oleic acid - a component which is known to be readily incorporated in mycobacterial lipids (35). Consequently, elevated amounts of oleic acid are found in bacteria grown on this medium compared with, e.g., Proskauer-Beck and Sauton agar media (31). When using a chemically complex, non-defined medium possibly containing lipid material care must be taken to avoid contamination during harvesting of cells. Best reproducibility of the chromatographic results is achieved when using a chemically defined medium. The incubation time does not seem to influence the profiles significantly provided that mycobacteria are taken from the logarithmic phase of growth (31), i.e. cells are harvested as soon as enough material is available for GC analysis.

Processing. Cellular fatty acids are transferred to methyl esters prior to GC analysis. A very simple technique utilizing direct injection of tetramethylammonium hydroxide digested mycobacteria enabled several species to be distinquished (24). However, other workers found this technique to be unsatisfactory due to poor reproducibility and occasional presence of extraneous, unidentified peaks in the chromatograms (28). For more reliable results, either acid methanolysis (heating in methanolic HCl) or alkaline saponification followed by methylation of the liberated acids(heating in NaOH or KOH in methanol, esterification through heating in BF_3 ot BCl_3 in methanol) is recommended. For convenience, whole mycobacterial cells rather than lipid extracts thereof are subjected to the transesterification or saponification procedures. Autoclaving of the mycobacteria before hydrolysis does not seem to affect the fatty acid profiles. Both saponification and methanolysis will

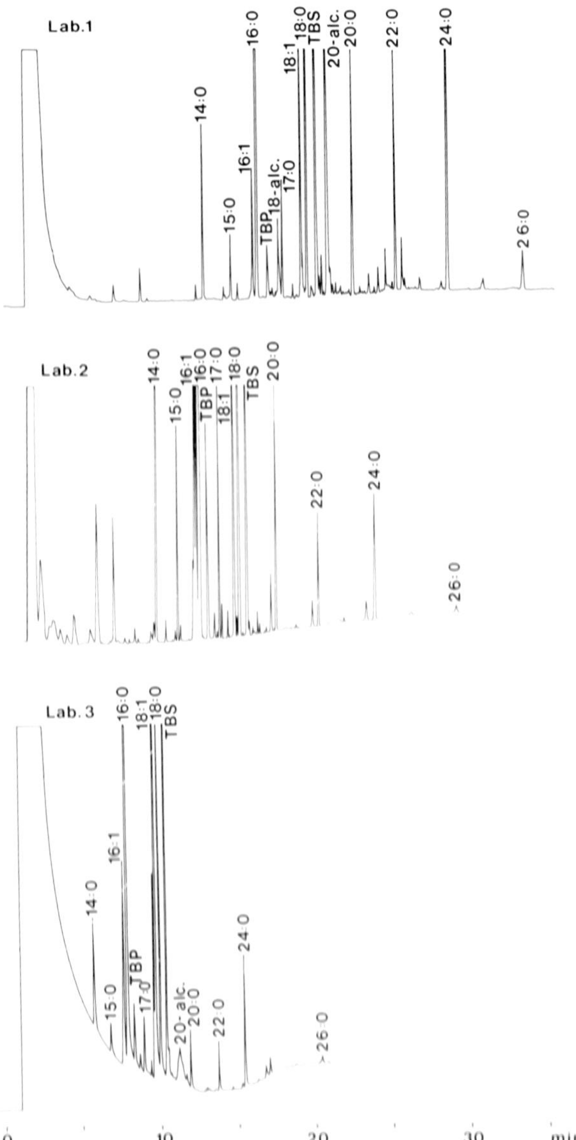

Fig. 4. Fatty acid profiles (methyl ester) representing *Mycobacterium avium* cultivated in Proskauer-Beck medium and processed and analyzed in three laboratories. Designations of the fatty acid peaks refer to number of carbon atoms and double bonds. TBP=10-methylhexadecanoic and TBS=10-methyloctadecanoic acid methyl ester, 18-alc=2-octadecanol, 20-alc=2-eicosanol. From ref. 33 with permission.

release secondary alcohols (mainly 2-octadecanol and 2-eicosanol) from wax-ester mycolates. To obtain reproducible chromatographic separation of these diagnostically important alcohols it is necessary to remove traces of free acids which might be present in the extracts used for GC analysis, for example by washing with an alkaline buffer solution (34).

Gas chromatography. The GC conditions used should be adjusted as to allow separation of methyl esters of cellular fatty acids from 14 to 26 carbon atoms, methyl esters (C 20 - C 26) resulting from pyrolytic degradation of mycolates (provided a high temperature of the injector of the chromatograph - at least 300°C - is used), and the two secondary alcohols described above. Capillary columns provide results compatible with those of packed columns, although their superior performance regarding separating efficiency results in more informative chromatograms and improved reproducibility of peak areas when calculated by electronic integrators (31). In addition, when using splitless injection onto a capillary column the sensitivity is often enhanced because the attenuation of the FID signal is typically lower, perhaps by one order of magnitude, and the peaks narrower and higher. Mycobacteria contain several (e.g., methyl branched) acids which elute very closely; a capillary column coated with a non-polar or medium polar stationary phase is preferred for their separation. If the lipid alcohols are adsorbed in the column, or elute as tailing peaks, they may be analysed after trifluoroacetylation (34). Use of a stationary phase chemically bonded to the capillary column wall is sometimes advantageous because acids and other material accumulated in the column can be removed by flushing the column with various organic solvents. This procedure will sometimes restore peak shape of the alcohols (34). Fig. 4 shows chromatograms representing one reference strain of *M. avium* after cultivation, processing and GC analysis in three different laboratories. This (and other) comparisons between chromatograms from different laboratories indicate possibilities of interlaboratory reproducibility of this GC technique for mycobacterial characterization (33).

SCOPE OF GC AND GC/MS ANALYSIS IN MYCOBACTERIAL DIAGNOSIS

GC and GC/MS have a wide range of application areas in clinical microbiology (for a review, see ref. 36). Clearly, GC analysis of mycobacterial fatty acids is useful in routine species identification because it is rapid, informative and requires minute (less than one mg) of cells. Another advantage is that the principles used for processing and analysis of the mycobacteria may be used for studies of other microorgansims. Thus, several clinical microbiological laboratories utilize fatty acid profiling in routine identification of e.g. non-fermentative bacteria such as *Alcaligenes*, *Legionella*, *Pseudomonas* etc (36).

GC/MS is by tradition considered a more sophisticated technique than merely GC and was in the past used only in a very restricted number of laboratories of physics and chemistry. However, along with the increased capacity and decreased costs particularly of the computer moiety of the instruments, intense developments of small, reliable mass spectrometers and the so-called mass selective detectors are in progress. Several such systems designed to solve a wide range of biological/chemical problems are already commercially available and are now rapidly moving into the various types of research laboratories including those of microbiology. The role of quantitative MS in the rapid diagnosis of mycobacterial infections will be further evaluated and reported in the near future.

ACKNOWLEDGEMENTS

Grants from the Swedish National Association against Heart and Chest Diseases are gratefully acknowledged.

REFERENCES

1a. Maller AI, Minnikin DE, Dobson G (1984) Biomed Mass Spec 11:79-86

1b. Larsson L, Odham G (1984) J Microbiol Meth 3:77-82

2. Odham G, Larsson L, Mårdh P-A (1984) In: Odham G, Larsson L, Mårdh P-A (eds) GC/MS Applications in Microbiology. Plenum, New York, pp 303-333

3. Larsson L, Mårdh P-A, Odham G, Westerdahl G (1981) Acta path microbiol scand Sect B 89:245-251

4. Mårdh P-A, Larsson L, Hoiby N, Engbaek HC, Odham G (1983) Lancet 8320: 367

5. Odham G, Tunlid A, Westerdahl G, Larsson L, Guckert JB, White DC (1985) J Microbiol Meth 3:331-344

6. Grubb R, Ursing B, Larsson L, Odham G, Olsson B, Åkerman M (1985) Läkartidningen 82:2349-2351

7. Larsson L, Mårdh P-A, Odham G, Westerdahl G (1980) J Chromatogr Biomed Appl 182:402-408

8. Odham G, Larsson L, Mårdh P-A (1979) J Clin Invest 63:813-819

9. Bergman R, Larsson L, Odham G, Westerdahl G (1983) J Microb Meth 1:19-22

10. Larsson L, Mårdh P-A, Odham G (1979) J Chromatogr Biomed Appl 153:221-223

11. Reiner E, Kubica GP (1969) Am Rev Respir Dis 99:42-49

12. Reiner E, Beam RE, Kubica GP (1969) Am Rev Respir Dis 99:750-759

13. Reiner E, Hicks JJ, Beam RE, David HL (1971) Am Rev Respir Dis 104:656-660

14. Larsson L, Mårdh P-A (1976) J Clin Microbiol 3:81-85

15. Alvin C, Larsson L, Magnusson M, Mårdh P-A, Odham G, Westerdahl G (1983) J Gen Microbiol 129:401-405

16. Larsson L, Bergman R, Mårdh P-A (1979) Acta path microbiol scand Sect B 87:205-209

17. Valero-Guillén PL, Martin-Luengo F (1983) Tubercle 64:283-290

18. Valero-Guillén PL, Pacheco F, Martin-Luengo F (1985) J Appl Microbiol. In press.

19. Saxegaard F, Andersen O, Jantzen E (1983) Acta Vet Scand 24:225-237

20. Andersen O, Jantzen E, Closs O, Harboe M, Saxegaard F, Fodstad F (1982) Ann Microbiol 133B:29-37

21. Julák J, Turecek F, Miková Z (1980) J Chromatogr 190:183-187

22. Daffé M, Laneélle MA, Asselineau C, Lévy-Frébault V, David H (1983) Ann Microbiol 134B:241-256

23. Mayall BC (1985) Pathology 17:24-28

24. Ohashi DK, Wade TJ, Mandle RJ (1977) J Clin Microbiol 6:469-473

25. Chomarat M, Flandrois JP, Viallier J (1981) Zbl Bakt Hyg I Abt Orig C2:21-32

26. Thoen CO, Karlson AG, Ellefsen RD (1971) Appl Microbiol 22:560-563

27. Thoen CO, Karlson AG, Ellefsen RD (1971) Appl Microbiol 21:628-632

28. Tisdall PA, Roberts GD, Anhalt JP (1979) J Clin Microbiol 10:506-514

29. Tisdall PA, DeYoung DR, Roberts GD, Anhalt JP (1982) J Clin Microbiol 16:400-402

30. Guerrant GO, Lambert MA, Moss CW (1981) J Clin Microbiol 13:899-907

31. Larsson L, Valero-Guillén P, Martin-Luengo F, Pacheco F (1985) Acta path microbiol immunol Sect B. In press

32. Larsson L, Draper P, Portaels F (1985) Int J Lepr. In press

33. Larsson L, Jantzen E, Johnsson J (1985) Eur J Clin Microbiol. In press

34. Larsson L (1983) Acta path microbiol immunol scand Sect B 91:235-239

35. Weir MP, Langridge WHR, Walker RW (1972) Am Rev Respir Dis 106:450-457

36. Larsson L, Mårdh P-A, Odham G (1983) In: Lab Management, April, 38-45

A NEW FLUORESCENCE TECHNIQUE OF MAJOR SENSITIVITY IN THE DIAGNOSIS OF TUBERCULOSIS

M. CASAL and J. CLEMENTE

Department of Microbiology, School of Medicine, University of Cordoba, 14004 Cordoba, (Spain)

INTRODUCTION

Today microscopy is a rapid, easy and orientative method, is important in the early diagnosis and treatment of tuberculosis.

There are as many different techniques of fluorescence microscopy (Truant, Blair, Smithwick, etc.) as there are of optic microscopy (Ziehl-Neelsen, Kinyoun, etc.).

In this study we have tried to compare various known microscopic techniques with a new technique developed in our laboratory, by means of conventional parameters (sensitivity, specificity, predictive and concording values).

MATERIAL AND METHODS

We studied 1,045 clinical samples of different origin. These samples came from new cases, relapses of old cases and from patients undergoing medical treatment.

The samples were decontaminated by the following methods: Tacquet and/or Kubica or Lowenstein (1). The samples were spread on the glass slides and when prepared they were fired by heat and dyed using the Ziehl-Neelsen and Kinyoun methods, the "big-drop" method, Smithwick fluorescence and big drop fluorescence. The glass slides were examined with an optic Nikon microscope -031 No.76.205, with an objective of 100x increasing 10x and a Nikon FL No.45.234 fluorescence microscope with a blue ray objective 20x and an increase of 8x.

To characterize the mycobacteria an objective of 40x and an increase of 8x were used. The smears were considered positive if one or more BAAR were found. They were quantified according to the usual procedure.

The samples, once processed, were cultivated in Lowenstein-Jensen and Lowenstein plus pyruvic acid. This culture was evaluated and quantified when positive (1).

From the results the sensitivity, specificity, predictive value and the correlation were studied (2)(8).

RESULTS

The results of this study are shown in tables I, II and III.

TABLE I

MICROSCOPIC VISUALISATION OF 1.045 SAMPLES

Techniques	Real Positives	Real Negatives	False Positives	False Negatives
Ziehl-Neelsen	151	797	9	88
	14.44%	76.26%	0.86%	8.42%
Kinyoun	148	799	7	91
	14.16%	76.45%	0.66%	8.70%
Big Drop	169	796	10	70
	16.17%	76.17%	0.95%	6.69%
Smithwick	185	784	22	54
	17.70%	75.02%	2.10%	5.16%
Fluorescence Big Drop	202	767	39	37
	19.33%	73.39%	3.73%	3.54%

TABLE II

COMPARISON OF TECHNIQUES OF MICROSCOPIC VISUALISATION OF TUBERCULOSIS IN 1.04 SAMPLES

Techniques	Sensitivity (%)	Specificity (%)	Predictive Value (%)	Correlation (%)
Ziehl-Neelsen	63.17	98.88	94.37	62.05
Kinyoun	61.92	99.13	95.48	61.05
Big drop	70.17	98.75	94.41	69.46
Smithwick	77.40	97.27	89.37	74.67
Fluorescence Big drop	84.51	95.16	83.81	79.26

The real positives, real negatives, false positives and false negatives are shown in Table I. These results are from a total of 1,045 samples using the five microscopic techniques mentioned. There were real positives with 14.44% with Ziehl-Neelsen, 14.16% with Kinyoun, 16.17% with the big drop technique, 17.70% with Smithwick and 19.33% with fluorescence big drop; 8.42% false negatives with Ziehl-Neelsen, 8.70% with Kinyoun, 6.69% with the big drop technique, 5.16% with Smithwick and 3.54% with fluorescence big drop. In Table II these techniques are compared with conventional parameters. There was a significant degree of sensitivity of 63.17% with Ziehl-Neelsen, 61.92% with Kinyoun, 70.71% with big drop, 77.40% with Smithwick and 84.51% with fluorescence big drop, and a high degree of specificity with the Kinyoun method (99.13%).

Table III shows the number of false negatives found using these techniques when compared with the number of cultures. It will be seen that the number of false negatives decreases using fluorescence big drop, and in a greater number of colonies in comparison with the Ziehl-Neelsen technique.

TABLE III

COMPARISON OF FALSE NEGATIVES IN VISUALISATION TECHNIQUES WITH THE NUMBER OF CULTURES IN 1.045 TUBERCULOSIS SAMPLES

Techniques	Less than 5 Colonies	5-20 Colonies	20-50 Colonies	50-100 Colonies	100-300 Colonies	Surface almost covered	Surface totally covered
Ziehl-Neelsen false negatives	26	6	15	21	11	9	
Kinyoun false negatives	25	6	17	20	14	9	
Big-Drop false negatives	25	4	13	13	9	6	
Smithwick false negatives	25	3	8	10	6	2	
Fluorescence Big-Drop false negatives	18	2	6	7	4		

DISCUSSION

In the microscopic diagnosis of tuberculosis, the Ziehl-Neelsen technique, when accurately performed and repeated 3 times, is reliable.

The disadvantage of this technique is that the visualisation process is slow. That is why more studies are performed to discover microscopic techniques so as to obtain the same or greater sensitivity and greater rapidity. Studies using the fluorescence microscope have shown a sensitivity of 78% and a specificity of 49.6% (7) and in our study a sensitivity of 77,40% with Smithwick and 84,51% with fluorescence big drop, and a specificity of 97,27% with Smithwick and 95,16% with fluorescence big drop (Table II). Some authors who centrifuged the samples before extending the smears obtained an even greater sensitivity (6). Other authors (4) in a comparative study of the Truant fluorescence microscope using the Kinyoun technique obtained the same results as in our study, showing greater sensitivity with the Truant technique. In a comparison of the fluorescence microscope with the number of cultures (5), similar results to our own were also obtained.

In conclusion, the fluorescence microscope big drop technique shows greater sensitivity and correlation in cultures with more than 300 colonies, i.e. with a surface almost covered, no false negatives were found using this technique in contrast to the other four techniques studied.

REFERENCES

1. Casal, M. (1983) Bacteriologia de la tuberculosis y micobacteriosis 1ª ed. Editorial AC. Madrid.

2. Laven, George T. (1971) Diagnosis of tuberculosis in children using fluorescence microscopic examination of gastric washings. American Review of Respiratory Disease. 115:743-749.

3. Nyboe, J. (1978) Etudes cooperatives des commissions scientifiques. Commissions des methodes diagnostiques. Rapports preliminaire de l´etude sus les lectures multiples d´etalements colores d´expectoration. Bull. Union Internation. Tuberc. 53:111-117.

4. Patrick R. Murray, Ph. D.; Carlene Elmore; and Donald J. Krogstad, M.D. (1980) The acid-fast stain: a specific and predictive test for Mycobacterial disease. Annals of International Medicine. 92:512-513.

5. Pollock, H.M. and Wieman, E.J. (1977) Smear results in the diagnosis of mycobacterioses using blue light fluorescence microscopy. Journal of Clinical Mycrobiology. 5:329-331.

6. Rickman, Thomas W. and Moyer, Nelson P. (1980) Increased sensitivity of acid-fast smears. Journal of Clinical Mycrobiology. 11:618-620.

7. Strumpf, Ira Jeffry.; Tsang, Anna. Y. and Sayre, James W. (1979). Re-evaluation of sputum staining for the diagnosis of Pulmonary Tuberculosis. American Review of Respiratory Disease. 119:599-602.

8. Toman, K. (1980) Tuberculosis deteccion de casos y quimioterapia. 1ª ed. OPS. Mexico.

Mycobacteria of Clinical Interest. M. Casal, editor

DETECTION OF TUBERCULOSIS BACILLI BY FLUORESCENCE MICROSCOPY

ARMANDO ALBERTE
Microbiologia, Hospital de la Seguridad Social, 47010 Valladolid (Spain)

INTRODUCTION

As the World Health Organization has indicated (1), controlling the spread of tuberculosis depends upon the earliest possible detection of new cases. Undoubtedly, such detection involves the field of microbiology.

Both microscopy and cultures of specimens from suspect patients have proven to be of great sensitivity and specificity (2,3) at the time of diagnosing tuberculosis. Microscopy, a simple and easily-managed technique, has been recommended throughout the world as a diagnostic method, as an epidemiologic tool, and for monitoring purposes during the patient's treatment. The technique's usefulness is especially evident in the developing countries, where their massive populations necessitate the most accessible studies possible, to overcome the lack of more complex diagnostic techniques. Various authors (4,5,6) have demonstrated, although with limitations, the reliability of microscopy in the examination of specimens with variable quantities of acid-fast bacilli.

On the other hand, there are clinical situations, such as in tuberculous meningitis, where anti-tuberculosis treatment must be started without waiting for the slow growth of the mycobacterium tuberculosis. In these cases microscopic observation, in conjunction with clinical and radiological data, plays an important role. Recently, automized culture methods have lessened the time-lag problem, allowing a more rapid diagnosis, although many centers cannot afford them (7).

Today numerous laboratories that have to examine a large number of specimens each day routinely employ fluorescence microscopy, together with the Ziehl-Neelsen stain, precisely because of the advantages already described (8,9). However, in the past few years there has been a great deal of controversy about the usefulness of acid-fast stains. For this reason, the purpose of this paper is to report the performance of acid-fast stains as compared to cultures of Koch's bacillus in a general hospital with 600 beds, based on 8,931 specimens processed in the past 5 years.

MATERIAL AND METHODS

Specimens

All specimens received for detection of Koch's bacillus between January of 1980 and December 1984 were studied, mostly samples of bronchopulmonary origin, followed by genitourinary, pleural, adenopathic, osteo-articular, LCR, etc. Only specimens which were cultured were counted, those contaminated being excluded. The specimens were from 3 types of tuberculosis cases: new, reactivated and under treatment. All strains were mycobacterium tuberculosis; other species without clinical significance were excluded.

Digestion and Concentration of Sputum

Specimens were digested and decontaminated by the lauryl sulfate sodium method (Tacquet and Tison) (10).

Microscopy

Smear concentrates were fixed on a hot plate at 50°C for one hour and stained with auramine O. The slides were examined with a Zeiss 143 standard using a 25x or 40x objective and a HBO 50 W blue light source. The ocular magnification was 10x. Interpretation and reports of the results were made in accordance with already-established methods (9). The smear was considered positive if three or more acid-fast bacilli were seen per smear.

Culture

Two Lowenstein-Jensens, one containing pyruvate, were inoculated with concentrated sediments and incubated at 35°C and observed for growth over an eight-week period.

Identification

Acid-fast organisms were identified by the methods already described (9).

Paired Specimens

We used methods described by Kubica.

RESULTS

Table I shows the results obtained in our laboratory when smears were compared with cultures. Of the 626 culture-positive specimens, 282 were detected by fluorescence staining (45.04%); and of 8,305 culture-negative specimens, 116 acid-fast smears were falsely positive (1.39%).

As can be seen in Table II, other results were as follows: 98.60% specificity, 70.85% true positives, 95.96% true negative, 29.14% relative false positive, 4.03% relative false negative and 54.95% absolute false negative. Overall correlation was 94.84%.

A summary of our findings as compared to those of previous investigations is presented in Table III.

DISCUSSION

These results must be considered in the context of a general community hospital which receives specimens for microbacteriological study from out-patient and hospital clinics, both for screening purposes and for confirmation of clinical suspects. This means that of every 100 specimens processed, only 8.3 show a positive result in either smear or culture. That is to say, 1 of every 18 specimens registers some type of positiveness. The study also includes

TABLE I
COMPARISON OF ACID-FAST SMEARS WITH CULTURES

	Smear Result		
Culture Result	# Positive	# Negative	Total Specimens
Positive	282	344	626
Negative	116	8,189	8,305
Total	398	8,533	8,931

TABLE II
ANALYSIS OF SMEAR-CULTURE COMPARISON

Factor	%
Sensitivity	45.04
True positive	70.85
Specificity	98.60
True negative	95.96
Relative false-positive	29.14
Absolute false-positive	1.39
Relative false-negative	4.03
Absolute false-negative	54.95
Overall correlation	94.84

out-patients undergoing therapy, who were given control examinations every 2 months until the 15th month.

With respect to the results, approximately one half of the culture-positive specimens also gave a smear-positive reading, which indicates a sensitivity of 45.04% and a specificity of 98.59%.

Obviously our percentage of false smear-positive readings (29.14%) can have different causes, but we believe that the follow-up control in our hospital of patients with therapy and isoniazid, together with the utilization of fluorescence microscopy, is the principal one.

TABLE III

COMPARATIVE ANALYSIS OF PAIRED ACID-FAST SMEARS AND CULTURES

Note that the columns of investigators' results are arranged in chronological order by publication dates, from earliest to latest, reading from left to right. See Reference Section for details.

	% Obtained by Investigator								
Analysis Factor	Boyd & Marr	Marraro et al	Burdash et al	Pollock & Wieman	Murray et al	Rickman* et al	Urbanczik Lab 1	Urbanczik Lab 2	Present Report
Sensitivity	22	24.1	42.7	49.7	38.91	82.4	37	54	45.04
True positive	44.8	58.3	93.3	92.5	95.18	80.7	-	-	70.85
Specificity	99.3	99.5	99.9	99.8	99.9	97.4	-	-	98.6
True negative	97.96	97.8	97.9	97.1	97.59	97.7	-	-	95.96
Relative false-positive	55.2	41.7	6.7	7.5	4.8	19.3	-	-	29.14
Absolute false-positive	0.7	0.5	0.1	0.2	0.079	2.6	-	-	1.39
Relative false-negative	2.02	2.2	2.1	2.9	2.40	2.3	-	-	4.03
Absolute false-negative	78	75.9	57.3	50.3	61.08	17.6	-	-	54.95
Overall correlation	97.3	97.3	97.8	98.3	97.63	95.7	-	-	94.84

*RCF 3800 x g

Taking into account the references reviewed, our sensitivity result can be compared to those achieved by Burdash et al (11) and Pollock and Wieman (12), with 42.7% and 49.7% respectively. The result obtained by Rickman et al (13) stands out for its sensitivity of 82.4%. This can be explained by their use of an optimal relative centrifugal force (RCF) of 3800 x g, which increased the sensitivity of the acid-fast smears. With a 54% sensitivity, Urbanczik's figures for his Laboratory 2 (14) come closest to theirs.

On the other hand, our relative false positive rate of 29.14% was exceeded only by Boyd and Marr (15) and Marraro (16), with 55.2% and 41.7% respectively. Although our result is high for this item, our earlier comments explain its possible causes. Likewise, as the presence of only three bacilli was sufficient to consider a smear positive in our study, smear-positive and culture-negative results were accumulated for patients under treatment.

Finally, fluorescence microscopy has the advantages of rapidity and high sensitivity when utilized by trained staff. Its use affords fast diagnoses, facilitating early treatment of new tuberculosis cases and thus, ultimately, a reduction of the disease itself.

TABLES

See the following two pages for Tables I, II and III.

REFERENCES

1. OMS, Serie de Informes Técnicos, No. 552, 1974 (Noveno Informe del Comité de Expertos de la OMS en Tuberculosis, págs. 15 y 26)
2. Nagpaul, D R et al Bulletin of the World Health Organization 43:17, 1970
3. Nagpaul, D R et al Proceedings of the 9th Eastern Region Tuberculosis Conference and 29th National Conference on Tuberculosis and Chest Diseases, Delhi, November 1974. Delhi, Indian Tuberculosis Association/International Union against Tuberculosis, 1975
4. David, H L Bacteriology of mycobacterioses. Atlanta, US Department of Health Service, Communicable Disease Center, 1976
5. Carvalho, A Zeitschrift für Tuberkulose und Erkran-Kungen der Thoraxorgane 63:305, 1932
6. Cruickshank, D B In: Sellors, T H and Livingston, J L ed. Modern practice of tuberculosis. London, Butterworth, 1952
7. Middlebrook, G, Reggiardo, Z and Tigerit, W D, 1977. Automatable radiometric detection of growth of mycobacterium tuberculosis in selective media. Am Rev Respir Dis 115:1066-1069
8. Meyer, L and David, H, 1980. Mycobacteriologie en fanté Publique Institut Pasteur, Paris
9. CommTech 16. Laboratory Diagnosis of the mycobacterioses. H M Sommers and J K McClatchy. Coordinating Editor: J A Morello. American Society for Microbiology. Washington, DC. March 1983

10. Tacquet, A and Tison, F, 1961. Nouvelle technique d'isolement des mycobactéries par le lauryl sulfate de sodium. Ann Inst Pasteur, 100, 676-680

11. Burdash, M M, Manes, J P, Ross, D and Bannister, E R, 1976. Evaluation of the acid-fast smear. J Clin Microbiol 4:190-191

12. Pollock, H M and Wieman, E J, 1977. Smear results in the diagnosis of mycobacterioses using blue light fluorescence microscopy. J Clin Microbiol 5:329-331

13. Rickman, T W and Moyer, W P, 1980. Increased sensitivity of acid-fast smears. J Clin Microbiol 11:618-620

14. Urbanczik, R, 1983. Present position of microscopy and culture in diagnostic mycobacteriology. Paper of Symposium "The TB Challenge--Where Do We Stand Today?", Brussels.

15. Boyd, J C and Marr, J J, 1975. Decreasing reliability of acid-fast smear techniques for detection of tuberculosis. Ann Intern Med 82:489-492

16. Marraro, R V, Rogers, E M and Roberts, T H, 1975. The acid-fast smear: fact or fiction? J Am Med Technol 37:277-279

17. Murray, P R, Elmore, C and Krogstad, D J, 1980. The acid-fast stain: a specific and predictive test for mycobacterial disease. J Clin Microbiol 92:512-513

TAXONOMY

TWO NEW POSSIBLE SENSITIVITY TESTS FOR QUINOLINES TO DIFFERENTIATE *M.FORTUITUM* FROM *M.CHELONEI*

M.CASAL, F.RODRIGUEZ, M.C. BENAVENTE and M. RUBIANO
Department of Microbiology, School of Medicine, University of Cordoba (Spain)

INTRODUCTION

Mycobacterium fortuitum and *Mycobacterium chelonei* are two species of the *M.fortuitum* complex at present considered to be pathogenic for man (1). The differential diagnosis between the two species is usually based on two biochemical tests: nitrate reduction (2) and iron uptake (3).

Casal et al (4) and Wallace et al (5) have proposed susceptibility test with pipemidic acid and polymixin respectively for differentiation of *M.fortuitum* from *M.chelonei*. The value of the pipemidic acid test has been confirmed by Levy-Frebault (6).

In this communication we describe a possible new differentiating test based on the susceptibility of each strain to ciprofloxacin and ofloxacin.

MATERIAL AND METHODS

122 strains of *M.fortuitum* and 80 strains of *M.chelonei* were included in the study. The cultures used came from our department and from various international collections: the National Collection of Thype Cultures (NCTC), London, U.K.; the Trudeau Mycobacterial Culture Collection (TMC), Sarnac Lake, N.Y., U.S.A. and the Czechoslovak Collection of Microorganisms (CCM), Brno, Czechoslovakia. Cultures also came from collections of investigators in the field (L. Eidus and A. Laszlo, Laboratory Centre for Disease Control, Ottawa, Ontario, Canada; R. Gordon, Rutgers University, New Brunswick, N.J., U.S.A.; P.A. Jenkins, Tuberculosis Reference Laboratory, Wales, U.K.; J. Viallier Hospital J.Courmont, P.Benite, Lyon, France; I.Tarnok, Forschungsinstitut, Borstel, Germany; H.David, Institut Pasteur, Paris, France; G.Sabater, Hospital Militar, Valencia, Spain; H.Saito, Shimane Medical College, Isumo, Shimane, Japan; E.Mankiewicz, Montreal Centre, Quebec, Canada.) and from our own laboratory.

M.fortuitum ATCC 6841 and *E.coli* ATCC 25922 were used as controls.

Mycobacteria for susceptibility testing were grown for 7 days at 28°C. on Dubos oleic agar base (Difco). Aqueous suspensions of the cultures were prepared and diluted with distilled water to a final concentration of about 10^7 CFU/ml. In each experiment, the initial concentration of organisms was determined by titration and plating in duplicate.

Agar dilution testing was performed with Mueller-Hinton agar (Difco). After autoclaving, the agar was cooled to 56°C. before the addition of quinolines to final concentrations that ranged from 0.25 to 128 μg/ml. The agar containing quinolines was poured into plates and allowed to solidify overnight. The plates were inoculated with 0.001 ml per spot with a Steers replicator.

Plates inoculated with *E.coli* and *M.fortuitum* were incubated at 37°C. and examined after 24 and 72 h. respectively. Plates inoculated with *M.chelonei* were examined after 72 h. of incubation at 28°C. The MIC was considered as the lowest concentration that completely inhibited visible bacterial growth.

RESULTS

The results are shown in Table I. All strains of *M.chelonei* were resistant to ciprofloxacin and ofloxacin. In contrast, all strains of *M.fortuitum* were susceptible with a CMI of 5 μg/ml. for ciprofloxacin and ofloxacin.

The results of this study show that ciprofloxacin and ofloxacin have the additional advantage of helping to differentiate *M.fortuitum* from *M.chelonei*. The susceptibility to these agents is markedly different in these two species. Using a CMI of 1 μg/ml. of ciprofloxacin and ofloxacin, *M.fortuitum* complex strains that are sensitive can be presumed to be *M.fortuitum* while resistant strains are likely to be *M.chelonei*. Differentiation of the two species is important because of the increasing incidence of the *M.fortuitum* complex in human infections (7)(8).

We believe that it is a valuable test that can be used in clinical microbiology laboratories that handle mycobacteria.

TABLE I

Activity of Norfloxacin and Ofloxacin against the *M.fortuitum* Complex

Antimicrobials	Species studied	No. of souches	C.M.I. μg/ml								
			0,25	0,5	1	2	4	8	16	32	64
Ciprofloxacin	*M.fortuitum*	122	104(85)	122(100)							
	M.chelonei	80				25(31)	31(39)	50(62)	74(92)	80(100)	
Ofloxacin	*M.fortuitum*	122	98(80)		122(100)						
	M.chelonei	80						12(15)	25(31)	80(100)	

ACKNOWLEDGEMENTS

We thank the investigators who provided the cultures used in the present study. We thank Bayer S.A. and Hoechst S.A. for providing ciprofloxacin and ofloxacin.

REFERENCES

1. Kubica GP,I Baes,RE Gordon,PA Jenkins,JBG Kwapinski,C McDurmont,SR Pattyn, H Saito,V Silicox,JL Stanford,K Takeya and M Tsukamura. (1972) A cooperative numerical analysis of rapidly growing mycobacteria. J. Clin. Microbiol. 73:55-70
2. Virtanen S (1960). A study of nitrate reduction by mycobacteria. Acta Tyberc.Scand.Suppl. 48:1-119
3. Wayne LG and JR Doubek (1968). Diagnostic key to mycobacteria encountered in clinical laboratories. Appl. Microbiol. 16:925-931
4. Casal M and FC Rodriguez (1981). Simple new test for rapid differentiation of the Mycobacterium fortuitum complex. 13:989-990
5. Richard J Walace JR,Jana M Swenson,Vella A Silicox and Robert C Good.(1982) Disk diffusion testing with Polymixin and Amikacin for differentiation of **Mycobacterium fortuitum** and **Mycobacterium cheloney**. Journal of Clinical Microbiology. 16.1003-1006
6. Veronique Levy-Frebault,Elie Rafidinarivo,Jean-Claude Prome,Jeannine Grandry,Henri Boisvert and Hugo L David.(1983). Mycobacterium fallax sp. nov. International Journal of Systematic Bacteriology.33:336-343
7. Grange JM (1980).Mycobacterial diseases.Edward Arnold,London.
8. Wolinsky E.(1979) Non tuberculous mycobacteria and associated diseases. Am. Rev. Respir. Dis. 119:107-159

Mycobacteria of Clinical Interest. M. Casal, editor

ENZYME CHARACTERIZATION OF MYCOBACTERIAL IMMUNOPRECIPITATES

MALIN RIDELL, RONNY ÖHMAN, GUN WALLERSTRÖM, VIVIANNE SUNDAEUS

Department of Medical Microbiology, University of Göteborg, Guldhedsgatan 10, S-413 46 Göteborg (Sweden)

INTRODUCTION

The antigenic composition of mycobacteria has been analysed by immunoelectrophoresis and immunodiffusion in many investigations (1, 2). Comparatively little is, however, known concerning the chemical structures of the revealed precipitinogens. Öhman and Ridell therefore initiated the use of selective enzyme staining procedures for characterization of mycobacterial antigens (3). In a recent study precipitation patterns of *Mycobacterium intracellulare* and *M. phlei* obtained by two-dimensional immunoelectrophoresis (2D-IE) were analysed (3). It was shown that one single precipitate of each pattern correspond to the enzyme malate dehydrogenase.

Antigen preparations from strains of the species *M. avium*, *M. bovis* BCG, *M. kansasii* and *M. marinum* were analysed by tandem 2D-IE using the *M. intracellulare* and *M. phlei* reactants as reference systems. All these four test strains also formed precipitates which were stained by the malate dehydrogenase staining procedure (3).

The precipitation patterns of *M. intracellulare* and *M. phlei* were furthermore analysed concerning the enzymes glutamic oxaloacetic transaminase, leucine aminopeptidase and glucose phosphate isomerase. All three of these enzymes were revealed in *M. phlei*, while only leucine aminopeptidase was revealed in *M. intracellulare* (3).

In the present study mycobacterial precipitinogens were analysed concerning isocitrate dehydrogenase and leucine aminopeptidase activity.

MATERIALS AND METHODS

Bacterial strains. *M. intracellulare* Battey-Boone TMC 1403, *M. phlei* NCTC 8151/ATCC 19249, *M. smegmatis* NCTC 8159/ATCC 19240.

Antigen preparations. The mycobacterial strains were cultivated, harvested and sonicated as previously described (3).

Antisera. The serum against *M. intracellulare* was produced in burros while the sera against *M. smegmatis* and *M. phlei* were produced in rabbits (3).

Immunological technique. A modification of the 2D-IE technique was employed. This modification has been developed by Samuelson (4) and was also used in the previous study (3).

Enzyme staining procedure. The 2D-IE plates were incubated at 37°C for about

1 h in the following mixtures. For isocitrate dehydrogenase: 40 ml 0.2 M Tris HCl (pH8.0), 0.3 ml 0.25 M $MgCl_2$, 4 ml 0.1 M isocitric acid, 15 mg NADP, 4 mg dimethylthiazolyl-diphenyltetrazolium and 5 mg phenazine methosulphate. For leucine aminopeptidase: 50 ml 0.1 M K_2HPO_4 (pH5.5), 1 ml 0.1 M $MgCl_2$, 30 mg L-leucine-β-napthylamide-HCl and 30 mg Fast Black K salt.

RESULTS AND DISCUSSION

The precipitation pattern obtained by 2D-IE of *M. intracellulare* consisted of at least 60 precipitation bands, the one of *M. phlei* of about 40 bands and the one of *M. smegmatis* of about 35 bands.
When the isocitrate dehydrogenase staining procedure was applied one precipitate in each pattern of *M. phlei* and *M. smegmatis* was coloured while all the other precipitates of these two organisms remained uncoloured. None of the *M. intracellulare* precipitates reacted, however, at the staining procedure for isocitrate dehydrogenase.

The *M. smegmatis* precipitation pattern was subjected to the staining procedure for leucine aminopeptidase and one single precipitation band reacted being coloured. This enzyme has previously been shown to correspond to one precipitate of the *M. intracellulare* precipitation pattern and to one precipitate of the *M. phlei* pattern.

TABLE 1

ENZYME ACTIVITY OF MYCOBACTERIAL ANTIGENS REVEALED BY SELECTIVE STAINING REACTIONS OF 2D-IE PRECIPITATION PATTERNS

	Mycobacterial precipitation system		
Enzyme	*M. intracellulare*	*M. phlei*	*M. smegmatis*
Malate dehydrogenase	+	+	–
Isocitrate dehydrogenase	–	+	+
Leucine aminopeptidase	+	+	+

Table 1 summarizes the results of the present and previous analyses (3) concerning malate dehydrogenase, isocitrate dehydrogenase and leucine aminopeptidase activity of precipitinogens from *M. intracellulare*, *M. phlei* and *M. smegmatis*.
In Fig 1 the precipitate corresponding to malate dehydrogenase is seen in the *M. phlei* precipitation pattern.

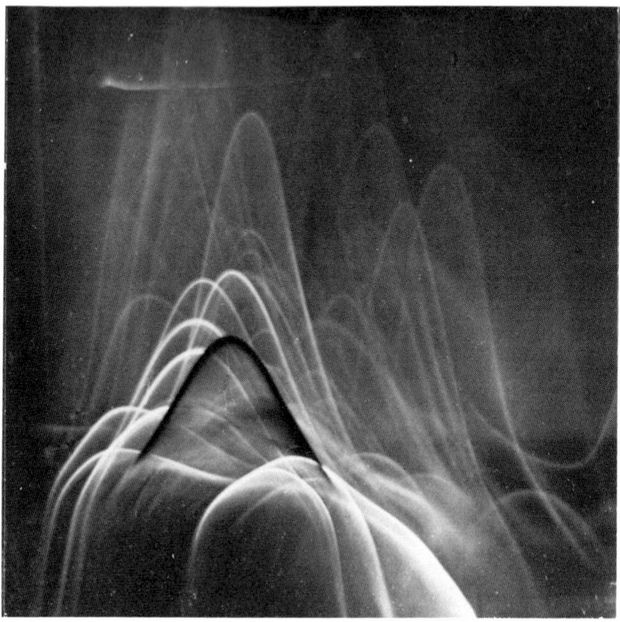

Fig 1. Precipitation pattern of *M. phlei* supplemented by selective staining for malate dehydrogenase.

The results of the present study demonstrate that selective enzyme staining procedures are useful in characterizing precipitates of multilinear precipitation patterns of mycobacterial origin. Such characterization is important in many fields e.g. for the possible development of serodiagnostic methods for mycobacterial diseases.

ACKNOWLEDGEMENTS

This investigation was supported by the Swedish National Association against Heart and Chest Diseases and the Ellen, Walter and Lennart Hesselman Foundation.

The authors would like to thank Dr M.F. Thorel, Laboratoire de Recherche Vétérinaire, Maisons-Alfort, France, for providing the *M. intracellulare* antigen preparation and Dr S. Chaparas, Mycobacterial and Fungal Branch, Food and Drug Administration, Bethesda, Maryland, USA, for providing the *M. intracellulare* antiserum (these reagents were produced on the initiative of the International Working Group on Mycobacterial Taxonomy).

REFERENCES

1. Daniel TM (1984) Soluble mycobacterial antigens In: Kubica GP, Wayne LG (eds) The Mycobacteria. Marcel Dekker Inc, New York, pp 417-465

2. Lind A, Ridell M (1984) Mycobacterial species: immunological classification: immunodiffusion and immunoelectrophoresis In: Kubica GP, Wayne LG (eds) The Mycobacteria. Marcel Dekker Inc, New York, pp 67-82

3. Öhman R, Ridell M (1985) Selective enzyme staining procedures for characterization of mycobacterial immunoprecipitates. Int Archs Allergy appl Immunol (in press)

4. Samuelson L (1985) A thin gel layer technique for improved resolution in two-dimensional immunoelectrophoresis. Int Archs Allergy appl Immunol (in press)

Mycobacteria of Clinical Interest. M. Casal, editor

DIFFERENTIAL MYCOLIC ACID PATTERNS OF RAPIDLY GROWING NONCHROMOGENIC MYCOBACTERIA

V. AUSINA, M. LUQUIN, L. MARGARIT and G. PRATS
Departamento de Microbiologia, Hospital de la Sta, Creu i Sant Pau. Facultad de Medicina de la Universidad Autonoma de Barcelona (Spain)

INTRODUCTION

Rapidly growing group IV mycobacteria are now more frequently recognized as a cause of infection than in previous years (1). Recent studies have shown that differential tests are not always satisfactory in characterizing individual isolates (2-4). Further studies are required to evaluate differential tests and thin-layer chromatography (TLC) analysis of complex lipids.

MATERIAL AND METHODS

Type strains of rapidly growing nonchromogenic species of mycobacteria including some new species(*M.fortuitum* ATCC 6841, *M.chelonae* NCTC 946, *M.smegmatis* ATCC 19420, *M.chitae* ATCC 19627, *M.agri* ATCC 27406,'*M.fallax*' CIP 8139 and '*M.porcinum*' ATCC 33776)and 93 isolates from clinical specimens, water and soil samples were studied.

Acid methanolysates of dried bacteria were prepared as described by Minnikin et al. (5). Alkaline hydrolysis of bacteria and extraction of lipids were performed as described by Levy-Frebault et al. (4); ether extracts were filtered, concentrated by evaporation and then methylated with diazomethane. For laboratory preparation of diazomethane from N-methyl-N-nitroso-p-toluene sulfonamide (Diazald, Merck), all safety precautions necessary were strictly observed.

Analytical two-dimensional TLC was done with glass plates (20x20 cm) coated with layers (0.25 mm) of silica gel (Kieselgel 60F254; Merck). A triple development with petroleum ether (b.p. 60-80°C)/acetone (95:5, v/v) in the first direction was followed in the second direction by a single development with toluene/acetone (97:3, v/v) as described by Minnikin et al. (5). The presence of separate components was revealed by spraying with 10% (w/v) molybdophosphoric acid in ethanol followed by charring at 120°C for 15 minutes. In addition, 57 differential tests were studied. The selection of differential tests was done as described in previous studies (2,3,4,6-9).

RESULTS

All strains in this study formed α-mycolates and were devoid of ketomycolates. Patterns composed of α-, α'- and epoxymycolates were found in reference strains of *Mycobacterium fortuitum*,'*M.peregrinum*', *M.smegmatis*, *M.chitae* and

'*M.porcinum*'. The presence of epoxymycolates was confirmed by degradation by acid methanolysis of the characteristic pairs of polar mycolates. The other species had distinct and characteristic patterns of mycolates: α- and α'-mycolates, *M.chelonae;* α-mycolates only, '*M.fallax*'; and α-, α'- and methoxymycolates, *M.agri* (Figure 1).

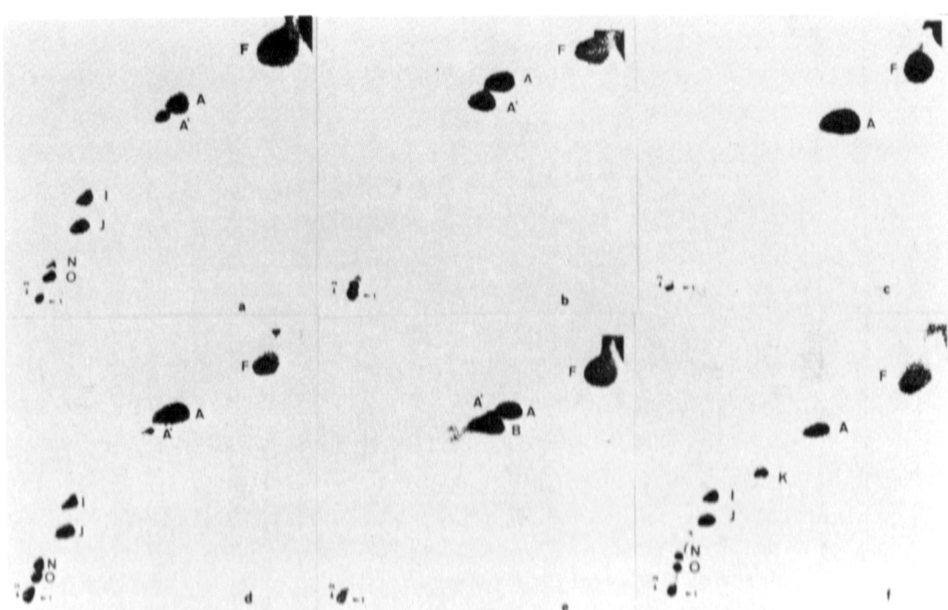

Figure 1. Two-dimensional TLC of acid methanolysates of a)*M.fortuitum*, b)*M.chelonae*, c)'*M.fallax*',d) *M.chitae*, e)*M.agri*, f) '*M.porcinum*'.

Abbreviations (Figure 1 and 2)

F: non-hydroxylated fatty acid methyl esters; A and A': α- and α'-mycolates; B: methoxymycolate; C: kethomycolate; D: ω-carboxymycolate; E: alcohols homologous with 2-eicosanol; I, J, N, O : components derived from epoxymycolates by acid methanolysis; X: unknown component (mycolic acid?).

Isolates of *M.fortuitum* (28 strains), *M.chelonae* (16 strains) and '*M.fallax*' (9 strains)were clearly distinguished by biochemical differential key tests and by analysis of their respective mycolic acids. The remaining 40 strains were distinct from the above and formed three distinct groups: very homogeneous in morphological growth, biochemical characteristics and susceptibility to antituberculous drugs and other antimicrobials. The isolates

TABLE 1

KEY TESTS FOR DIFFERENTIATING MEMBERS OF THE M. FORTUITUM COMPLEX, M. FALLAX AND GROUPS 1, 2 AND 3.

Group	1	2	3	4	5	6	7	8	9	10	11	12	13	14	15
M. fortuitum															
biovar. *fortuitum*	+	+	+	+	+	+	-*	-	-	+	+	-	+	+	+*
biovar. *peregrinum*	+	+	+	+	+	+	-*	+	-	+	+	+*	+	+	+*
biovar. 3	+	+	+	+	+	+	-	+	+	+	+	+	+	+	
M. chelonae															
subsp. *chelonae*	+	+	+	-	-	-	+	-	-	+	-	-	-	+	-
subsp. *abscessus*	+	+	+	-	-	+	-	-	-	+	-	-	-	+	-
M. fallax	-	+*	-	+	-	-	-	-	-	+	+	-	-	-	+*
Group 1	+*	+	+*	-*	-	-	-*	+	-	+	+	+	-	-	±
Group 2**	+*	+	+	-*	-	-*	-	-	-	-	-	-	-	-	-
Group 3	+	-	-	+	-	-	-	-	-	+	+	-	-	-	+

1. Arylsulfatase (3 days). 2. Growth in the presence of: NH_2OH (500 μg/ml).
3. Growth on MacConkey agar without crystal violet. 4. Nitrate reductase. 5. Iron uptake, 6. Growth in the presence of 5% NaCl. 7 to 9. Use as sole carbon source: sodium citrate, mannitol, inositol, respectively. 10 to 13. Acid from glucose, fructorse, mannitol and inositol, respectively.
14. Growth in presence of ethambutol (2 μg/ml). 15. Inhibition by pipemidic acid (20 μg disk).

* Ocassional variations.

**Moderate growth rate (7 days at 30ºC and 37ºC).

which are designated as group 1 (24 strains) and group 2 (11 strains) had α-, and ω-carboxymycolates and long-chain alcohols, homologous with 2-eicosanol. The isolates designated as group 3 (5 strains) had α-, α'-mycolates and a major uncharacterized acid-stable component which may have chemotaxonomic value.

Differential reactions of 93 isolates are compared in Table 1.

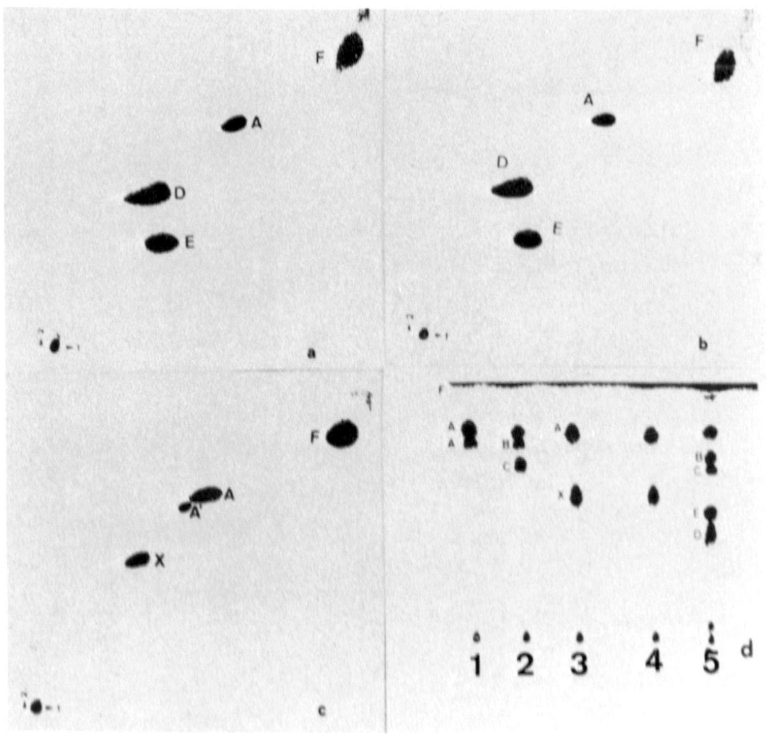

Figure 2. a, b and c) Two-dimensional TLC of acid methanolysates of group 1, 2 and 3 strains. d) Unidimensional TLC of acid methanolysates of 1. *M.chelonae*, 2. *M.kansasii*, 3 and 4. Strains of group 3. 5. *M.intracellulare*.

CONCLUSIONS

The results reported in the present paper reinforce the value of mycolic acid analysis in mycobacterial systematics.

Representatives of rapidly growing nonchromogenic mycobacteria show the presence of six characteristic patterns of mycolic acid methyl esters.

The isolates which are designated as groups 1, 2 and 3 are different from other rapidly growing nonchromogenic mycobacteria in biochemical properties, mycolic acid pattern and susceptibility to several antimicrobial agents.

The isolates which are designated as group 3 show a major acid-stable component which could correspond to a new type of mycolic acid.

REFERENCES

1. Wallace, J.R.Jr., Swenson, J.M., Silcox, V.A., Good, R.C., Tschen, J.A., Stone, M.S. (1983). Rev. Infect. Dis. 5: 657-679.
2. Kubica, G.P. et al. (1972). J. Gen. Microbiol. 73: 55-70.
3. Silcox, V.A., Good, R.C., Floyd, M.M. (1981). J. Clin.Microbiol. 14: 686-691.
4. Lévy-Frébault, V., Daffé, M., Sen Goh, K., Lenéelle, M.A., Asselineau, C., David, H.L. (1983). J. Clin. Microbiol. 17: 744-752.
5. Minnikin, D.E., Hutchinson, I.G., Caldicott, A.B. and Goodfellow, M. (1980). J. Chromatogr. 188: 221-233.
6. Tsukamura, M. (1967). Tubercle. 48: 311-338.
7. Saito et al. (1977). Int. J. Syst. Bacteriol. 27: 75-85.
8. Casal, M.J., Rodríguez, F.C. (1981). J. Clin. Microbiol. 13: 989-990.
9. Wallace, R.J.Jr., Swenson, J.M., Silcox, V.A., Good, R.C. (1982). J. Clin. Microbiol. 16: 1003-1006.

Mycobacteria of Clinical Interest. M. Casal, editor

LIPID PROFILES OF MEMBERS OF THE *MYCOBACTERIUM TUBERCULOSIS* COMPLEX

D.E. MINNIKIN[1], J.H. PARLETT[1,2], G. DOBSON[1,2], M. GOODFELLOW[2], M. MAGNUSSON[3] AND M. RIDELL[4]

Departments of [1]Organic Chemistry and [2]Microbiology, The University, Newcastle upon Tyne, (U.K.). [3]Tuberculin Department, Statens Seruminstitut, Amager Boulevard 80, Copenhagen (Denmark). [4]Department of Medical Microbiology, University of Gothenburg, Guldhedsgatan 10, Gothenburg (Sweden)

INTRODUCTION

The species *M. tuberculosis, M. bovis, M. africanum* and *M.microti* are genetically very similar and together they comprise the 'tuberculosis complex' (1). The genetic similarity has raised the question whether they should remain as four separate species or be considered as subspecies of *M. tuberculosis*. Phenetic differences are apparent, however, in the clinical manifestations, biochemical activity and chemical composition. Data on the lipid composition indicates a discontinuous distribution of lipid types between species and also among variants within the species *M. tuberculosis* and *M. bovis* (2-4). In the present communication, lipid profiles of representatives of the *M. tuberculosis* complex will be compared.

MATERIALS AND METHODS

The strains investigated are shown in Table I. Cultivation conditions have been described previously for *M. microti* (5), *M. tuberculosis* H37Ra and H37Rv (3) and the MNC (Statens Seruminstitut) strains (3,6). Free lipids were extracted by a two stage procedure (2) designed to produce separate non-polar and polar fractions. Each of these fractions was then taken up in a known amount of solvent and aliquots applied to thin-layer chromatographic plates which were analysed with five developing systems (2) (see Fig. 1). Defatted cells were subjected to alkaline hydrolysis followed by phase-transfer catalysed esterification with iodomethane, the resulting long-chain compounds being analysed by thin-layer chromatography (TLC) (2).

RESULTS AND DISCUSSION

An example of the free lipid profiles is shown in Fig. 1 and a summary of the distribution of characteristic lipids given in Table I. All strains produced α-, keto- and methoxymycolates, excepting one group of BCG strains which lacked the latter component (3). The glycosyl phenolphthiocerol (M, Table I) is not found in *M. tuberculosis* and certain BCG strains lacked both this lipid and significant amounts of dimycocerosates of the phthiocerol family (A-C). Trehalose dimycolates (CF) were always present in small or

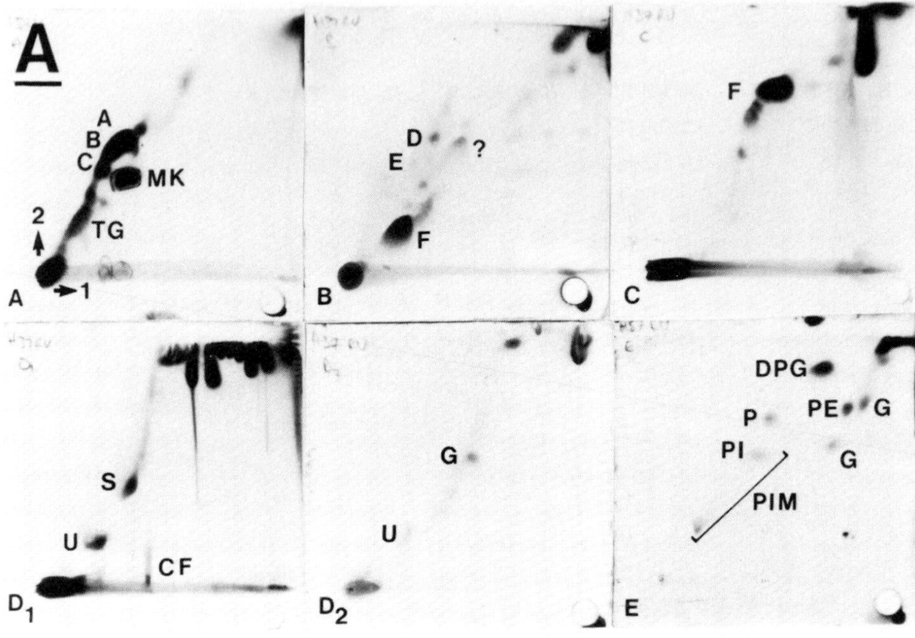
A
A
B
C
MK
2
TG
1
A
D
?
E
F
B
F
C
D1
S
U
CF
D2
G
U
DPG
P
PE
G
PI
G
PIM
E

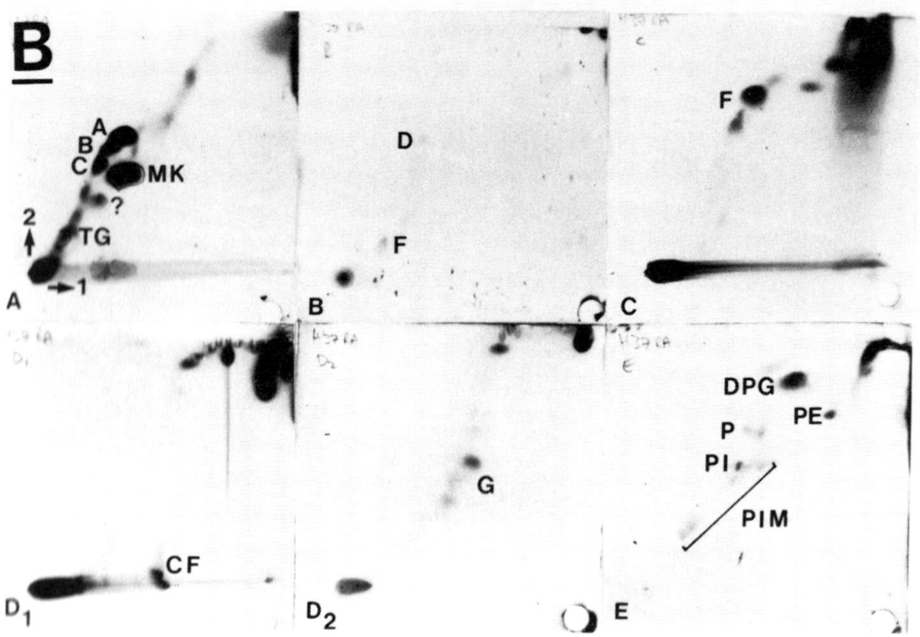
B
A
B
C
MK
?
2
TG
1
A
D
F
B
F
C
CF
D1
G
D2
DPG
PE
P
PI
PIM
E

TABLE I

DISTRIBUTION OF LIPIDS IN THE 'TUBERCULOSIS' COMPLEX

Taxon		Methoxy-mycolate	A–C[1]	D, E[1]	U[1]	S[1]	M[2]
M. tuberculosis	H37Rv	+	+	+	+	+	–
	H37Ra	+	+	+	–	–	–
	MNC 3, 57	+	+	+	+	+	–
M. africanum	MNC 322	+	+	–	–	–	+
M. microti	OV 254	+	+	–	–	–	+
M. bovis	MNC 8	+	+	+	(+)[3]	+	+
	MNC 27	+	+	–	(+)	(+)	+
M. bovis BCG	a)	+	+[5]	–	–	–	+[5]
	b)	–	+[5]	–	–	–	+[5]

1. See Fig. 1 for explanation of abbreviations.
2. 2-methylrhamnosyl phenolphthiocerol dimycocerosate, chromatographing just below F in TLC system C (Fig. 1) (see Ref. 2).
3. Brackets represent small amounts not positively identified.
4. a) Mcreau, Swedish, Russian, Japanese; b) Pasteur, Glaxo, Prague, Danish, Chinese substrains.
5. Absent in Glaxo and Moreau substrains.

trace amounts but the trehalose-based sulpholipids (S) and mycolipenate-containing glycolipids (D, E, U) showed variations with virulence in *M. tuberculosis* and were also detectable in certain *M. bovis* strains. The polar mycolipenates/mycosanoates of trehalose (U) are lipid antigens (8). Variations in the content of sulpholipids (S) and a methyl ether of phenolphthiocerol with virulence in *M. tuberculosis* have been noted (7). Mycolipenates (phthienoates) have been previously detected in certain strains of

Fig. 1. TLC analysis of free lipids of *M. tuberculosis* H37Rv (A) and H37Ra (B). A, B, C and D_1 are for non-polar and D_2 and E for polar extracts. Developing systems: A i) Petroleum ether (b.p. 60–80°C)/ethyl acetate 98:2 x 3, ii) petroleum ether/acetone 98:2 x 1; B i) petroleum ether/acetone 98:2 x 3, ii) toluene/acetone 95:5 x 1; C i) chloroform/methanol 96:4, ii) toluene/acetone 80:20; D_1 and D_2 i) chloroform/methanol/water 100:14:0.8, ii) chloroform/acetone/methanol/water 50:60:2.5:3; E i) chloroform/methanol/water 60:30:6 ii) chloroform/acetic acid/methanol/water 40:25:3:6. Detection, ethanolic molybdophosphoric acid for all lipids, in combination with other specific sprays (2). Abbreviations: A–C, dimycocerosates of phthiocerols A, B and phthiodiolone A; CF, 'cord factor' (trehalose dimycolate); D and E, non-polar mycolipenates of trehalose; DPG, diphosphatidylglycerol; F, free fatty acids; G, glycolipid; P, phospholipid, PE, phosphatidylethanolamine; PI, phosphatidylinositol; PIM, phosphatidylinositol mannosides; S, sulphated trehalose phthioceranates; U, polar mycolipenates/mycosanoates of trehalose.

M. africanum (4).

These preliminary results require consolidation by the analysis of more strains. The present small-scale procedure allows the quantitative interrelation of the whole range of mycobacterial lipids to be rapidy and reliably visualized (2,9). Having established a general lipid profile, sophisticated analytical procedures such as gas chromatography - mass spectrometry (10, 11) and high performance liquid chromatography (12) would allow precise quantitative data to be exploited, particularly for rapid and sensitive identification.

ACKNOWLEDGEMENTS

Grants: D.E.M. and M.G., LEPRA and M.R.C. (G8216538); M.R., Swedish National Association against Heart and Chest Diseases. *M. microti* provided by P. Draper.

REFERENCES

1. Wayne LG (1984) In : Kubica GP, Wayne LG (eds) The Mycobacteria : a sourcebook. Marcel Dekker, New York, pp 25-65
2. Dobson G, Minnikin DE, Minnikin SM, Parlett JM, Goodfellow M, Ridell M, Magnusson M (1985) In : Goodfellow M, Minnikin DE (eds) Chemical Methods in Bacterial Systematics. Academic Press, London, pp 237-265
3. Minnikin DE, Parlett JH, Magnusson M, Ridell M, Lind A (1984) J Gen Microbiol 130 : 2733-2736
4. Daffé M, Lanéele MA, Asselineau C, Lévy-Frébault V, David H (1983) Ann Microbiol 134B : 241-256
5. Davidson LA, Draper P, Minnikin DE(1982) J Gen Microbiol 128 : 823-828
6. Minnikin DE, Minnikin SM, Parlett JH, Goodfellow M, Magnusson M (1984) Arch Microbiol 139 : 225-231
7. Grange JM, Aber VR, Allen BW, Mitchison DA, Goren MB (1978) J Gen Microbiol 108 : 1-7
8. Minnikin DE, Dobson G, Sesardic D, Ridell M (1985) J Gen Microbiol 131 : 1369-1374
9. Minnikin DE, Dobson G, Parlett JH (1985) In : Habermehl KC (ed) Rapid Methods and Automation in Microbiology and Immunology. Springer, Berlin, pp 274-282
10. Larsson L (1985) In : Habermehl KO (ed) Rapid Methods and Automation in Microbiology and Immunology. Springer, Berlin, pp 248-254
11. Mallet AI, Minnkin DE, Dobson G (1984) Biomed Mass Spectrom 11 : 79-86
12. Minnikin DE, Dobson G, Goodfellow M, Draper P, Magnusson M (1985) J Gen Microbiol 131 : 2013-2021

DNA RESTRICTION ENDONUCLEASE ANALYSIS OF SEVERAL STRAINS OF RAPIDLY GROWING MYCOBACTERIA

M.J. GARCIA and E. TABARES
Departamento de Medicina Social y Preventiva, Facultad de Medicina, Universidad Autonoma, c/ Arzobispo Morcillo n^{0} 4, 28029 Madrid (Spain)

INTRODUCTION

Characterization of bacterial DNA has been used to study their taxonomy (8). Hybridization and base composition have mainly been used so far, but restriction endonuclease analysis has recently been applied successfully to the typing of bacteria (1,3), and it may be useful in the future for genetic description of species.

This method has great interest since it allows comparison of clinical of biological isolates over a genotypic base, instead of on a phenotypic base as in usual typing methods.

MATERIAL AND METHODS

Bacterial strains and growth media

M.fortuitum ATCC 6841 (type strain), *M.gadium* ATCC 27726 (type strain), *M. gilvum* Stanford 132 and *M.phlei* IMRU 500 were grown in Lowenstein-Jensen medium (Difco) at 37°C until colonies were observed.

Extraction and purification of DNA

Strains were cultured in 200 ml Dubos-Albumin medium (Difco), at 37°C in a gyrotory shaker for 3-8 days.

The cells were lysated by Lysozyme and Pronase (Sigma) treatment (9) and DNA was purified using M.A.K. (Methylated-Albumin-Kieselguhr) chromatography (11).

Restriction endonuclease analysis

Samples of mycobacterial DNA (2 μg) were digested with about 20 U of each of the following restriction endonucleases type II (7), using the temperature and buffer specified by the supplier: *Bgl II*, *Cla I*, *Eco RI* and *Hind III*.

The restriction enzymes were obtained from B.R.L. except *Hind III* which was obtained from New England Biolabs.

Digests were analyzed by gel electrophoresis on 25 cm long horizontal slabs of 0.5 % (w/v) agarose gel (Bio-Rad) in Loening's buffer (6). After electrophoresis (20 hours at 2 V/cm), the gels were stained for 30 minutes in 1 mg/L ethidium bromide and placed on an Ultraviolet Transilluminator and photographed using a Kodak wratten 23 A gelatin filter.

RESULTS

Analysis of mycobacterial DNA with Bgl II, Cla I and Eco RI restriction endonucleases

The gel electrophoretograms of the DNA of the investigated strains digested by *Bgl II*, *Cla I* and *Eco RI*, showed characteristic banding patterns (Figure 1) easily distinguishable from one another. These patterns could be consistently reproduced.

Higher molecular weight fragments were obtained after *M.phlei* DNA had been digested by *Eco RI* and *Cla I* restriction endonucleases.

Analysis of mycobacterial DNA with Hind III restriction endonuclease

Analysis of DNA of the four mycobacterial strains with *Hind III*, showed a low degree of digestion with high molecular weight fragments (Figure 1). Under the conditions of our experiments the DNA of the Herpes Simplex Virus-2 was completely digested (Figure 2).

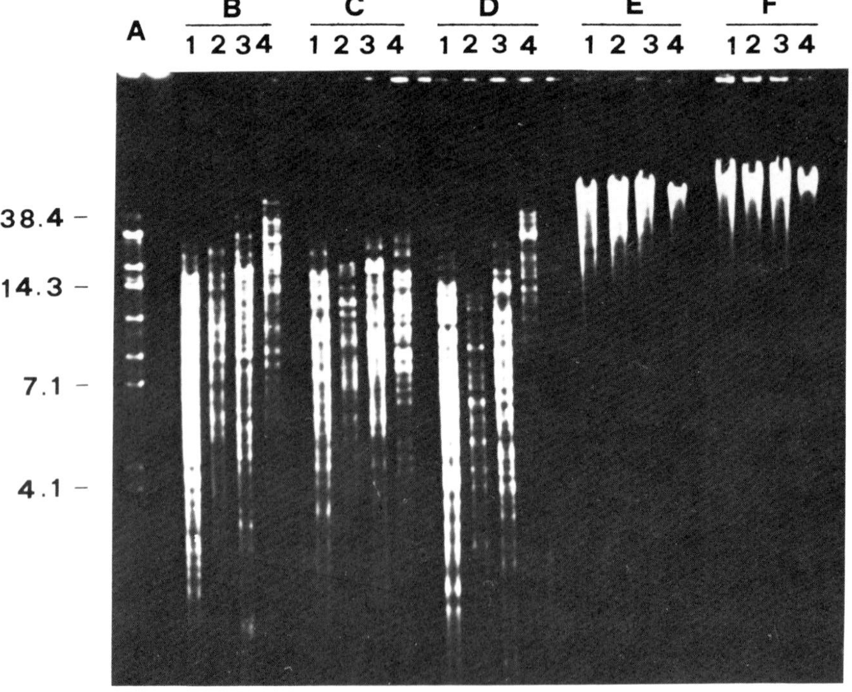

Fig. 1. A - Herpes Simplex Virus-2 DNA *Bgl II* restriction fragments (fragment sizes (in Kilobase pairs)(5) are indicated on the left); B - Agarose gels of *Eco RI* restriction fragments; C - Agarose gels of *Bgl II* restriction fragments; D - Agarose gels of *Cla I* restriction fragments; E - Agarose gels of *Hind III* restriction fragments; F - Agarose gels of undigested mycobacterial DNA. Slots 1- *M.fortuitum* ATCC 6841; Slots 2- *M.gadium* ATCC 27726; Slots 3- *M.gilvum* Stanford 132; Slots 4- *M.phlei* IMRU 500.

The patterns can be consistently reproduced with different enzyme concentrations and incubation periods. This mycobacterial DNA pattern is very different from that obtained with *Bgl II*, *Cla I* and *Eco RI* endonucleases (Figure 1), or that described for other bacterial DNAs (1,2).

Densitometry of the electrophoretic patterns of undigested and *Hind III* digested DNAs, showed that part of the DNA is resistant to digestion. The quantity is similar in all cases.

Methylation of the terminal Adenine of the recognition sequence inhibits the enzymatic activity of *Hind III* (7). The mycobacterial DNA contains N^6-methyladenine (10) which may prevent enzymatic activity; N^6-methyladenine may be involved in the restriction-modification systems in mycobacteria (4), and therefore may be a factor in the recognition sequence of *Hind III*.

Further studies are in progress to investigate whether this resistance to *Hind III* digestion is a characteristic of all mycobacteria or is restricted to rapidly growing mycobacteria, since this resistance has not been reported in slowly growing mycobacterial DNA (3).

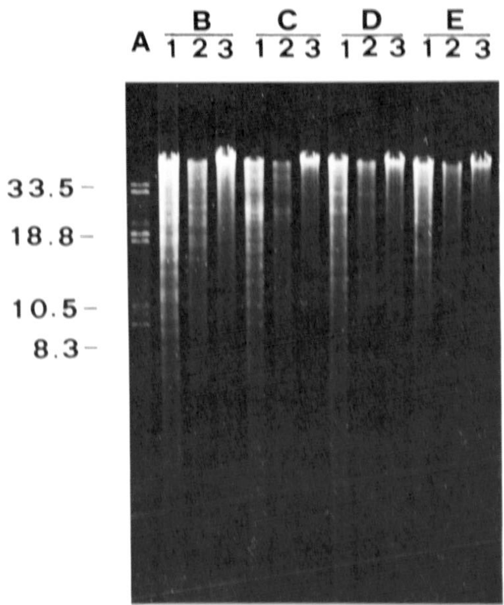

Fig. 2. A - Herpes Simplex Virus-2 DNA *Hind III* restriction fragments (fragment sizes (in Kilobase pairs)(5) are indicated on the left); B - *M. fortuitum* ATCC 6841; C - *M. gadium* ATCC 27726; D - *M. gilvum* Stanford 132; E - *M. phlei* IMRU 500. Slots 1- 1 μg of *Hind III* digested DNAs of different strains; Slots 2- 0.5 μg of *Hind III* digested DNAs of different strains; Slots 3- 0.5 μg of undigested DNAs of different strains.

REFERENCES

1. Bjorvatn B, Lund V, Kristiansen B-E, Korsnes L, Spanne O, Lindqvist B (1984) J Clin Microbiol 19:763-765

2. Bradbury WC, Pearson AD, Marko MA, Congi RV, Penner JL (1984) J Clin Microbiol 19:342-346

3. Collins DM, De Lisle GW (1984) J Gen Microbiol 130:1019-1021

4. Crawford JT, Cave MD, Bates JH (1981) J Gen Microbiol 127:333-338

5. Morse LS, Buchman TG, Roizman B, Schaefer PA (1977) J Virol 24:231-248

6. Post LE, Conley AJ, Mocarski ES, Roizman B (1980) Proc Nat Acad Sci 77:4201-4205

7. Roberts RJ (1983) Nucl Acids Res 11:r135-r167

8. Schleifer KH, Stackebrandt E (1983) An Rev Microbiol 37:143-187

9. Simpson DK, Wayne LG (1977) Biol Actin Rel Org 12:36-44

10. Somogyi PA, Masobel M, Földes I (1982) Acta Microbiol Acad Sci Hung 29:181-185

11. Tabarés E, Valladares Y, Alvarez Y (1973) Rev Esp Onc XX:7-18

AUTOMATED METHODS OF LABORATORY DIAGNOSIS

RADIOMETRIC DIAGNOSIS OF *MYCOBACTERIUM TUBERCULOSIS*

A. LASZLO

National Reference Centre for Tuberculosis, Laboratory Centre for Disease Control, Health and Welfare Canada, Ottawa, Ontario K1A 0L2, Canada

INTRODUCTION

Automated and radiometric methods for culturing and for antimicrobial susceptibility testing have been in general use in diagnostic bacteriology for more than a decade; so much so , that this technology is now of critical importance to many segments of microbiology. In mycobacteriology however, despite the fact that the protracted doubling time of *M. tuberculosis* should have made it a prime target, the development of rapid diagnostic procedures has only occurred very recently.

Beginning in 1975, Cummings and co-workers (1) used (U-^{14}C) glycerol and (U-^{14}C) acetate as substrates for the detection of mycobacterial growth. A special 7H12 medium containing (1-^{14}C) palmitic acid as a substrate was introduced by Middlebrook and co-workers (2) in 1977. A year later Kertcher and co-workers (3) used the radiometric growth detection system to test drug susceptibility of *M. tuberculosis* after addition of an anti-tuberculosis drug to a medium containing (^{14}C) formate. The indirect drug susceptibility testing technique for *M. tuberculosis* was carried a step further by adjusting the inoculum size in such a way that semi-quantitative results could be achieved with a 1% threshold as determinant of resistance. Finally, in 1984 Laszlo and Siddiqi (4) evaluated a radiometric technique for the differentiation of *Mycobacterium tuberculosis* complex from the other mycobacteria using p-nitro-α-acetylamino-β-propiophenone (NAP) as a selective inhibitory agent. Their results showed that a concentration of 5 μg/ml NAP in 7H12 medium effectively separates the *M. tuberculosis* complex from 35 other mycobacterial species in 4 to 6 days.

The results of large scale independent evaluations of the rapid radiometric method (BACTEC) applied to diagnostic clinical mycobacteriology are beginning to build a body of evidence difficult to ignore. In 1983, G.D. Roberts and co-workers (5) evaluated this method in the recovery of mycobacteria from acid-fast smear positive specimens and found that the mean recovery time for *M. tuberculosis* is shortened by more than 10 days when compared to conventional methods. M. Morgan and co-workers, in 1983 (6) compared the recovery of mycobacteria from smear negative specimens by BACTEC and conventional methodologies. They found that 7H12 medium alone detects more mycobacteria than any other single conventional medium.

In 1981 D.E. Snider and co-workers (7) reported on a large scale comparison of radiometric and conventional drug susceptibility testing of M. tuberculosis. They found that radiometric methods were more reliable in determining resistance to Rifampin than to Isoniazid, Steptomycin and Ethambutol. In 1983, Laszlo and co-workers (8) compared the BACTEC radiometric drug susceptibility testing to the conventional proportion method, the absolute concentration and the resistant ratio methods and showed that the radiometric method does not differ significantly in reliability from conventional methodologies.

This current study was designed to measure the impact of some of these new methodologies on the operations of our National Reference Centre for Tuberculosis. We therefore introduced the radiometric NAP differentiation test and the radiometric drug susceptibility test in our laboratory routine in parallel with conventional methodology. The NAP test was evaluated as an early screening test for the M. tuberculosis complex which could allow selective drug susceptibility testing of presumptive M. tuberculosis isolates only. Results were analyzed in terms of diagnostic accuracy and speed.

MATERIALS AND METHODS

Cultures

Over a period of several months, a total of 250 non-pigmented clinical mycobacterial isolates referred to the National Reference Centre for Tuberculosis, L.C.D.C., Health and Welfare Canada, were examined radiometrically without prior subculture, the conventional identification and drug susceptibility testing were performed as usual.

Stock Solutions

NAP was synthesized in our chemistry laboratory, starting from p-nitroacetophenone by the procedure of Long and Troutman (9). NAP was dissolved in absolute alcohol (1 mg/ml) at 55°C in a water bath and then diluted with distilled water. The solution was filter sterilized (pore size, 0.2 μm). Small amounts of stock solution of 100 μg/ml were stored at room temperature for up to 3 weeks or at -20°C for up to 2 months. 0.1 ml of this stock introduced in 7H12 vials yields a final concentration of 5 μg/ml. Similar procedures were utilized to prepare stocks of the other antituberculosis drugs (10).

Radiometric Methods. The methodology used has been described elsewhere (10,4). Briefly, bacterial suspensions approximating the McFarland No. 1 Standard were prepared from the referred subculture in a solution of 0.2% fatty acid free albumin (Sigma Chemical Co., St. Louis, Mo.) and 0.02% Tween 80 in deionized water. 0.1 ml aliquots of this suspension were inoculated into each of two Middlebrook 7H12 vials containing NAP at concentrations of 0 and 5 μg/ml. The inoculated vials were incubated at 37°C and read daily for 1 to 6 days on a BACTEC 460 instrument. Both instrument and medium vials were purchased from Johnson Laboratories, Towson, Md.

When isolates were identified as belonging to the *M. tuberculosis* complex, the control vial was utilized to inoculate properly diluted aliquots into Middlebrook 7H12 vials containing antituberculosus drugs. The drug concentrations used were Isoniazid (INH) 0.2 μg/ml, Dihydro streptomycin sulfate (DSM) 4.0 μg/ml, Rifampin (RIF) 2.0 μg/ml, and Ethambutol (EB) 3 μg/ml. One control vial was inoculated with 0.1ml of a 1:00 dilution of the standard inoculum. The vials were then incubated at 37°C and read daily until the results could be interpreted. A Ziehl Neelsen (Z.N.) smear was prepared from bacterial growth from the control vial and examined for acid-fastness and bacterial morphology.

Conventional Methodology

The identification of the mycobacterial isolates was confirmed by standard methods (11, 12). Drug susceptibility of *M. tuberculosis* isolates was determined by the proportion method (13) performed on L.J. medium. The critical concentrations used in this standard procedure were INH 0.2 μg/ml DSM 4.0 μg/ml, RIF 40.0 μg/ml and EB 2.0 μg/ml. Critical proportion for resistance was 1%.

RESULTS

Growth detection

Of 250 non-pigmented mycobacterial cultures, 242 grew both on conventional L.J. and on radiometric Middlebrook 7H12 media. 3 cultures failed to grow on either media and 5 cultures grew only in 7H12 medium. Of the 5 cultures which did not grow on L.J. medium, 4 were *M tuberculosis* and one was an atypical mycobacterium. These cultures were excluded from further analysis.

Identification. 242 cultures recovered from L.J. medium were identified by conventional procedures. *M. tuberculosis* accounted for 61.6% or 149 of the referred cultures, the remaining cultures belonged to 9 different species of mycobacteria other than tuberculosis (M.O.T.T.) (See Table 1).

Radiometric screening by NAP selective inhibition.

Table 1 shows that selective inhibition test performed by the BACTEC radiometric method identified 150 cultures as belonging to *M. tuberculosis* complex and 92 cultures as belonging to M.O.T.T. When these results were compared to the conventional identification, this identification test had failed to recognize 1 culture of the *M. tuberculosis* complex and 2 cultures of M.O.T.T. This represents an overall agreement of about 99.0%

TABLE 1 RADIOMETRIC SCREENING BY NAP SELECTIVE INHIBITION

Species	Number of Cultures	NAP (5 μg/ml) Number Susceptible	Number Resistant
M. tuberculosis	149	148	1
M. avium or *M.intracellulare*	65	0	65
M. fortuitum	10	0	10
M. terrae	6	1*	5
M. Kansasii	5	0	5
M. gordonae	2	0	2
M. marinum	2	0	2
M. gastri	1	0	1
M. chelonae subsp. *chelonae*	1	0	1
M. chelonae subsp. *abscessus*	1	1*	0
TOTAL	242	150	92

*When tested at 30°C culture, is NAP resistant.

TABLE 2
ANALYSIS OF DISAGREEMENTS BETWEEN RADIOMETRIC (RAD) AND CONVENTIONAL (CONV) RESULTS

Drug	CONV-S RAD-R	CONV-R RAD-S	% Agreement
DSM	1	0	99.25
INH	3	0	97.8
RIF	0	0	100.0
EB	2	1	97.6

The two false M. tuberculosis cultures, were identified as M. terrae and M. chelonae subsp. chelonae by conventional means. Both cultures showed optimum growth at 30°C. When further tested radiometrically for drug susceptibility, the 1% control of the M. terrae culture failed to grow at 37°C, whereas M. chelonae subsp chelonae showed resistance to all four drugs tested. When retested at 30°C both were resistant to NAP inhibition. Two cultures of M. marinum, both isolated from skin lesions were originally tested for NAP inhibition at both 30°C and 37°C. In this case also, they were found susceptible at 37°C but resistant at 30°C. This confirms our earlier observations (4).

Presumptive identification of about 93.0% of M.O.T.T. cultures was reportable one day after inoculation, all M. tuberculosis were reportable four days after inoculation.

Radiometric Drug Susceptibility Test. Following the screening by the NAP inhibition test, cultures presumptively identified as belonging to the M. tuberculosis complex were tested for susceptibility to four major antituberculosis drugs. Radiometric versus conventional comparison was made in 134 cultures. Conventional methodology found 2 (1.5%) cultures resistant to EB, 7 (5.2%) resistant to RIF, 16 (11.9%) resistant to INH and 16 (11.9%) resistant to DSM. The overall agreement between conventional and radiometric results was 98.5%.

Table 2 shows the analysis of disagreements between the results of radiometric and conventional drug susceptibility testing. This table readily shows that overall agreement for each of the 4 drugs is high. A more meaningful measurement of agreement between methods can be obtained by calculating the sensitivity of the test method ie. its capacity to distinguish correctly resistant cultures and the specificity ie. its capacity to distinguish susceptible cultures.

Table 3 shows the sensitivity and specificity values of the radiometric data are very similar to those published previously (8). The low sensitivity values for EB can be explained by the fact that only 2 cultures were found to be resistant to this drug by the conventional method.

Where A is the number of cultures susceptible by both methods, B is the number of cultures susceptible by the conventional method and resistant by the radiometric method, C is the number of cultures resistant by the conventional method and susceptible by the radiometric method, and D is the number of cultures resistant by both methods.

Reporting. 53 or 39.6% of the results were reportable in 4 days, 101 or 75.4% in 5 days, 126 or 94.0% in 8 days and 134 or 100.0% in 11 days. The average reporting time was 5.2 days.

TABLE 3

SENSITIVITY SPECIFICITY AND PREDICTIVE VALUES OF RAD METHOD AS COMPARED TO THE CONVENTIONAL

Drug	Sensitibity	Specificity	PVR	PVS
DSM*	1.0	.99	.94	.99
INH*	1.0	.97	.84	1.0
RIF**	1.0	1.0	1.0	1.0
EB**	.66	.98	.50	1.0

*Based on 134 comparisons

**Based on 126 comparisons

Sensitivity = D/(C+D) Specificity = A/ (A+B)

PVR = Predictive Value (Resistance) = D/ (B+D)

PVS = Predictive Value (Susceptibility) = A/ (A+C)

DISCUSSION

The results of this study confirm that rapid radiometric diagnostic tests such as the NAP selective inhibition test for the M. tuberculosis complex followed by the radiometric drug susceptibility tests are extremely reliable and compare favourably with conventional methodologies. This study also shows that referred cultures growing on solid medium can be processed by radiometric procedures without prior subculture. This circumstance by itself shortens the time needed for reporting.

The NAP selective inhibition test used as an early screening test identifies members of the M. tuberculosis complex in four days and differentiates most non-pigmented MOTTs in 24 hours, only about 1% of all cultures tested were misidentified. Erroneous diagnosis by this test alone does not necessarily lead to erroneous reporting e.g. in this study two false M. tuberculosis were tested for drug susceptibility, in one case the 1% control would not grow at 37°C, and when retested at 30°C it clearly showed resistance to NAP. Conventional identification procedures then showed that this isolate was a M. terrae with an optimal growth at 30°C. The other culture, later identified as M. chelonae subsp. chelonae, showed resistance to all four drugs tested. A simple recommendation of confirmation of all presumptive multiple drug resistant M. tuberculosis isolates would avoid the mislabeling of similar cultures. One culture of M. tuberculosis was not recognized by the radiometric procedure alone but the presence of cording in the Z.N. smear would be a valid reason for reconsideration of the diagnosis. Conventional media on the other hand failed to support the growth of 3 isolates of M. tuberculosis.

The NAP screening procedure allows for selectivity in the performance of drug susceptibility tests. In a time of ever increasing diagnostic workload due to MOTT, this fast screening procedure will prevent the performance of frequently unnecessary work such as drug susceptibility testing of atypical mycobacterial isolates.

Finally, when one takes into account the time needed for presumptive identification, and the time needed for the interpretation of drug susceptibility testing it can easily be shown that a complete radiometric report on M. tuberculosis isolates is obtainable in as early as 8 days, the average reporting time being 9 days after receipt of the cultures.

ACKNOWLEDGEMENTS

I thank Mrs. V. Handzel for excellent technical assistance.

REFERENCES

1. Cummings, DM, Ristroph, D, Camargo, EE, Larson, SM, Wagner, HN (1975) Radiometric detection of the metabolic activity of Mycobacterium tuberculosis. J. Nucl. Med. 16:1189-1191

2. Middlebrook, G, Reggiardo, Z, Tigertt, WD (1977) Automatable radiometric detection of growth of Mycobacterium tuberculosis in selective media. Am. Rev. Respir. Dis. 115:1067-1069

3. Kertcher, JA, Chen, MF, Charache, P, Hwangbo, CC, Camargo, EE, McIntyre, PA, Wagner, HN Jr (1978) Rapid radiometric susceptibility testing of Mycobacterium tuberculosis. Am. Rev. Respir. Dis. 117:631-637

4. Laszlo, A, Siddiqi, SH (1984) Evaluation of a rapid radiometric differentiation test for the Mycobacterium tuberculosis complex by selective inhibition with p-nitro-α-acetylamino-β-Hydroxypropiophenone. J. Clin. Microbiol. 19:694-698

5. Roberts, GD, Goodman, NL, Heifets, L, Larsh, HW, Linder, TH, McClatchy, JK, McGinnis, MR, Siddiqi, SH, Wright, P (1983) Evaluation of the BACTEC radiometric method for recovery of mycobacteria and drug susceptibility testing of Mycobacterium tuberculosis from acid fast smear positive specimens J. Clin. Microbiol. 18:689-696

6. Morgan, MA, Horstmeier, C.D., DeYoung, D.R., Roberts, G.D. (1983) Comparison of a radiometric method (BACTEC) and conventional culture media for recovery of mycobacteria from smear-negative specimens. J. Clin. Microbiol. 18:384-388

7. Snider, DE Jr, Good, RG, Kilburn, JO, Laskowski, LF Jr, Lusk, RH, Marr, JJ, Reggiardo, Z, Middlebrook, G (1981) Rapid susceptibility testing of Mycobacterium tuberculosis. Am. Rev. Respir. Dis. 123: 402-406

8. Laszlo, A, Gill, P, Handzel, V, Hodgkin, MM, Helbecque, DM, (1983) Conventional and radiometric drug susceptibility testing of Mycobacterium tuberculosis complex. J. Clin. Microbiol. 18:1335-1339

9. Long, LM, Troutman, HD, (1949) Chloramphenicol (chloromycetin) VII. Synthesis through p-Nitro-acetophenone. J. Am. Chem. Soc. 71:2473-2475

10. Siddiqi, SH. Libonati, JP, Middlebrook, G. (1981) Evaluation of a rapid radiometric method for drug susceptibility testing of Mycobacterium tuberculosis. J. Clin. Microbiol. 13:908-912

11. Tuberculosis Laboratory Methods - National Reference Centre for Tuberculosis, Laboratory Centre for Disease Control, Health and Welfare Canada (1985)

12. Vestal, AL (1975) Procedures for the isolation and identification of mycobacteria. Center for Disease Control (DHEW publication No. (CDC) 76.8230)

13. Canetti, G, Fox, W, Khomenko, A, Mahler, H, Mehon, NK, Mitchison, DA, Rist, N, Smelev, NA (1969) Advances in techniques of testing mycobacterial drug sensitivity, and the use of sensitivity tests in tuberculosis control programs. Bull. W.H.O. 41:21-43

Mycobacteria of Clinical Interest. M. Casal, editor

RADIOMETRIC INVESTIGATION OF SUSCEPTIBILITY OF MYCOBACTERIUM TUBERCULOSIS

GIOVANNI FADDA, GIACOMO FODDAI, LUCIA MUREDDU, FRANCA SANCIU and STEFANIA ZANETTI

Chair of Science Microbiology and Institute of Medical Microbiology, University of Sassari, 07100 Sassari (Italy)

INTRODUCTION

Conventional procedures for studying the drug susceptibility of *Mycobacterium tuberculosis* isolates are time-consuming, and require a minimum of 3 or 4 weeks to complete. During that time patients with resistant organisms may be receiving treatment with an inappropriate or ineffective drug regimen, which can result in the infected organisms becoming resistant to additional drugs and/or failure to control the disease (22).

For these reasons the development of automated and more rapid techniques for determining drug susceptibility of *M. tuberculosis* isolates is desirable (22). Only in the last few years have automated methods been employed for culture of microorganisms and for antimicrobial susceptibility tests in bacteriology. Among these procedures, and one of the most widely used, is the Bactec radiometric system (Johnston Laboratories, Inc., Cockeysville, MD.). This method uses media containing ^{14}C-labeled substrates which, when metabolized by bacteria, yield detectable levels of 14 CO^2, which are read as growth index (G.I.) readings (6). The Bactec radiometric method has been applied successfully to blood culturing (19), to radiometric detection of antibiotic effects on bacterial growth (5), to the differentiation of Neisseria by substrate metabolism (19), and for serum assay of aminoglycoside antibiotics (19).

Cummings and coworkers in 1975 (3) carried out preliminary work to show that the same principle could be used to detect growth in pure culture of *M. tuberculosis*. Middlebrook et al. (18) introduced a special liquid medium, 7H12, which contains ^{14}C palmitic acid as the labeled substrate.

Several reports on the detection and recovery of mycobacteria from sputa and from other extrapulmonary specimens by the radiometric procedure have been published (1,4,8,9). These studies indicate that the radiometric method used in conjunction with conventional media appears to maximize the recovery of mycobacteria and significantly reduces the time required to report positive results (9).

The possibility of using the same principle for testing drug susceptibility of mycobacteria by adding an antituberculous drug to the 7H12 liquid medium has also been demonstrated. Siddiqi et al. (21) developed a technique of indirect susceptibility testing, using 7H12 liquid medium, adjusting the inoculum size in such a way that semiquantitative results could be achieved with a 1% threshold as a determinant of resistance (21,22). Siddiqi et al. (21) and ourselves, (8,9), have obtained 98% agreement between conventional and radiometric methodologies for the common primary antituberculous drugs with several strains of *M.tuberculosis*. Other reports showed that this is also true for second-line drugs (ethionamide, cycloserine, capreomycin, kanamycin and pyrazinamide) (11-12).

Spiro-piperydil rifamycins are members of a new class of ansamycin antibiotics produced synthetically (Farmitalia Carlo Erba Research Laboratory, Milan, Italy). One member of this class (4-N-isobutyl-spiro-piperydil-rifamycin S) whose code number is LM 427 (ANS) showed remarkable activity against *M.tuberculosis in vitro* and *in vivo*. The *in vitro* susceptibility to ANS was also determined in 7H12 broth in the radiometric Bactec system, and there was good agreement between the results with radiometric and conventional procedures (10, 13).

Since we now have 4 years of experience with the automated Bactec 460 TB instrument, the following report gives an evaluation of the system for radiometric study of susceptibility testing and MIC determinations of *M.tuberculosis* with regard to common primary drugs and to the new rifamycin-derivative, LM 427.

TABLE 1
CONDITIONS OF SUSCEPTIBILITY TESTING

Materials and Methods		Proportion method	Radiometric method
Middlebrook medium		7H11 agar	7H12 broth
Inoculum from 7H12 GI between 300-400 or MacFarland 1 suspension.		0.1 ml and dilutions	0.1 ml in drug vials 0.1 ml (1:100) in control vial
Drug final concentration (mg/l)			
Streptomycin	(SM)	2.0-10.0	4.0
Isoniazid	(INH)	0.2- 1.0	0.2
Ethambutol	(EMB)	5.0-10.0	10.0
Rifampin	(RM)	1.0	2.0*
LM 427	(ANS)	1.0	1.0
Incubation		5-10% CO2 37°C	36-38°C
Reading		at day 21	when GI control ≥30

* when compared with ANS at concentration of 0.1.

MATERIAL AND METHODS

Drug susceptibility testing

The 170 strains of *M.tuberculosis* used in this study were isolated from specimens received for culture of mycobacteria. The rapid radiometric drug susceptibility procedure employed (Tab.1), described by Siddiqi et al. (21), is a modified version of the proportion method.

0.1 ml suspension of each isolate was inoculated into rubber-sealed vials containing two ml. Middlebrook 7H12 liquid culture medium and included anti-tuberculous drugs. Each isolate was also inoculated into a control vial, without drugs, at a dilution of 1:100 in special diluting fluid containing fatty-acid-free albumin (0.2%) and Tween 80 (0.002%). After inoculation, each vial was tested on a Bactec 460 TB instrument to provide CO_2 in the headspace. All of the bottles were incubated at 36° - 38°C and tested daily by Bactec 460 TB at intervals of about 24 hours. At each reading the instrument aspirated gases out of the vials for readings of $^{14}CO_2$ and introduced fresh air into the vial (5% CO_2). When the growth index of the control read at least 30, the results were interpreted by comparing the increase in GI (ΔGI) from the previous day in the control with that in the drug vial. If the variation in GI of the test vial was equal to, or greater than the ΔGI of the control, 1% or more of the bacterial population was considered to be resistant to the drug. The strain was then reported as resistant (21). Typical variation in GI with respect to time is shown in Table 2.

TABLE 2
RADIOMETRIC SUSCEPTIBILITY TESTING
Typical variation of GI with respect to time

Drug	Day 1	2	3	4	5	Results
Control (1:100)	5	10	20	35	70	
Streptomycin	38	62	65	56	31	Susceptible
Isoniazid	32	64	82	98	129	Resistant
Ethambutol	208	300	496	380	224	Susceptible

A susceptibility test by the radiometric procedure was done either from a positive primary 7H12 liquid isolation medium, when the GI was between 300 and 400, or from the growth on solid Lowenstein-Jensen slant media. In the latter case, individual strains were suspended in the special diluting fluid. The suspensions were adjusted to an opacity equivalent (MacFarland n.1), and inoculated into 7H12 liquid medium. The vials incubated at 36° - 38°C were checked on a Bactec 460 TB instrument until the GI became between 300 and 400. The drugs tested were the four first-line antitubercular drugs and the rifamycin derivative LM 427 (ANS). The final concentrations (mg/l) used were: streptomycin (SM),4; isoniazid (INH),0.2; rifampin (RM),2; and ethambutol (EMB),10. The in vitro activity of ANS was also compared with RM by testing *M. tuberculosis* isolates at a concentration of one mg/l for each drug. The susceptibility patterns of the strains to test drugs were also tested by the proportion method (2) modified by Wayne and Krasnow (26), with paper disks containing standardized amounts.

The final concentrations (mg/l) used in 7H11 agar medium were: SM, 2-10; INH, 0.2-1; EMB, 5-10; RM, 1; ANS, 1. Resistance was indicated when the number of colonies exceeded 1% of the controls. For quality controls, we periodically used the *M. tuberculosis* strain H-37 RV which is sensitive to the drugs tested.

Minimal inhibitory concentrations (MICs)

Minimal inhibitory concentrations of drugs tested (SM,INH,EMB,RM,ANS) were studied for 84 *M. tuberculosis* isolates, using radiometric and conventional procedures. The conditions of MIC analysis are given in Table 3. The MICs were determined by growth on 7H12 radiometric vials and on 7H11 agar plates to which serial two-fold dilutions of stock drug solution were added. One vial and one plate were used for each concentration of the drug, respectively for the radiometric and conventional method. In the latter case, the inoculum consisted of about 10^6 cells per ml. Incubation was at 37°C and the MIC was defined as the lowest concentration of the drug which resulted in no growth after 21 days.

MIC determinations by the radiometric procedure were made from a positive primary 7H12 liquid medium when the GI was between 300 and 400. All bottles were incubated at 36° - 38°C and were checked daily with a Bactec 460 TB instrument. The MIC was defined as the lowest concentration of the drug which resulted in no growth when the GI of the control read at least 30. We did not distinguish between the bacteriostatic and bactericidal activity of the drugs.

RESULTS

Table 4 shows the results of the 170 *M. tuberculosis* isolates for which susceptibility tests were done by the conventional and rapid radiometric methods. There was 98.8% agreement for determinations with SM, 98.2% with INH, and 100% agreement with EMB, RM and ANS.

34, 50, 22, 44 and 29 strains were noted to be resistant respectively to SM, INH, EMB, RM and ANS by both methods.

TABLE 3
CONDITIONS OF MIC DETERMINATION

Materials and Methods	Conventional method	Radiometric method
Middlebrook medium	7H11 agar	7H12 broth
Inoculum from suspension 10^6 cells/ml from 7H12 GI 300-400	0.1 ml	0.1 ml in drug vials 0.1 ml (1:100) in control vial
Drug serial two-fold dilutions; final concentration (mg/l) from-to		
Streptomycin (SM)		16.0 0.12
Isoniazid (INH)		6.4 0.01
Ethambutol (EMB)		80.0 0.15
Rifampin (RM)		16.0 0.001
Lm 427 (ANS)		16.0 0.001
Incubation	5-10% CO_2 37°C	36-38°C
Reading	at day 21	when GI ≥ 30

TABLE 4

RESULTS OF TWO SUSCEPTIBILITY METHODS

	Drug*										Results**	
Number of	SM		INH		EMB		RM		ANS		Methods	
strains	N	%	N	%	N	%	N	%	N	%	C	B
170	134	78.8	117	68.8	148	87.0	126	74.1	141	82.9	S	S
	0	-	1	0.5	0	-	0	-	0	-	S	R
	2	1.2	2	1.2	0	-	0	-	0	-	R	S
	34	20.0	50	29.4	22	13.0	44	25.9	29	17.1	R	R
Agreement	98.8		98.2		100		100		100			
Average (time/day)											22	5.4
Range (time/day)											18-25	3-7

* SM,streptomycin; INH,isoniazid; EMB,ethambutol; RM,rifampin; ANS,LM 427

** C,conventional method; B,Bactec radiometric method.

Of the five cultures for which disagreements were noted, four were observed to be resistant by the conventional method and susceptible by the radiometric procedure. Only one strain was susceptible to INH by the conventional method and resistant by the Bactec radiometric system.

The average time to report the results was 22 days and 5.4 days respectively by conventional and radiometric methods, with a range of 18-25 days and 3-7 days.

The susceptibility of RM and ANS data indicates that all strains susceptible to RM by both methods at a concentration of 1 mg/l were also inhibited by the same concentration of ANS. However, of the 44 strains resistant to RM, 15 were susceptible to ANS.

The minimal inhibitory concentrations for the five drugs tested, determined on 84 *M. tuberculosis* strains by the conventional and radiometric methods, are listed in Table 5. These results show complete correlation between the two methodologies and were reported respectively in an average time of 21 and 6 days respectively. The MICs of the various drugs showed a fairly wide range: SM 0.5-6 mg/l, INH 0.05-6.0 mg/l, EMB 2.5-80.0 mg/l, RM 0.008-16.0 mg/l, and ANS 0.004-8.0 mg/l.

TABLE 5

RESULTS OF MIC DETERMINATIONS BY TWO METHODS

Number of strains	MIC(mg/l) Drug	Range		MIC-50	MIC-90	Methods* C	Methods* B
	Streptomycin	0.5	6.0	1.0	0.2		
	Isoniazid	0.05	6.0	0.1	0.2		
	Ethambutol	2.5	80.0	5.0	10.0		
	Rifampin	0.008	16.0	0.008	0.016		
	LM 427	0.004	8.0	0.004	0.008		
Agreement						100%	
Average (time/day)						21	6

* C, conventional method; B, Bactec radiometric method.

The in vitro activity of ANS in comparison with RM against *M. tuberculosis*, tested on 77 isolates, is shown in Table 6.

ANS (MIC = 0.004-0.008 mg/l) showed about twice the potency of RM (0.008-0.016 mg/l) against sensitive strains and a higher activity against RM-resistant isolates (>16 mg/l). Strains resistant to streptomycin, isoniazid and ethambutol were usually susceptible to both ANS and RM.

TABLE 6

IN VITRO ACTIVITY OF LM 427 IN COMPARISON WITH RIFAMPIN ON 77 M. TUBERCULOSIS STRAINS

Number of strains	resistant to*	MIC (mg/l) range ANS	MIC (mg/l) range RM
44	-	0.004-0.008	0.008-0.016
8	RM	8.0	>16.0
5	SM	0.004-0.008	0.008-0.016
5	INH	0.004-0.008	0.008-0.016
5	EMB	0.004-0.008	0.008-0.016
5	RM-INH	8.0	>16.0
2	RM-SM-INH	8.0	>16.0
3	RM-SM-INH-EMB	8.0	>16.0

* SM, streptomycin; INH, isoniazid; EMB, ethambutol; RM, rifampin; ANS, LM 427

DISCUSSION

The main disadvantage of conventional methods for testing susceptibility of drugs to *M. tuberculosis* is the long waiting period before results can be obtained. This period, about three weeks, minimizes the usefulness of the results (21,22).

In this study the susceptibility to common first-line antituberculous drugs and to ansamycin LM 427, of 170 *M. tuberculosis* isolates was determined by comparing conventional and radiometric procedures. Overall, results were reported respectively in an average time of 22 and 5.4 days, with a time-saving of 17 days. Our data have also confirmed the good degree of correlation between the two techniques. Only some discrepancies were found. The disagreements between the two methods were mainly resistance by conventional procedures and susceptibility by the Bactec radiometric system. This type of disagreement can be attributed to imperfect dispersion of the bacteria in the inoculum in the conventional plate method (21,25), or in any case, can depend on what each technique measures. The conventional procedures in fact use bacterial growth as end-point, whereas the Bactec radiometric system measures the metabolism of palmitic acid (22).

The radiometric method has also been conveniently applied to MIC determinations of the five antituberculous drugs tested. Results showing complete agreement between the two techniques were obtained within 6 days by the radiometric method as compared to the usual 21 days required by conventional procedures.

An antibiogram based on minimum inhibitory concentrations of antituberculous drugs, using the less time-consuming radiometric method as suggested by Loder and Gruft (17), could be a useful tool, perhaps in combination with phage typing, in the epidemiology of *M. tuberculosis*.

Our data on in vitro activity of LM 427 compared with rifampin by testing *M. tuberculosis* isolates are in agreement with those obtained by others (7,10, 13,24,27) and showed a higher activity of ansamycin, and incomplete cross-resistance between the two drugs. The susceptibility of rifampin-resistant *M. tuberculosis* strains to LM 427 may be attributed to the increased lipophilic nature of the spiro-piperidyl rifamycins which are inhibitory at concentrations that prevent rifampin from penetrating the cells (7,24,27).

SUMMARY

This study reports in vitro activity of several *M. tuberculosis* isolates tested for drug susceptibility and MIC against the new rifamycin derivative LM 427 (ANS) and against the four primary antitubercular drugs, streptomycin (SM), isoniazid (INH), ethambutol (EMB) and rifampin (RM) by the rapid radiometric procedure using 7H12 liquid medium with the ^{14}C labeled substrate.

The results indicate that all strains susceptible to RM, at a concentration of 1 mg/l were also inhibited by the same concentration of ANS. However, ANS (MIC = 0.004-0.008 mg/l) show about twice the potency of RM (MIC = 0.008-0.016 mg/l) against sensitive strains and somewhat higher activity (MIC = 8 mg/l) against resistant isolates (MIC 16 mg/l). Strains resistant to SM, INH and EMB were susceptible to both ANS and RM. These data are in good agreement with those obtained by the conventional 7H11 plate method. The results were obtained more rapidly in 21 days and 5.4 days respectively, for drug-susceptibility testing, and within 6 days compared to the usual 3 weeks required for conventional procedures of MIC determination.

REFERENCES

1. Ausina V.,Matas J.,Luquin M.,Coll P.,Carbo´ Ll.,Prats G.(1984) Enf.Infec. y Microbiol. Clin. 2:236-241

2. Canetti G.,Rist N.,Grosset J.,(1963) Rev.Tuber.Pneumol. 27:217-272

3. Cummings D.M.,Ristroph D., Camargo E.E., Larson S.M.,Wagner H.N.Jr. (1975) J. Nucl. Med. 16:1189-1191

4. Damato J.J.,Collins M.T.,Rothland M.V.,McClatchy J.K. (1983) J.Clin.Microbiol 17:1066-1073

5. De Blanc H.J.,Charache P.,Wagner H.N.Jr.(1971) Antimicrobial Agents and Chemioterapy 2:360-364

6. De Land F.H.,Wagner H.N.Jr. (1969) Radiology 92:154-155

7. Della Bruna C.,Schioppacassi G.,Ungheri D.,Jabes D.,Morvillo E.,Sanfilippo E.(1984) J. Antibiotics 36:1502-1506

8. Fadda G.,Roe S.L.(1984) J. Clin. Microbiol. 19:720-721

9. Fadda G.,Cossellu S.,Roe S.L.,Rubattu L.,Zanetti S.(1984) Clin. Chem. Newsletter 4:72-77

10. Fadda G.,Mureddu L.,Sanciu F.,Zanetti S.(1985) Abstracts Ann. Meet. American Society for Microbiology (Abstract C99) Las Vegas,Nevada

11. Von Graevenitz A.,Koch E.,Selfinger M.(1984) J. Microbiol. Meth.3:95-100

12. Heifets L.B.,Iseman M.D.(1985) J. Clin. Microbiol. 21:200-204

13. Heifets L.B.,Liudholm P.,Iseman M.B., (1985) Abstracts Ann.Meet.American Society for Microbiology (Abstract U 59) Las Vegas, Nevada

14. Kertcher J.A.,Chen M.F.,Charache P.,Hwangbo C.,Camargo E.F.,McIntyre P.A., Wagner H.N.Jr.(1978) Am.Rev.Resp.Dis. 117:631-637

15. Laszlo A.,Gill P.,Mandzel V.,Hodgkin M.M.,Helbecque D.M. (1983) J. Clin. Microbiol. 18:1335-1339

16. Laszlo A.,Michaud R. (1984) Bulletin of the international union against tuberculosis 59:185-187

17. Loder A.B.,Gruft L. (1985) Abstracts Ann.Meet.American Society for Microbiology (Abstract U51) Las Vegas, Nevada

18. Middlebrook G.,Reggiardo Z.,Tigertt W.D. (1977) Am. Rev. Resp. Dis. 115:1066-1069

19. Randall E.L. (1975) In Schlessinger D. (ad), Microbiology AMS Washington D.C. pp.39-40

20. Roberts G.D.,Goodman N.L.,Heifets L.B.,Larsh H.W.,Lindner T.H., McClatchy J.K.,McGinnis M.R.,Siddiqi S.H.,Wright P. (1983) J.Clin.Microbiol. 18:689-696

21. Snider D.E.,Good R.C.,Kilburn J.O.,Laskowski L.F.Jr.,Lask R.H.,Marr J., Reggiardo Z.,Middlebrook G.(1981) Am.Rev.Respir.Dis.123:402-406

22. Siddiqi S.H.,Libonati J.P.,Middlebrook G.(1981) J.Clin. Microbiol. 13:908-912

23. Takashi H.,Foster V.(1983) J.Clin. Microbiol.17:380-381

24. Ungheri D.,Morvillo E.,Sanfilippo A.(1983) G.Ital.Chemioter.30:97-100

25. Vincke´ G.,Yegers O.,Vanacter H.,Jenkins P.A.,Butzler J.P.(1982) J.Antimicrob.Chemother.10:351-354

26. Wayne L.G.,Krasnow I.(1966) Am.J.Clin.Pathol.45:769-771

27. Woodley C.L.,Kilburn J.O.(1982)Am.Rev.Respir.Dis.126:585-589

COMPARISON OF A RADIOMETRIC (BACTEC) AND CONVENTIONAL CULTURE METHOD FOR RECOVERY AND IDENTIFICATION OF MYCOBACTERIA FROM SMEAR-NEGATIVE SPECIMENS

L. MATAS, M. LUQUIN, M.J. CONDOM, P. COLL and V. AUSINA
Departamento de Microbiologia, Hospital de la Sta. Creu i Sant Pau,
Facultad de Medicina de la Universidad Autonoma de Barcelona (Spain)

INTRODUCTION

Rapid laboratory diagnosis of mycobacterial disease is essential for medical and epidemiological purposes. The use of rapid radiometric methods which use BACTEC instrumentation (Johnston Laboratories Cokeysville, Md.) for both recovery and identification of mycobacteria and drug-susceptibility testing of *Mycobacterium tuberculosis* (TB) has been previously reported (1-6,8).

Previous studies reported the BACTEC system to be as successful as conventional culturing methods in detecting mycobacterial growth in clinical specimens and also revealed that the BACTEC system detected mycobacterial growth more rapidly that did conventional cultural methods (1,2,5).

A specimen with a negative acid-fast smear and a positive culture is not an unusual occurrence in a clinical laboratory; approximately 35% of positive mycobacterial cultures in our laboratory have negative direct acid-fast smears.

This study was carried out to evaluate the BACTEC system and conventional methods for detecting mycobacteria in specimens with a negative acid-fast smear and a positive culture. In addition, the study evaluates p-nitro-a-acetyl-amino-b-hydroxypropiophenone (NAP), 8-azaguanine (8-AZG) and para-nitrobenzoic acid (PNB) inhibition for the rapid differentiation of TB from MOTT bacilli.

MATERIAL AND METHODS

Acid-fast smear-negative specimens submitted to our laboratory from November 1982 to December 1984 for routine mycobacterial culture were included in this study. Of 6086 specimens, 223 had negative direct acid-fast smears, but mycobacteria were recovered by either the conventional or the radiometric method. Processing of specimens for preparation of inocula, digestion and decontamination were carried out according to standard procedures (7).

Two Lowenstein-Jensen (LJ) slants were each inoculated with 0.2 ml of the processed specimen. In addition, 0.1 ml of the specimen was inoculated into a bottle containing 2 ml of Middlebrook 7H12 broth medium (12A medium, Johnston Laboratories, Inc.) containing ^{14}C-palmitic acid. To reduce non-mycobacterial contamination, 0.1 ml of an antimicrobial mixture (PACT supplement, Johnston Laboratories, Inc.) was added to each broth. The resulting concentration of

PACT in each culture bottle was 50 U polymixin B, 5 μg amphotericin B, 25 μg carbenicillin, and 2.5 μg trimethoprim per ml of culture medium.

A group of 1751 specimens was also inoculated into additional bottles each containing 4 ml of 7H12 broth medium. These 4 ml 7H12 bottles contained 0.2 ml of inoculum and 0.2 ml of PACT supplement.

The LJ slants were incubated at 35°C and examinated for appearance of growth at weekly intervals for 8 weeks. A carbon dioxide-air mixture was introduced into 7H12 vials by the BACTEC instrument. The 7H12 bottles were incubated at 37°C and screened for $^{14}CO_2$ output from ^{14}C-palmitic acid at 3-day intervals for 8 weeks with the BACTEC. Middlebrook 7H12 broth media producing a growth index (GI) of >20 were considered to be positive. Positive bottles were examined daily until a reading of 100 was reached, and then samples of each medium were used to prepare acid-fast stains and subcultures on LJ medium.

Isolates from LJ slants were identified by conventional methods (7). NAP (5 μg/ml) inhibition testing was performed according to the manufacturer's instructions; PNB (500 μg/ml) and 8-AZG (250 μg/ml) inhibition testing was performed as described by Park et al. (4).

RESULTS

A total of 223 clinically significant mycobacteria were recovered from 6086 acid-fast smear-negative specimens submitted for culture (211 *M.tuberculosis*, 2 *M.avium-intracelllulare*, 5 *M.kansasii*, 1 *M.scrofulaceum*, 2 *M.fortuitum* and 2 *M.chelonae*).

Mycobacteria were recovered from 191 specimens (85.6%) by the radiometric method with 7H12 medium and from 190 specimens (85.2%) by the conventional method with LJ medium. Although neither the radiometric nor the conventional method alone was able to recover all cultures, the data (Tables I and II) suggest that improved isolation rates are possible if 7H12 and LJ are used together.

The average time to report the culture results was 18.6 days (range, 13.6 to 28.5 days) and 32.4 days (range, 25.5 to 44.5 days) for the BACTEC and conventional methods, respectively. The use of 7H12 medium significantly reduced the time required to report positive culture results in specimens with negative smears.

The contamination rates in the 7H12 and LJ media were 2.8 and 4.4 respectively. In this study the addition of PACT supplement to the 7H12 medium reduced contamination without affecting overall recovery. In addition the BACTEC system offers the advantage of a liquid medium into which selected antimicrobial agents, other than PACT can be added (4).

Doubling of medium volume with increased inoculum size yielded the highest

TABLE I

MYCOBACTERIA ISOLATED FROM CLINICAL SPECIMENS BY RADIOMETRIC AND CONVENTIONAL METHODS

Mycobacteria	No.recovered with: 7H12	LJ	Total no. recovered
M. tuberculosis	181	182	211
MOTT*	10	8	12
Total no.(%)	191(85.6)	195(85.2)	223(100.0)

*Excluding *M. gordonae* and other nonsignificant isolates.

TABLE II

MYCOBACTERIA ISOLATED FROM 6086 ACID-FAST SMEAR-NEGATIVE SPECIMENS

Specimens	Total no.	No.recovered with:* 7H12	LJ	Total no.* recovered
Sputa	3331	72	65	80
Bronchial washings	768	27	22	28
Gastric aspirates	210	21	21	26
Organic fluids	931	40	44	48
Urines	465	3	6	6
Surgical tissues	160	22	24	27
Pus and others	221	6	8	8
Total no.(%)	6086	191(85.6)	190(85.2)	223(100.0)

*Including TB and significant MOTT isolates.

recovery rate (93.4%) in the BACTEC system (Table III). However, the increased inoculum size did not decrease significantly the detection time of TB by the BACTEC procedure; the average time to report the culture results was 23.1 and 21.0 days for the 0.1 and 0.2 ml inocula, respectively.

Fresh clinical isolates were identified correctly as TB or MOTT, in an average of 3 to 5 days by the BACTEC radiometric procedure. Identification of TB by NAP, 8-AZG and PNB inhibition showed 100% agreement with results obtained by customary procedures. Growth of all isolates of TB were inhibited by NAP, 8-AZG and PNB while all the MOTT bacilli tested were inhibited.

TABLE III
EFFECT OF INCREASED INOCULUM SIZE ON RECOVERY RATE OF *M.TUBERCULOSIS* IN BACTEC 7H12 MIDDLEBROOK MEDIUM

Specimens	Total no.	No. recovered with:			Total no. recovered
		LJ	2ml 7H12*	4ml 7H12*	
Sputa	244	16	18	19	19
Bronchial washings	723	42	40	46	49
Gastric aspirates	38	1	2	1	2
Organic fluids	640	16	10	17	19
Others	106	1	1	2	2
Total no.(%)	1751	76(83.5)	71(78.0)	85(93.4)	91(100.0)

*2ml bottles received 0.1 ml of inoculum and 4 ml bottles received 0.2ml of inoculum.

The BACTEC system for recovery of mycobacteria, although more costly than conventional procedures using slanted media, is a valuable improvement in clinical microbiology laboratory performance. The benefits include, shorter time for recovery, ready separation of TB from MOTT bacilli (8) and susceptibility testing of TB (9), in addition to a higher yield of positive cultures.

REFERENCES

1. Takahashi, H., Foster,F. (1983). J.Clin.Microbiol. 17:380-81

2. Damato J.J., Collins, M.T., Rothlauf, M.V., McClatchy, J.K. (1983). J.Clin.Microbiol. 17:1066-1073.

3. Morgan, M.A., Hortsmeier, C.D., De Young, D.R., Roberts, G.D. (1983). J.Clin.Microbiol. 18:384-388.

4. Park, C.H., Hixon, D.L., Ferguson, C.B., Hall, S.L., Risheim, C.C., Cook, C.B. (1984). Am.J.Clin.Pathol. 81:341-345.

5. Ausina, V., Matas, L., Luquin, M., Coll, P., Carbó, LL., Prats, G. (1984) Enf.Infec. y Microbiol.Clin. 2:16-25.

6. Fadda, G., Roe, S.L. (1984). J.Clin.Microbiol. 19:720-721.

7. Vestal, A.L. Department of Health Education and Welfare publication nº (CDC) 76-8230. Centers for Disease Control, Atlanta, Ga.

8. Siddiqi., S.H., Hwangbo, C.C., Silcox, V., Good, R.C., Snider, D.E., Jr., Middlebrook, G. (1984). Am. Rev. Respir. Dis. 130: 634-640.

9. Siddiqi, S.H., Libonati, J.P., Middlebrook, G. (1981). J.Clin. Microbiol. 13:908-912.

SUSCEPTIBILITY TESTING OF M.TUBERCULOSIS TO STREPTOMYCIN, INH, RIFAMPICIN AND ETHAMBUTOL BY A RADIOMETRIC METHOD

L. MADARIAGA, J.M. ZUBIAUR and R. CISTERNA
Department of Microbiology, Hospital Civil de Bilbao, Faculty of Medicine, U.P.V./E.H.U. (Spain)

INTRODUCTION

The basic principle of radiometric bacterial growth detection is the use of a ^{14}C labeled substrate that is metabolized by bacteria. $^{14}CO_2$ is released into the atmosphere above the medium in the vial, and the radioactivity can be measured quantitatively with the help of a Bactec instrument. The quantity of $^{14}CO_2$ produced is directly proportional to the rate and amount of growth of bacteria. For mycobacteria, vials contain 7H12 Middlebrook culture medium.

This method permits a reduction in the time needed for drug susceptibility testing of *M.tuberculosis* in comparison with conventional methods.

We report here the results obtained in our hospital after testing 69 strains of *M.tuberculosis*.

MATERIAL AND METHODS

The 69 strains tested were isolated from clinical specimens in Lowenstein-Jensen medium.

For the susceptibility test, index of 300 or more; 0.1 ml from an initial BACTEC 12A vial with a growth, was inoculated into four vials containing streptomycin 4 μg/ml, INH 0 2 μg/ml, rifampicin 2 μg/ml and ethambutol 10 μg/ml respectively (final concentration). A control vial was inoculated with the initial inoculum diluted 100 times in the recommended diluting fluid. All the vials were read on a BACTEC instrument immediately after inoculation to introduce a 5% CO_2 atmosphere and were incubated at 37°C. Vials were read every 24 hours.

Once the growth index (GI) in the control vial was 30 or more, the results were interpreted as follows: Δ GI of control vial greater than Δ GI of the drug vial, susceptible; Δ GI of the control less than Δ GI of the drug vial, resistant.

RESULTS

The time to interpret the results ranged between 2 and 14 days, the mean value being 6.05 days and the standard deviation 2,31.

Only in one case was less than 3 days needed, and in seven cases 9 or more were necessary.

All the strains, except three, were susceptible to the four tested drugs. Two strains were resistant to INH and the other to INH and streptomycin.

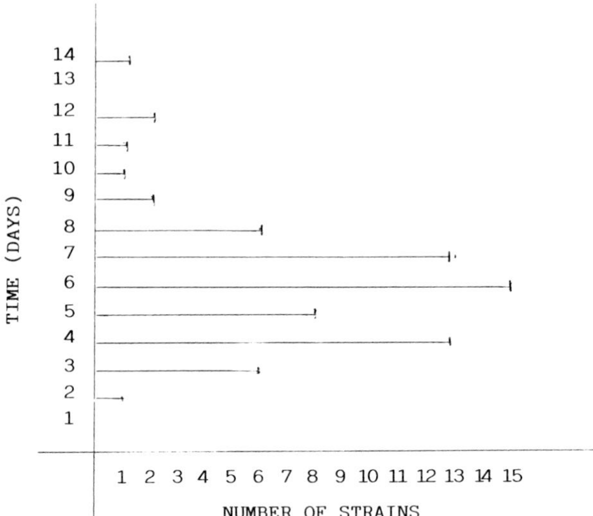

Fig. 1. Time needed to interpret the susceptibility test.

CONCLUSION

The mean value of days that we needed in our study is very close to that obtained by other authors. Since there is agreement between the radiometric and the conventional method, as has been shown in other papers, the BACTEC radiometric method seems to have important advantages, particularly for clinical purposes.

A time of less than three days must be considered, due to a too heavy inoculum, which may cause some problems in interpreting resistance. On the other hand, a time of more than 9 days may be due to a very light inoculum or to low viability. In this case the interpretation may not be affected.

Mycobacteria of Clinical Interest. M. Casal, editor

USE OF THE RADIOMETRIC METHOD FOR ISOLATION OF MYCOBACTERIA

J.M. ZUBIAUR, L. MADARIAGA, G. MARTIN and R. CISTERNA
Department of Microbiology, Hospital Civil de Bilbao, Faculty of Medicine, U.P.V./E.H.U. (Spain)

INTRODUCTION

The long time needed for *M. tuberculosis* to grow has always presented a difficulty for accurate diagnosis, when microscopic examination of stained smears is negative, and symptoms are not specific.

A new method, recently introduced, for decreasing the time, is the radiometric technique. In this method, mycobacteria metabolize the ^{14}C substrate and release $^{14}CO_2$ into the atmosphere. The radioactivity of $^{14}CO_2$ is measured with the help of a BACTEC 460 instrument.

We present here our results after testing 897 specimens by means of the radiometric and the conventional method.

MATERIAL AND METHODS

The 897 tested specimens were obtained from hospitalized and external patients: 524 from the respiratory tract (sputum, bronchial brushing and aspiration and translaryngeal aspiration), 168 from serosa (pleura, sinovial and pericardium), 32 CRL, 77 urine and 96 of other origin.

Among the specimens from the respiratory tract, 31 were positive, as shown by microscopic examination of direct smears. Sputa were only accepted when they were highly suspected of harboring infectious diseases.

Specimens that required decontamination were treated by both the following methods in parallel:

Conventional method: inoculation in L-J medium after digestion with N-acetyl-L-cysteine, concentration and decontamination with sodium dodecylhydrogen-sulphate; neutralization was done with H_3PO_4.

For this procedure, decontamination was done with Na_3PO_4 (10%) and neutralization with H_3PO_4. Prior to inoculation, 0.1 ml of PACT were added to each 12A vial (7H12 Middlebrook medium). PACT is an antimicrobial solution containing polymyxin B 1000 units/ml, amphotericin B 100 µg/ml, carbenicillin 500 µg/ml and trimethoprim 50 µg/ml.

Vials were read with a BACTEC instrument immediately after inoculation and successively on the 3rd, 6th, 9th and 14th days, and weekly thereafter for 8 weeks.

RESULTS

Mycobacteria were isolated from 48 specimens (5.35% of the total): 34 sputum (9.21%), 4 bronchial aspiration and brushing (2.87%), 4 pleural effusions (2.96%), 1 urine (1.29%), 1 CRL (3.125%), 1 pericardial effusion (10%), 1 sinovial effusion (4.34%) and 2 pleural biopsies (11.76%) allowed the isolation of *M. tuberculosis*.

Table I shows the number of positive cultures in 7H12 Middlebrook medium and Lowenstein-Jensen medium.

TABLE I

	7H12 + L-J +	7H12 - L-J +	7H12 + L-J -	7H12 cont.* L-J +
Sputum	28	0	1	5
Bronchial brushing and aspiration	3	1	0	0
Pleural effusion	4	0	0	0
Urine	1	0	0	0
CRL	1	0	0	0
Pericardial effusion	1	0	0	0
Sinovial effusion	0	1	0	0
Pleural biopsy	0	2	0	0

*The culture was contaminated.

CONCLUSION

In our results the sensitivity of both methods was very similar, and the use of the radiometric method offers no important advantage as regards the number of isolations.

The most important advantage relates to the mean time required for growth of *M. tuberculosis* with both methods. The mean number of days necessary to obtain a positive culture in 7H12 Middlebrook medium was 13.44, but was 25 days in L-J medium.

ISOLATION OF *MYCOBACTERIUM* spp. BY CONVENTIONAL AND RADIOMETRIC METHODS: A COMPARATIVE STUDY

M.A. VITORIA, C. RUBIO-CALVO, J. GIL, A. REZUSTA and M.P. EGIDO
Department of Microbiology, School of Medicine, University of Zaragoza, 50009 Zaragoza (Spain)

INTRODUCTION

An automated and radiometric method was developed by DeLand and Wagner (2) for detection of microbial growth. Cummings et al. (1) showed that the same principle could be utilized to detect growth of *M. tuberculosis*. Middlebrook et al. (5) further developed the technique and introduced a special 7H12 medium which contained (^{14}C)palmitic acid as the labeled substrate.

Several reports have been published on the detection and recovery, as well as susceptibility testing of mycobacteria from clinical specimens by the radiometric method (3,6,7,8).

We studied the capacity of the radiometric method to recover *M. tuberculosis* or mycobacteria other than tubercle (MOTT) bacilli, since awareness of infection caused by MOTT bacilli is increasing (10). We compared the radiometric and conventional methods, and their effectiveness on recovery of mycobacteria from clinical specimens.

MATERIAL AND METHODS

From October 1983 to May 1985 we examined 2600 specimens for mycobacteria at the University Hospital of Zaragoza. In all cases, the specimen collection, digestion and decontamination were performed according to Vestal (9) and Lennette (4). One Lowenstein-Jensen (LJ) slant and one Lowenstein with pyruvate (LP) slant were each inoculated with 0.1 ml of processed specimen. In addition 0.1 ml of the specimen was inoculated into a vial with Middlebrook 7H12 broth containing (^{14}C)palmitic acid. To reduce non-mycobacterial contamination, 0.1 ml of PACT solution (Johnston laboratories), consisting of polymixin B, amphotericin B, carbenicillin and trimethoprim, was added to each 7H12 medium.

Both LJ and 7H12 cultures were incubated in 5% CO_2 at 37°C. The LJ and LP slants were examined weekly for 8 weeks for appearance of growth. 7H12 vials were screened for production of (^{14}C)palmitic acid at 3-day intervals for the first 14 days, then weekly for 4 weeks with BACTEC 460 (Johnston laboratories). When the BACTEC culture vial registered a growth index (GI) of 50 or more, a small volume of the medium was removed and stained for acid-fast bacilli (AFB) by the Zhiel-Neelsen technique. If the FAB stain was positive, the specimen was checked for bacterial contamination by subculturing onto a blood agar plate and observed for growth after 18 hours of incubation. Identification of mycobacteria was carried out by conventional methods (4,9).

RESULTS

Of the 2600 specimens cultured, we recovered mycobacteria from 172 specimens (6.6%). *M. tuberculosis* was recovered from 116 specimens (4.4%) and MOTT bacilli from 34 specimens (72.3%) by the conventional methods with LJ and LP. Using the radiometric method with 7H12 medium *M. tuberculosis* was recovered from 115 specimens (99.1%) and MOTT bacilli from 31 specimens (65.9%) (Tables 1, 2). The average time required to detect positive cultures in LJ or LP media was 29 days versus 19 days in the 7H12 medium. The contamination rates in the 7H12 medium was 5.8%. The range of MOTT bacilli recovered during the evaluation is shown in Table 3.

22 strains grew only in the BACTEC system, whereas 37 strains grew only in conventional media.

DISCUSSION

Although it did not prove possible to recover all cultures by the conventional or the radiometric method alone, the 7H12 medium used in conjunction with classic media appears to maximize the recovery of mycobacteria.

Our data suggest that improved isolation rates are possible if radiometric and LP are used together for recovery of *M. tuberculosis*, since only 1 specimen was positive in LJ alone.

Of the total 47 MOTT bacilli isolates, 34 (72.3%) were recovered by the conventional method, and 40 (85.1%) were recovered by the radiometric method. A total of 13 MOTT isolates (27.6%) were not recovered by the classical method, and 17 (36.1%) were not recovered by the radiometric method. When the radiometric method was used together with the conventional method, recovery increased by 13 (27.6%).

Our results are in agreement with those obtained by Takahashi (8) and Fadda (3).

The use of the radiometric method reduces the time required to report positive results.

TABLE 1. Mycobacteria isolated from clinical specimens

	n°total of isolates	Conventional L.J.	Conventional L.P.	radiometric
M. tuberculosis	125	102(81,2%)	100(80%)	115 (92 %)
MOTT	47	23(48,9%)	22(46,8%)	40 (85,1%)

TABLE 2. Relationship between presence of Acid Fast Bacilli (AFB) and recovery of Mycobacterium tuberculosis from several media.

No AFB found: 33 smears BACTEC	L.J.	L.P.	n° (%)	Presence of FAB: 92 smears BACTEC	L.J.	L.P.	n° (%)
+	+	+	14(42,4)	+	+	+	77(83,6)
-	-	+	5(15,1)	-	-	+	6(6,5)
+	-	-	4(12,1)	+	-	-	5(5,4)
-	+	+	4(12,1)	-	+	+	4(4,3)
+	-	+	3(9)				
+	+	-	2(6)				
-	+	-	1(3)				

TABLE3. Recovery of 47 MOTT isolated from clinical specimens

MOTT	Total n°of isolates	BACTEC(%)	L.J.(%)	L.P.(%)	AFB
M. gordonae	15	9 (60)	8 (53,3)	6 (40)	2
M. chelonei	13	8 (61,5)	8 (61,5)	7 (53,8)	0
M. xenopi	9	7 (77,7)	3 (33,3)	3 (33,3)	2
M. fortuitum	3	2 (66,6)	0	2 (66,6)	0
M. terrae	3	0	2 (66,6)	1 (33,3)	0
M. avium complex	1	1 (100)	1 (100)	1 (100)	1
M. dierhoferi	1	1 (100)	0	1 (100)	0
M. nonchromogenicus	1	1 (100)	1 (100)	0	0
M. scrofulaceum	1	1 (100)	0	1 (100)	1

REFERENCES

1. Cummings DM, Ristroph D, Camargo EF, Larson SM, Wagner HN.(1975). Radio metric detection of metabolic activity of Mycobacterium tuberculosis. J. Nucl. Med. 16: 1189-91.

2. DeLand FH, Wagner HN. (1969). Early detection of bacterial growth with carbon 14 labeled glucose. Radyilogy: 92-154.

3. Fadda G, Roe SL. (1984). Recovery and susceptibility testing of Mycobacterium tuberculosis from extrapulmonary specimens by the BACTEC radiometric. J. Clin. Microb. 19: 720-21.

4. Lennette EH, Balows A, Hausler WJ, Truant JP,(ed) (1980). Manual of clinical microbiology 3rd ed. American Sosiety for Microbiology, Washington D.C.

5. Middebrook G. (1977). Automatable radiometric detection of growth of Mycobacterium tuberculosis in selective media. Am. Rev. Resp. Dis. 115: 1066-69.

6. Morgan Ma, Doerr KA, Hempel HO, Goodman NL, Roberts GD. (1985). Evaluation of the p-nitro-α-acetylamino-β-hydroxypropiophenone differential test for identification of Mycobacterium tuberculosis complex. J. Clin. Microb. 21: 634-35.

7. Siddiqi SH, Libonati JP, Middlebrook G. (1981). Evaluation of a rapid radiometric method for drug susceptibility testing of Mycobacterium tuberculosis. J. Clin. Microb. 13: 903-12.

8. Takahashi H, Foster V. (1983). Detection and recovery of mycobacteria by a radiometric procedure. J. Clin. Microb. 17:380-81.

9. Vestal AL. (1975). Procedures for the isolation and identification of mycobateria. U.S. Public Health Service Publication n°75- 8230. Center for Disease Control, Atlanta, Ga.

10. Wollinky E. (1979). Nontuberculousis mycobateria and associated diseases. Am. Rev. Resp. Dis. 119: 107-59.

Mycobacteria of Clinical Interest. M. Casal, editor

RAPID AND PRECISE IDENTIFICATION OF MYCOBACTERIAL SPECIES BY GAS CHROMATOGRAPHY AND GAS CHROMATOGRAPHY-MASS SPECTROMETRY OF MYCOLIC ACIDS

KENJI KANEDA, SADAO IMAIZUMI, SEIKO MIZUNO, IKUKO TOMIYASU, MICHIO TSUKAMURA and IKUYA YANO
Department of Bacteriology, School of Medicine, Niigata University, Niigata, and National Chubu Hospital, Obu, Aichi (Japan)

Identification of mycobacterial species is usually performed by a combination of many biochemical tests, but this requires a lot of work and differentiation between species is often not clear. Minnikin et al. (1) reported several subclass patterns of mycolic acids, 2-alkyl-3-hydroxy long-chain fatty acids, of Mycobacteria on thin-layer chromatography (TLC), but this is insufficient for the identification of each species. Here, we further reveal the molecular composition of α-mycolic acids in different species using gas chromatography (GC) and gas chromatography-mass spectrometry (GC/MS).

MATERIAL AND METHODS

Thirteen rapidly growing species and twelve slowly growing ones were grown in glucose-peptone-yeast extract medium on a shaker at $30^{o}C$ and in Sauton medium at $35^{o}C$ for 3-4 w. respectively. The bacteria harvested were hydrolysed and then transmethylated. These methyl ester fatty acids were separated into several subclasses of mycolic acids on TLC. The α-mycolic acid methyl esters recovered from a plate were then prepared to form trimethylsilyl (TMS) ether derivatives, which were injected into GC (Hitachi 063) and GC/MS (Hitachi M-80B) equipped with a 0.3 m glass column coated with 1% OV-101, operated at 330 - $340^{o}C$.

RESULTS AND DISCUSSION

On TLC, each mycobacterial species revealed its characteristic subclass pattern of mycolic acids (Fig. 1). On gas chromatograms, the TMS ether derivatives of α-mycolic acid methyl esters were well separated according to the total number of carbon atoms, as shown in Fig. 2 for *M.intracellulare* and *M.avium*. The mass spectra at the main peak on the gas chromatogram of *M.avium* (Fig. 3) proved to be in accordance with $C_{80:2}$ α-mycolic acid with a $C_{56:2}$ β-unit and a $C_{24:0}$ α-unit, from the fragmentation pattern of α-mycolic acid as shown in Fig. 4 (2,3). On the other hand, $C_{80:2}$ α-mycolic acids of *M.tuberculosis* possessed a $C_{26:0}$ α-unit, as shown by (B-29)=482 (Fig.3). Mass chromatograms monitoring the mass number of the (M-15)ions of each carbon-numbered α-mycolic acid (dien) and the (B-29)ions of the $C_{22:0}$,

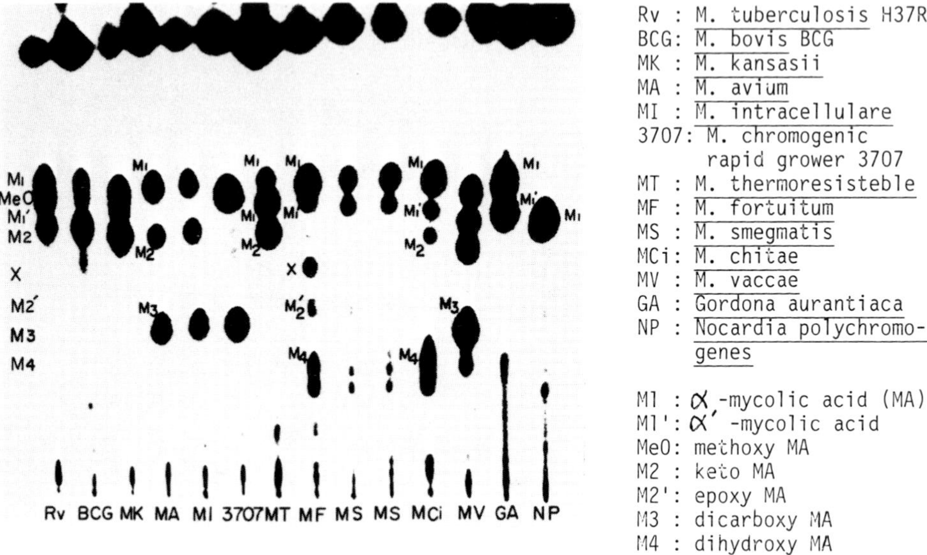

Fig. 1. Thin-layer chromatogram of the total fatty acid methyl esters from representative mycobacterial species and two related taxa. Developing solvent was n-hexane/diethyl ether 4:1, by vol.

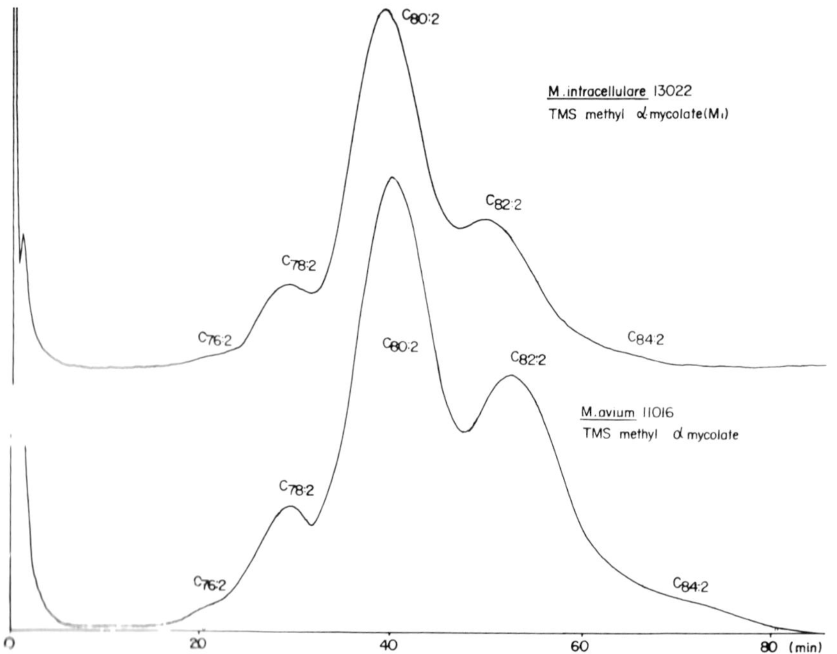

Fig. 2. Gas chromatograms of TMS ether derivatives of methyl α -mycolate of M. intracellulare (upper) and M. avium (lower). The two species show similar patterns. The chemical structure of each peak was determined by GC/MS.

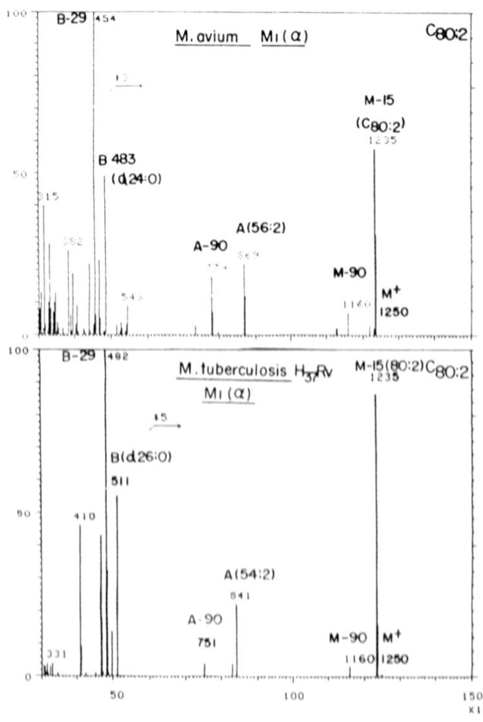

Fig. 3. Mass spectra of TMS ether derivatives of methyl α-mycolate of M. avium (upper) and M. tuberculosis (lower). From the mass number of the characteristic fragment ions generated as shown in Fig. 4, the upper is determined to be C80:2 with C56:2 (β) and C24:0 (α), and the lower is C80:2 with C54:2 (β) and C26:0 (α).

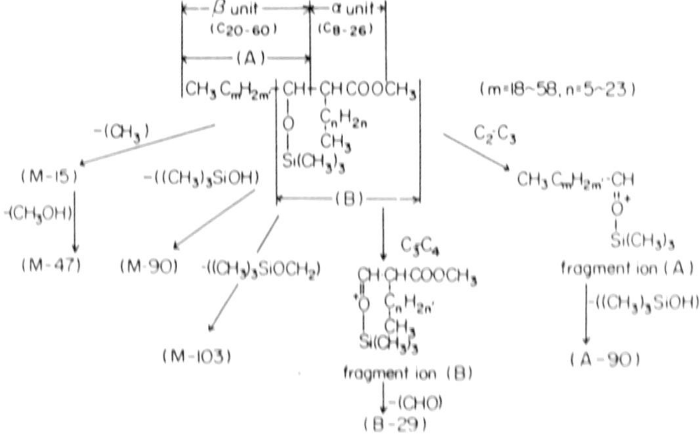

Fig.4. General structure and mass fragmentation pattern of TMS ether derivatives of methyl mycolate. By electron impact, several characteristic ions are generated: (M), (M-15) and (M-90) are derived from the whole molecule, and (A) and (A-90) from the straight portion or β-unit, and (B) and (B-29) from the branched portion or α-unit containing portion. From the mass number of these fragment ions the chemical structure of α-mycolic acid can be determined.

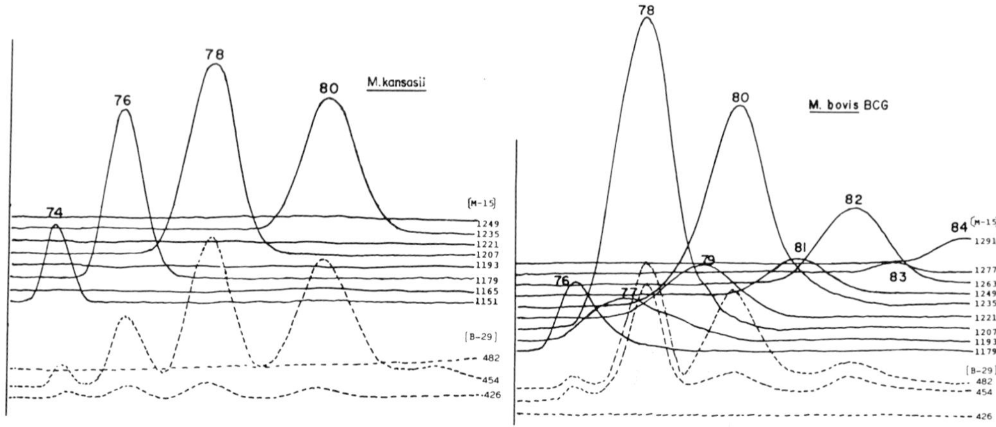

Fig. 5. Mass chromatograms of TMS ether derivatives of methyl α-mycolate of M. kansasii (left) and M. bovis (right). (M-15) ions of each carbon-numbered dien mycolic acids (solid lines) and (B-29) ions, m/z 482, 454 and 426, being corresponded to C26:0, C24:0, C22:0 α-unit (dotted lines) respectively were monitored.

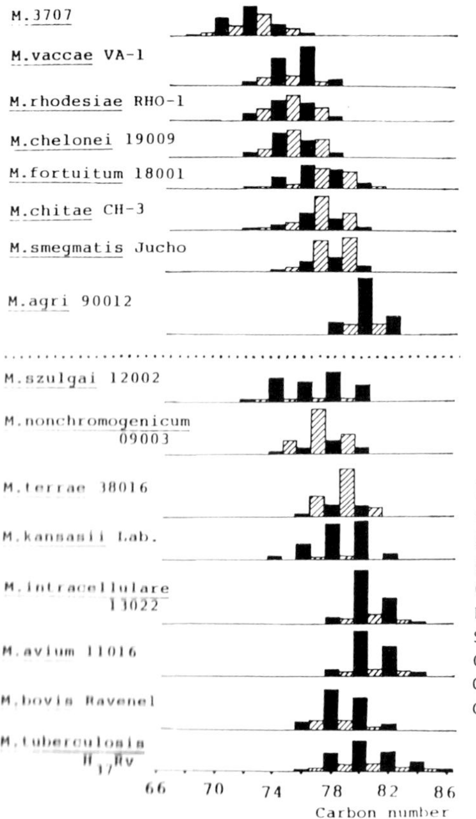

Fig. 6. α-mycolic acid molecular species composition of representative rapidly growing (above the dotted line) and slowly growing (under the line) mycobacteria. Generally even carbon-numbered molecules ( ) exist mainly, in some species like *M. chitae* and *M. terrae* odd ones ( ) are dominant. Distribution of the carbon chain length is also characteristic to each species.

$C_{24:0}$ and $C_{26:0}$ α-units containing α-mycolic acid enabled the quantification of each molecular species and the ratio of the homologues differing at the α-unit. In *M.kansasii* molecules with even numbers of carbon atoms with a $C_{24:0}$ α-unit predominated, while in *M.bovis* there were some with odd numbers of carbon atoms, and the α-unit was mainly $C_{26:0}$ (Fig. 5).
The molecular composition characteristic of each species, grown under the same cultural conditions calculated from the mass chromatograms is shown diagrammatically in Fig. 6. The grouping of species according to α-mycolic acid composition was similar to that of previous classifications (4): *M.parafortuitum* complex; C_{72-77}, $C_{22:0}$ α-unit; *M.fortuitum* complex; C_{72-79}, $C_{24:0}$ α-unit, containing an equal amount of molecules with even and odd numbers of carbon atoms; *M.terrae* complex; C_{75-81}, $C_{24:0}$ α-unit, mainly odd numbered; *M.avium-intracellulare-scrofulaceum* complex; C_{78-84}, $C_{24:0}$ α-unit, mainly even numbered *M.tuberculosis* complex; C_{76-84}, mainly $C_{26:0}$ α-unit.

In conclusion: (1) a single analysis of α-mycolic acids by GC/MS gives much information rapidly (up to 2 h) about the precise structure and the quantitative data of each molecular species. (2) the characteristic α-mycolic acid composition of each species is valuable as a finger print for the classification and identification of mycobacterial species, particularly of new isolates from clinical or natural sources.

REFERENCES

1. Minnikin D E (1982) In: Ratledge C, Stanford J (eds) The Biology of The Mycobacteria. Vol. 1. Academic Press, London, pp 95-184.
2. Yano I et al. (1978) Biomed Mass Spectrom 5: 14- 24.
3. Toriyama S et al. (1978) FEBS Letters 95: 111-115.
4. Goodfellow M and Wayne LG (1982) In: Ratledge C, Stanford J (eds) The Biology of The Mycobacteria. Vol. 1. Academic Press, London, pp. 471-521.

EPIDEMIOLOGY

DESCRIPTION OF NEW EPIDEMIOLOGICAL MARKER IN TUBERCULOSIS

MANUEL CASAL ROMAN, M[a] JOSE LINARES SICILIA and M[a] MANUELA MORALES SUAREZ-VARELA
Department of Microbiology, School of Medicine, Cordoba University, Cordoba (Spain)

INTRODUCTION

The epidemiological interest in finding a new, easy to use epidemiological marker for tuberculosis is understandable, given the difficulty of detecting tuberculosis in patients from the asymptomatic initial stages to the full disease, and also because of possible reinfection, either endogenic or exogenic. We should like to report our investigations of the relationship between the strains isolated and a few characteristic features of the patients.

In the intensive search for such an epidemiological marker, only phagotypy by Frodman, Will and Bogen (1) in 1954, mycobacteriocins by Tokina and Takena (2) in 1977 and enzymatic biovarieties by Casal and Linares (3,4) in 1984 have so far been described, and it is on the latter work which is still being studied that we have based our study.

MATERIAL AND METHODS

A total of 241 strains isolated from specimens from sputum, L.C.R., urine, bronchial aspiration, purulent exudate and pleural fluid were provided by the Mycobacteria Reference Centre of the Faculty of Medicine in Cordoba.

The strains were inoculated into Lowenstein-Jensen medium where they were kept for a minimum of 30 days at 37°C. We performed our test study at 40-50 days.

The API ZYM system (5) was used to study the enzymatic activity. Each kit has a plastic strip containing 20 cupules, 19 of which have an enzymatic substrate and one cupule is reserved as control.This system detects the presence of alkaline phosphatase, esterase (C4), lipase esterase (C8), lipase (C14), leucine arylamidase, trypsin, chymotrypsin, acid phosphatase, phosphoamidase, alpha-galactosidase, beta-glucosidase, beta-galactosidase, beta-glucuronidase, alpha-glucosidase, N-acetyl-beta-glucosaminidase, alpha-mannosidase, alpha-fucosidase, valine arylamidase and cystine arylamidase.

A suspension of *Mycobacterium tuberculosis* was made in sterile distilled water, with turbidity approximating that of McFarland 5-6 nephelometric standard. Two drops of the mycobacterial suspension were added with a Pasteur pipette to each cupule of the API ZYM strip. Each strip was placed in a moist chamber and incubated at 37°C for 5 hours. After incubation, 1 drop of reagent A (25 g of Tris, 110 ml of HCl, and 100 g of lauryl sulfatase in 1 liter of distilled water) and 1 drop of reagent B (3.5 g of fast blue BB in 1 liter of 2-methoxyethanol) were added to each cupule. The resulting colors which developed within 5 minutes were compared with the API ZYM color chart. The tests were repeated at least twice with each isolate to confirm the results. Each kit of the system was controlled by a study of enzymatic activity using a control *Mycobacterium tuberculosis* strain, of which we knew the enzymatic profile.

RESULTS

Table I shows the results according to enzymatic activity. The enzymes which gave constant results, remaining positive in all cases, were alkaline phosphatase, esterase (C4), lipase esterase (C8), lipase (C14), acid phosphatase and phosphoamidase. We did not find the following enzymes to be active: chymotrypsin, alpha-galactosidase, beta-galactosidase, beta-glucuronidase, N-acetyl-beta-glucosaminidase, alpha-mannosidase and alpha-fucosidase.

TABLE I

ENZYMATIC BIOVARIETY OF *MYCOBACTERIUM TUBERCULOSIS*

Biovariety	Number of strains (%)	Enzymatic activity in cupule				
		7	8	9	16	17
A	124 (51.45)	+	+	-	-	+
B	45 (18.67)	+	+	-	-	-
C	19 (7.88)	-	-	-	-	-
D	12 (4.97)	+	-	-	-	-
E	9 (3.73)	+	+	-	+	+
F	9 (3.73)	+	+	+	-	-
G	6 (2.48)	-	-	-	-	+
N.C.	17 (6.50)					

Total number of strains: 241

N.C.: Unclassified
7 - valine arylamidase; 8 - cystine arylamidase; 9 - trypsin;
16 - alpha-glucosidase; 17 - beta-glucosidase

The enzymes with differing activity were valine arylamidase, cystine arylamidase, trypsin, alpha-glucosidase, beta-glucosidase. If valine arylamidase, cystine arylamidase and beta-glucosidase were active they were considered as biovariety A. In this study 124 (51.45%) were active.

If only valine arylamidase and cystine arylamidase were positive, they were considered as group B (45 strains, 18.67%). If they remained positive, they were considered as group C (19 strains, 7.88%).

When valine arylamidase were active they were considered as biovariety D (12 strains, 4.97%). When all strains were positive, with the exception of trypsin, they were considered as biovariety E (9 strains, 3.73%) and biovariety F if there was no alpha-glucosidase and beta-glucosidase activity. The strains were considered as biovariety G when positive for beta-glucosidase (6 strains, 2.48%). There was also a group of 17 strains (6.50%) which we hope to classify when we have obtained a greater number.

REFERENCES

1. Froman S, Will D W , Bogen E (1954) Bacteriophage active against virulent **Mycobacterium tuberculosis**. Isolation and activity. Amer. J. Publ. Hlth. 44:1326-1333
2. Tokina H, Takeya T (1977) Typage par les mycobacteriocines du bacilli tuberculous de type humain. Kekkaku.HAP. 52:11-15
3. Casal M, Linares MJ (1984) Enzymatic Profile of **Mycobacterium tuberculosis** Eur. J. Clin. Microbiol. 3:155-156
4. Casal M, Linares MJ (1984) Preliminary investigation of **Mycobacterium tuberculosis** Biovars. J. Clin. Microbiol. 20:1015-1016
5. Humble MW, King A, Philips I (1977) API ZYM: a simple rapid system for detection of bacterial enzymes. J. Clin. Pathol. 30:275-277

Mycobacteria of Clinical Interest. M. Casal, editor

SCREENING FOR TUBERCULOSIS IN ELDERLY NURSING HOME RESIDENTS

S. GRZYBOWSKI, E. A. ALLEN, C. W. CHAO, E. DORKEN, D. A. ENARSON, J. L. ISAAC-RENTON, B. B. PUSELJA, W. A. BLACK

Divisions of Laboratories and Tuberculosis Control, Ministry of Health, British Columbia, Canada; Divisions of Medical Microbiology and Respiratory Diseases, University of British Columbia, Vancouver, British Columbia, Canada

INTRODUCTION

The incidence of tuberculosis in North America increases progressively with age, especially in men (1). This has prompted interest in the detection of tuberculosis in the elderly, particularly since this age group constitutes an ever increasing segment of the population in all economically advanced countries.

Special attention has been paid to the elderly in nursing homes since even a single undetected case in an institution may cause a potential mini outbreak (2, 3).

The optimal method for screening a nursing home situation is debatable and the lack of a generally accepted approach raises the question as to whether any single approach is feasible for all geographic areas. The present study, therefore, was designed to screen nursing home residents by chest X-ray, tuberculin skin testing and sputum bacteriology, and compare the merits of each of the methods.

MATERIALS AND METHODS

The study lasted from December 1981 - October 1983, and evaluated 625 patients in four residential care facilities. Since the predicted number of positives was small in this group, the patient files of all 218 recently-diagnosed cases of active tuberculosis aged 65 or more were analysed and used as "controls".

Chest X-ray was carried out using standard procedures. The films were read independently by two highly experienced physicians, and reviewed by a third doctor for accuracy, consistency and film quality.

Tuberculin skin testing was done using Tween-stabilised PPD bioequivalent to 5 TU. All tests were performed by a single highly experienced Registered Nurse and were read in 48 hours.

Two types of sputum specimen were obtained: i) an "on the spot" specimen in which one of the team waited while the specimen was collected, ii) a more traditional "pooled" sputum in which the specimen container was left at the bedside for a maximum of 24 hours. Specimens were processed using standard methods (4), which included staining by the Ziehl-Neelsen method and culturing on Lowenstein-Jensen with nalidixic acid, Lowenstein-Jensen with added pyruvate and a Tarshis blood agar. Media were incubated for a minimum of 8 weeks.

RESULTS

Not every nursing home resident was available for every test. Of a potential 625, only 584 were eligible for all three tests, and only 309 completed all three tests. The 584 patients included 361 females and 223 males. The "control" group of 218 included 142 males and 76 females.

Failure to comply among the 625 was noted in 34.4%, 19.7% and 7.2% respectively for sputum bacteriology, chest X-ray and tuberculin skin testing. The same trend was noted in the 584 sub-group.

Abnormal (positive) chest X-ray findings were noted in 170 (32.8%) of the 519 patients who had a chest X-ray. The abnormalities included one new active pulmonary TB, 4 inactive pulmonary TB, 19 apical scarring, 1 possible infection with MOTT, and 145 various other abnormalities.

Tuberculin skin test reactions of 10 mm or more were found in 30% of the nursing home population and 75% of the "control" group. Reactions of 10 mm or more were noted in 36% of males and 26% of females. In both males and females there was a clear trend towards a negative test with increasing age.

The ability to provide a 'pooled' sputum specimen was noted in 87.6% of the "controls" and 59.4% of the nursing home population. "On the spot" collection in the nursing home group yielded specimens from only 47.1% of the group. A higher percentage of males were able to produce a specimen and, in both males and females, the ability, and/or willingness to produce a sputum specimen declined with age.

Correlations between smear and culture results were obtained for 191 "controls". Positive cultures were obtained in 97.1% of smear-positives, and 82.7% of smear-negatives. Of 136 cases from whom 3 sputum samples had been obtained, all smear-positives yielded a positive culture with the first specimen; for smear-negatives, the yield of positive cultures increased with the number of samples cultured.

DISCUSSION

For this type of random survey in a nursing home population in an economically advanced country such as Canada, the tuberculin skin test is of little value.

With sputum bacteriology, only about 60% of the patients were able, and/or willing, to give a sputum sample. Elderly males were more likely to produce sputum than elderly females, a factor which might be related to heavy cigarette smoking. The 'pooled' method of collection was slightly more successful than the "on the spot" technique.

Chest X-ray produced abnormal results in 33% of the 519 persons tested. Among this group was one newly recognised case of active tuberculosis. The compliance rate with the chest X-ray was intermediate between that of sputum collection and the tuberculin skin test. A potential benefit of chest X-ray is the capability of detecting other abnormalities. However, in this survey less than 2% of the findings yielded results which prompted useful clinical intervention. Also, there are serious practical problems associated with the interpretation of chest X-rays in the elderly debilitated patient.

In our jurisdiction, therefore, a combination of chest X-ray and sputum bacteriology is optimal for this type of random survey. The use of either technique on its own may be appropriate for specific situations. Each of the methods used has its advantages and disadvantages (5, 6, 7), and local cost-benefits will determine which techniques(s) is used where.

REFERENCES

1. Stead WW (1983) Does the Risk of Tuberculosis Increase in Old Age? J Infect Dis 147:951-955
2. Stead WW (1981) Tuberculosis Among Elderly Persons: An outbreak in a Nursing Home. Ann Intern Med 94:606-610
3. Stead WW, Lofgren JP, Warren E, Thomas C (1985) Tuberculosis As An Endemic and Nosocomial Infection Among the Elderly in Nursing Homes. N Engl J Med 312:1483-1487
4. Vestal AL (1978) Procedures for the isolation and identification of mycobacteria. U.S. Department of Health, Education and Welfare publication CDC-79-8230, Centers for Disease Control, Atlanta

5. Snider DE, Anderson HR, Bentley SE (1984) Current Tuberculosis Screening Practices. Am J Public Health 74:1353-1356

6. Soo Hoo GW, Palmer DL, Sopher RL (1984) Reducing Tuberculosis Detection Costs. Chest 86:860-862

7. Welty C, Burstin S, Muspratt S, Tager IB (1985) Epidemiology of Tuberculous Infection in a Chronic Care Population. Am Rev Respir Dis 132:133-136

Mycobacteria of Clinical Interest. M. Casal, editor

FOLLOW-UP OF BCG VACCINATION THROUGH THE MANTOUX REACTION

M.L. VIDAL, J.G. HORTELANO and E. ROMAN
Servicio de Infecciosos, Hospital Infantil "Le Paz", Madrid (Spain)

INTRODUCTION

The interpretation of the Mantoux reaction in children immunized with BCG (1,2) may be problematic, especially when immunization is recent. The present study was undertaken to analyze the evolution of the tuberculin reaction in the years following BCG immunization.

MATERIAL AND METHODS

In a group of children immunized with BCG vaccine during their first days of life, a Mantoux test was performed at 3 months, 1 year and 2 years after immunization. This group originally comprised 750 children; the reaction was done at 3 months in 356 children, at 12 months in 187 children and at 24 months in 98 children, all of which were born at La Paz hospital and were immunized with BCG by their second day of life. The immunization took place in the months of February-March, 1980 and the vaccine was obtained from the Health Department. None of the children had a history of family members suffering from tuberculosis.

Tuberculin PPD RT 23 was obtained from the Copenhagen Serologic Institute, prepared at Ibys Laboratories, using 2 units in each Mantoux test, equivalent to 5 units of the international PPD S pattern. Readings were taken at 48 and/or 72 hours. The results according to the size of infiltration were evaluated in three groups:

- Group I: Mantoux negative "Negative"
- Group II: Mantoux 9 mm "Weakly positive"
- Group III: Mantoux 9 mm "Positive"

The statistical analysis of the results, after applying a percentage comparison test with matching data (3), showed a significant variation between the 3 month and 1 year groups as well as between the 1 year and 2 year groups. The ratio of those that turned positive at 3 months to those at 1 year is significant, as are the number of cases that turn negative during the period between 1 year and 2 years.

RESULTS

Group A

(negative results at 3 months of life control) — 34 cases

	1 year		2 years	
No change	10	52.63%	14	93.35%
Change to weakly positive	7	36.83%	0	0.00%
Change to positive	2	10.52%	1	6.14%
	19		15	

Group B

(Weakly positive at 3 months) — 190 cases

	1 year		2 years	
No change	66	43.70%	20	51.28%
Change to negative	17	11.25%	11	28.20%
Change to positive	68	45.03%	8	20.51%
	151		39	

Group C

(Positive at 3 months) — 57 cases

	1 year		2 years	
No change	13	100.00%	32	72.72%
Change to negative	0	0.00%	6	13.63%
Change to weakly positive	0	0.00%	6	13.63%
	13		44	

The statistical analysis of the results, after applying a percentage comparison test with matching data (3), showed a significant variation between the 3 month and 1 year groups as well as between the 1 year and 2 year groups.

The ratio of those that turn positive at 3 months to those at 1 year is significant. The number of cases that turn negative during the period between 1 year and 2 years is also significant.

OVERALL RESULTS

	Negative %	Weak positive %	Positive %
at 3 months	12.2	79.6	8.1
at 1 year	15.5	40.0	44.3
at 2 years	27.5	28.5	43.8

COMMENTS

In the group of children immunized with BCG vaccine during their first days of life, none developed a positive tuberculin test at 3 months after immunization; nevertheless, this same group showed a significant increase in the number of tuberculin-positive reactions at 1 year. However, those in the group with a negative tuberculin test at 1 year did not turn positive at the 2 year control.

The fact that the Mantoux reaction in immunized children takes several months, or even a year, to attain its final size may be related to the lack of cellular immunity in the first months of life. We believe that this may be relevant when evaluating the Mantoux reaction in a child under 1 year of age, since a positive result may be considered true; on the other hand, if it is not positive, the test may be considered conditional to another reading about 12 months of age.

Values for the Mantoux reaction taken about 1 year of age are more reliable, since no variations following BCG immunization have been observed in children of this age group (4, 5).

REFERENCES

1. Palmer CE, SHAW IW, Comstoch GW (1958) Comunity trials of BCG vaccination. Am Rev Tub 77:877-907
2. Abrahans EW (1979) Tuberculin hypersensitivity following BCG vaccination in Brislane School Children. Tubercle 60(2):109-113
3. Domenech JM I Massons. Bioestadística: Métodos estadísticos para investigadores
4. Mande P (1982) Le BCG en France aujourd'hui. Arch Fr Pediat 39:791-798
5. Hortelano JG, Vidal ML (Nov 1983) Reacción del Mantoux en una población escolar. Curso de Medicina y Sanidad Escolar, pag. 337-342

RADIOLOGICAL CONTROL OF PULMONARY TUBERCULOSIS USING A PLANIMETER

M. GARCIA LOPEZ, J.Ga. HORTELANO and M.L. VIDAL
Servicio de Infecciosos, Hospital Infantil "La Paz", Madrid (Spain)

It is well known that the radiological evolution of infantile tuberculosis and the clinical-analytical data may not run parallel. Clinical improvement is often coupled with much slower radiological evolution throughout treatment (6). The aim of the present investigation is to demonstrate this (1,2).

MATERIAL AND METHODS

Cases were children undergoing treatment for pulmonary tuberculosis with gangliar involvement, at the Department of Infectious Diseases, the Children's Hospital, "La Paz".

Control of the evolution was undertaken by analysing:

a. Basic clinical data: weight, activity, appetite, fever, cough and positive pulmonary auscultation.

b. Basic analytic data: leucocyte count, sedimentation rate, and bacilloscopy (gastric and bronchial aspiration).

c. Radiological data: simple anterior-posterior radiography of the thorax, the size of the lesions being determined by the use of a digital-electronic planimeter, Planix 7 (Tamaya). To measure the radiological lesions, 'stencils' were traced on transparent paper placed over the X-rays, and from these 'stencils' (Fig. 1) the area of the image recorded by X-ray was measured with the planimeter (Fig. 2).

All these tests of the clinical evolution together with the analytic and radiographic data were carried out at the commencement of treatment, and at the first, second, third, sixth and ninth months of treatment.

Statistical analysis was made using percentages: for testing of signs, the non-parametrical method was employed.

RESULTS

Number of cases studied: 14.

Sex: female 5; male 9.

Age: between 6 months and seven years.

Treatment received: similar in all cases, with the same 2-3 antibiotics for at least 9 months.

Adenopathy

Total number of adenopathies studied: 19

Increase after 1 month of treatment : 15 (78,94%)

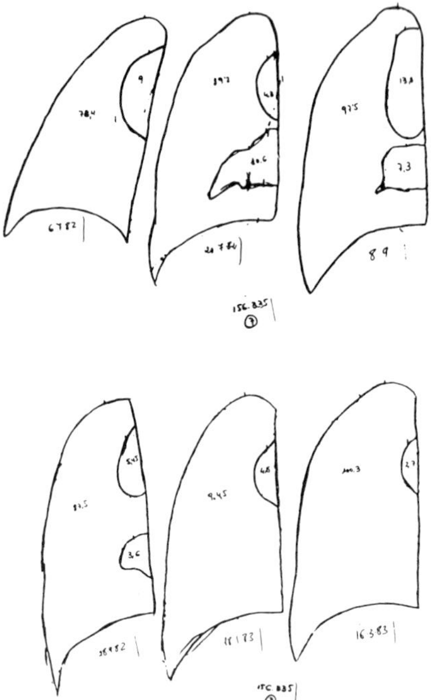

Fig. 1.

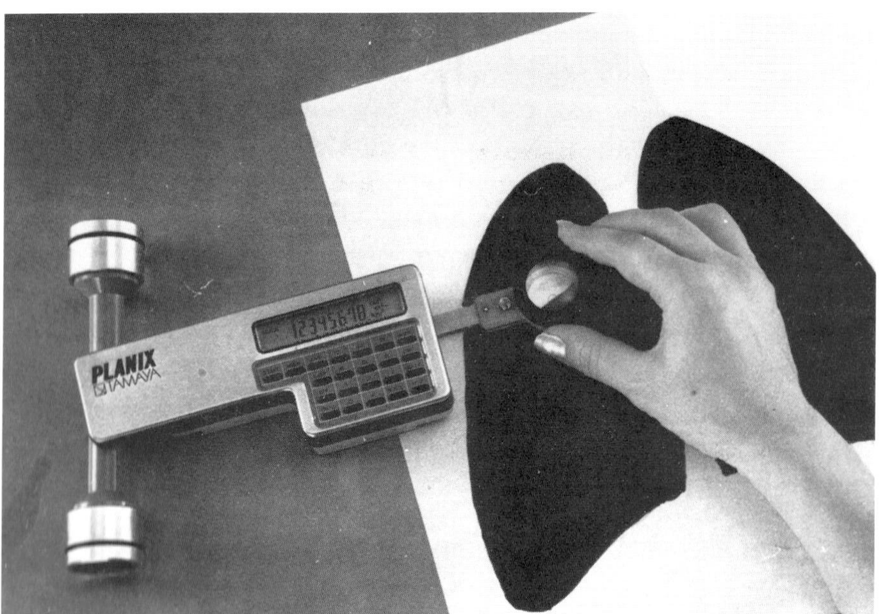

Fig. 2.

Decrease after 1 month of treatment : 3 (15.78%)

No variation after 1 month of treatment : 1 (5.26%)

Began to decrease after increasing:

in the 2nd month 5

in the 3rd month 5

in the 6th month 4

in the 12th month 1

Infiltrations

Total numbers of infiltrations studied 11

Increase after 1 month of treatment 8 (72.72%)

Decrease after 1 month of treatment 3 (27.27%)

Began to decrease after increasing:

in the 2nd month 3

in the 3rd month 3

in the 6th month 2

Statistical analysis

Data obtained after more than 1 month of treatment

Activity (improvement)	7/7	(100.00%, $p < 0.05$)
Appetite (improvement)	13/13	(100.00%, $p < 0.01$)
Weight (increase after 3 months)	11/12	(91.66%, $p < 0.01$)
Fever (improvement)	12/12	(100.00%, $p < 0.01$)
Sedimentation rate (decrease)	8/9	(88.88%, $p < 0.05$)
Adenopathy (increase)	15/19	(78.94%, $p < 0.05$)
Infiltrations (increase)	8/11	(72.72%, $p < 0.05$)

COMMENTS

Between the first and second month after beginning treatment there was a noticeable improvement in general clinical signs, in clinical signs of respiratory infection, and the data of the basic analyses.

The radiological lesions, however, continued to increase in size up to three months of treatment in 2/3 cases. After three months of treatment, some lesions began to regress, but others continued to grow (5).

The increase in size of the radiological lesions during treatment was more marked in cases with gangliar lesions than in those with parenchymatous lesions, most probably due to the 'abscessifying' tendency of gangliar tuberculosis and its lesser 'permeability' to antibiotics, compared with infiltrative lesions.

For these reasons, an improvement in clinical and analytical data during the first months of treatment may be taken as an index of the efficacy of treatment, whereas an increase in radiological thoracic lesions, especially of the gangliar type, does not necessarily indicate failure of the therapy. This development must be taken into account by the pediatrician to avoid any change in the treatment. (When is tuberculosis cured?) (3,4).

The size of the lesions is satisfactorily rendered by the digital planimeter, thus avoiding the inconvenience of subjective estimation, or calculation by measuring the diameter of the radiological image.

The use of the planimeter, however, would appear to be indicated only for investigation, or other very specific instances.

REFERENCES

1. Aderele W (1980) Radiological patterns of pulmonary tuberculosis childrens. Tubercle 61(3):157-163
2. Barret-Connor E (1980) The periodic chest roentgenogram for the control of tuberculosis in health care personnel. Amer Rev Resp Dis 122(1):153-155
3. Editorial (1982) When is pulmonary tuberculosis cured? The Lancet 22:1(8282):1163-1164
4. Font J, Coca A, Torres A (1982) Tuberculosis. Persistencia del síndrome febril a pesar de un correcto tratamiento antituberculostático. Rv Clin Esp 166(3-4):131-133
5. Gatner E, Burkhadt K (1980) Correlation of the results of x-Ray and sputum cuture in tuberculosis prevalence surveys. Tubercle 61(1):27-31
6. Vidal R, Villaplana R (1981) Correlaciones clínico radiológicas en 100 casos de tuberculosis miliar. Rev Clin Esp 161(2):109-112

Mycobacteria of Clinical Interest. M. Casal, editor

PULMONARY TUBERCULOSIS AND ASSOCIATED DISORDERS

L.M. ENTRENAS-COSTA, F. SANTOS-LUNA, L. MUNOZ-CABRERA, A. SALVATIERRA-VELAZQUEZ, J. MUNOZ-ALGUACIL, F. SEBASTIAN-QUETGLAS, A. COSANO-POVEDANO and J. LOPEZ-PUJOL
Respiratory Division, 'Reina Sofia' Regional Hospital, 14004 Cordoba (Spain)

INTRODUCTION

After the introduction of effective chemotherapy, the incidence of pulmonary tuberculosis has decreased in Spain as in other countries (1). Since then, the diagnosis and treatment of tuberculosis has been transferred to general hospitals (1,2). However, there is an increasing number of immunocompromised patients due to the use of very aggressive procedures against neoplastic and systemic diseases (3-5). Tuberculosis in immunocompromised patients is considered difficult to treat (3-6). We have made a study of these patients to determine differences in diagnosis, evolution and treatment response.

MATERIAL AND METHODS

An analysis was made of the results of a prospective study of pulmonary tuberculosis performed at the 'Reina Sofia' Hospital between January 1979 and June 1985; 144 patients, 106 males (74%) and 38 females (26%), were bacteriologically diagnosed to have pulmonary tuberculosis with a mean age of 41 ± 19 years (mean ± standard deviation). In all cases the tuberculosis bacillus was isolated in sputum and/or bronchial aspirate. Other disorders had previously been diagnosed in 32 patients (22%).

In all patients the antitubercular chemotherapy was a combination of rifampicin, isoniazid and ethambutol, used in generally accepted time and dosages (7). A change to drugs with low hepatoxicity was indicated if severe hepatic injury was detected (8).

RESULTS

The immunocompromised patient group comprised 19 males and 13 females with a mean age of 51 ± 15 years (not significant statistically).

Associated conditions are shown in Table I. Eight patients had two or more diseases.

High and longterm fever was the main clinical symptom at onset. The chest X-ray showed polymorphism. The radiological image at time of diagnosis is shown in Table II. The localisation of the lesions was atypical in 20 patients (65%).

The clinical evolution was favorable in all cases. No reactivation was found. The response to chemotherapy was no different in the noncompromised host. Sputum

TABLE I

ASSOCIATED CONDITIONS

Underlying disease	Number	(%)
Diabetes	12	8.5
Chronic liver disease	7	4.9
Corticotherapy	4	2.8
Hemopathy	4	2.8
Immunosuppressive therapy	4	2.8
Pneumoconiosis	3	2.1
Gastrectomy	3	2.1
Radiotherapy	3	2.1
Dialysis	3	2.1

TABLE II

RADIOLOGICAL PRESENTATION

Radiology	Number	(%)
Local exudative image	13	40.6
Cavity with/without lesion	8	25.0
Fibroproductive reaction	7	21.9
Bronchogenic dissemination	3	9.4
Miliary dissemination	1	3.1
Total	32	100.0

conversion was obtained in every case at the end of the eight weeks of treatment.

CONCLUSIONS

The association of pulmonary tuberculosis in immunocompromised patients is a usual finding in general hospitals (1-6) (22% in our series).

The association of febrile syndrome and atypical chest X-ray in the immunocompromised host should make one suspect tuberculosis (3,5,9). Initiation of treatment should be as soon as possible, even in the absence of bacteriological confirmation, since the course of tuberculosis can be rapid and fatal in these patients (3,5,6). However, we have not found this true in our experience.

The treatment was the same and for the same duration as for noncompromised patients (3-6) except when severe liver injury due to chemotherapy was detected (8). The clinical evolution, chest X-ray and mycobacteriology were also similar.

REFERENCES

1. Rodríguez Ramos S, Pascual Pascual T, Martínez González del Río J.Incidencia de la Enfermedad Tuberculosa en un Hospital General. Arch Bronconeumol 1983; 19: 80-84.

2. Dandoy S. Current Status of General Hospital Use for Patients with Tuberculosis in United States. Am Rev Resp Dis 1982; 126: 270-273.

3. Dautzenberg B, Grosset J, Bechner J, Lucciani J, Debre P, Herson S, Truffot C, Sors C. The Management of Thirty Immunocompromised Patients with Tuberculosis. Am Rev Resp Dis 1984; 129: 494-496.

4. Feld R, Bodey GP, Groschel D. Mycobacterioses in Patients with Malignant Disease. Arch Intern Med 1976; 136: 67-70.

5. Kaplan MH, Amstrong D, Rosen P. Tuberculosis Complicating Neoplastic Disease. A Review of 201 Cases. Cancer 1974; 33: 850-858.

6. Millar JW, Horne NW. Tuberculosis in Immunosuppressed Patients. Lancet 1979; 2: 1176-1178.

7. Brouet G, Roussel G. Trial 6, 9, 12 Overall Methods and Results. Rev Fr Mal Respir 1977; 5 (supp. 1: 5-13).

8. Blackman F, Lin D, Zwiel P, Costanzo P, Cohen M, Reichman LB. Hepatotoxic antituberculosis Drug Combinations are not Contraindicated in Patients with Pre-existing Liver Disease or Alcoholism. A Preliminary Report. Am Rev Resp Dis 1982; 125: 171.

9. Miller WT, Mac Gregor RR. Tuberculosis: Frecuency of Unusual Radiographic Findings. Amer Jour Roent 1978; 130: 867-875.

Mycobacteria of Clinical Interest. M. Casal, editor

BONE AND JOINT TUBERCULOSIS. UNUSUAL LOCALIZATIONS

N. MARTIN CASABONA*, A. ORCAU PALAU*, T. GONZALEZ FUENTE*, M. GARCIA GONZALEZ* and J. BAGO GRANELL**
Service of Microbiology*, Department of Traumatology**, Ciudad Sanitaria 'Vall d'Hebron', Barcelona, Generalitat de Catalunya, Departament de Sanitat i Seguretat Social (Spain)

INTRODUCTION

Osteoarticular infection is the third most often found form of extrapulmonary tuberculosis after renal and lymphatic forms. It is found in 73.6% of cases in the vertebral column, hip-joint or knee-cap and the time between appearance of symptoms and diagnosis may be weeks or even months. Other sites are less common and clinical symptoms are often confused with other processes. In this group the average time between the development of the process and diagnosis is even longer. In this chapter, we investigate the reasons behind this type of lesion.

MATERIAL AND METHODS

In the last 10 years, we received 437 samples of osteoarticular origin for mycobacterial investigation. These were obtained from our hospital, Department of Traumatology, and other hospitals and outpatient clinics. Over this same period, we went through 117 cases of osteoarticular tuberculosis (O.T.), clinically diagnosed in the Department of Traumatology, accepting for the study those fulfilling the following conditions: (1) culture and/or positive bacilloscope for osteoarticular specimens; (2) characteristic tuberculosis (TB) histology for the same specimens; (3) positive culture on non-osteoarticular specimens, with a clinical picture and radiology indicating O.T.; (4) clinically active TB without positive bacteriological tests, not localised but with a clinical picture and an osteoarticular lesion shown on radiography; (5) clinical history of TB with compatible osteoarticular radiographic pictures. Seventy-two cases were accepted of which 73.6% had lesions in the vertebral, hip-joint or knee-cap. The other 19 cases had lesions localised elsewhere which will be discussed here.

RESULTS

General microbiological tests. Of 437 samples received, 64 (14.6%) were positive of which 12 (18.7%) were diagnosed by bacilloscopy. Fifty-six strains were identified as *Mycobacterium tuberculosis*, 2 as *M.bovis*-BCG, 2 as *M. gordonae*, 1 as *M. fortuitum* and 3 as those of groups III and IV in the Runyon classification.

Patient history. Of the 72 patients, 33 (45.8%) had vertebral lesions, 10 (13.8%) had hip-joint lesions and 10 (13.8%) in the knee-cap. The remainder (26.3%) form the cases studied here (Table I).

There were equal numbers of male and female patients with ages between 22 and 81 years (mean of 48.5). The mean time between development of the process and establishing the diagnosis was 174 weeks. For vertebral lesions, this was 35 weeks, hip-joint 53 weeks and knee-cap 112 weeks. The time was less for sacro-iliac lesions, but for a patient with a pubic symphisis lesion and one with sternal cartilage lesion, the illness took 10 and 13 years respectively to develop.

TABLE I

CLINICAL DATA AND DIAGNOSTIC METHODS

Case Nr.	Site of lesion	Age/sex	Time for lesion to develop (weeks)*	Reasons for consultation	Histo-logy	Micro-biology B	C
1	Trochanter osteitis	72/F	260	Pain	+	-	+
2	Trochanter osteitis	67/M	208	Recurring gluteal abscess	N.D.	-	+
3	Trochanter osteitis	46/M	24	Thigh tumour	N.D.	+	+
4	Sternal cartilage joint	36/F	260	Pain,recurring fis-tula tumour	+	-	+
5	Sternal cartilage joint	34/F	676	Recurring fistula	+	N.D.	
6	Sternal cartilage joint	52/M	?	Tumour	N.D.	-	+
7	Wrist flexor	29/M	208	Pain, tumour	+	N.D.	
8	Wrist flexor	36/M	4	Carpal tunnel	+	N.D.	
9	Wrist flexor	52/M	52	Swelling	N.D.	-	+
10	Wrist flexor	49/M	?	Crepitant tendo-synovitis	+	N.D.	
11	Carpal radial joint	71/M	56	Swelling fistula	N.D.	-	+
12	Tibio tarsal joint	81/F	156	Fistula	+	+	+
13	Subastragalar joint	25/F	20	Pain, swelling	+	+	+
14	Calcaneocuboid joint	37/M	?	Pain, tumour	+	-	+
15	Pubic symphisis	48/M	520	Fistula	N.D.	-	+
16	Esternum	68/F	24	Fracture	N.D.	-	+
17	Elbow	66/M	156	Arthritis	+	-	+
18	Sacroiliac joint	32/F	1	Pain	N.D.	N.D.	
19	Sacroiliac joint	22/F	20	Pain	N.D.	N.D.	

F = female ; M = male ; ? = unknown ; B = bacilloscopy ; C = culture ; N.D. = not done ;
* time between appearance of symptom and diagnosis

Symptoms bringing the patient to consultation varied with the site of the lesion; in 3 cases, the lesions were multifocal (trochanter-hip-joint, sternal cartilage-vertebral and tibiotarsal-hip-joint). In 9 cases, there was a history of pulmonary tuberculosis and 1 patient had an active pulmonary infection. There was no radiological evidence for the sternal cartilage lesions or for 3 tendon-sheath cases.

Histology was positive in 10 patients (52.6%) and bacteriology in 13 (68.4%). On combining both techniques, 89.4% were diagnosed. In one patient (No. 17) the bacterium was isolated in sputum. Seventeen patients underwent medical treatment with 3 antituberculous drugs. Treatment of case No. 6 is unknown and case No. 4 did not receive antituberculous therapy. Initial techniques involving puncture or scraping were carried out in patients with tumours or abscesses; all those with tendon sheath infections underwent synovectomy. In 12 patients, the infection and functional sequellae were reduced considerably. In the sternal cartilage case, where no therapy was undertaken, the patient developed Pott's disease D_{12}-L_1 within a year. In 3 patients (2 with lesions in the wrist flexor and one at the elbow), there remained functional limitations and ankylosis appeared in the radiocarpal joint. In the patient with a pubic symphisis lesion, fistulas remained and in one case (No. 6) the follow-up remained unknown.

DISCUSSION

The clinical picture varies according to the site; however, in the group we studied 26.3% had fistulas, which suggested an advanced state of the disease process. In some cases, tuberculosis had not been suspected before, which considerably delayed the diagnosis (1).

Rehn-Graves (2) found that the trochanter major was affected in 1-2% of patients with OT; we found this to be greater at 4.1% (3 of 72 cases). The age of patients with lesions at the sternal cartilage joint was lower, which agrees with data from other authors (3,4). Radiological lesions could not be seen; only histological and bacteriological tests enabled diagnosis.

Tendon sheath TB infections may be confused with carpal tunnel syndrome (5, 6), as indeed occurred in one of our patients; in one case only, radiology demonstrated periarticular osteoporosis.

We have only found one case cited in the literature by Newton (7) which indicated pubic symphisis involvement. Kuntz (8) found that multi-sited infections comprised about 10% of cases of OT, whereas we found 5.6%. The same author reported a case of sternal osteitis which he hardly mentions in the overview since it was so rare. In our patient, the sternum fractured as a result of the bone lesion but this resolved under medical treatment.

In the patient with an elbow infection, medical treatment only was undertaken. This was successful apart from reduced motility, as occurred in 17 of 22 patients studied by Martini (9).

Since histological and radiological examinations give faster results for O.T., bacteriological tests are seldom performed. Since there are few bacteria in these lesions (10) bacilloscopy is of little use and culture is slow due to their slow metabolism. Moreover, to obtain specimens in closed lesions, surgical intervention is required. However, as Martini commented (11), radiology enables diagnosis of O.T. in 80% of cases. Late or false diagnosis occurs in the other 20%, especially in rare sites where isolation and identification of the bacteria are the only certain means of diagnosis. In these specimens, bacteriological tests should include tests for mycobacteria and associated histological examination which will increase the percentage of diagnoses (12,13).

REFERENCES

1. Uriel Latorre S, Muguerza Eraso I, Echevarria Albeniz F, López Roger R, Amor Azpeitia T (1979) Rev Clin Esp 155:277-281
2. Rehn-Graves S, Weinstein AJ, Calabrese LH, Cook SA, Boumphrey FRS (1983) Arthritis Rheum 26:77-81
3. Pardell Alantá H, Gispert Nicolau FJ, Mundet Surroca J, García Marcé I, Ruiz Pardo MJ (1982) Mee Clin (Barc) 79:472-474
4. Mall JC, Genant HK, Camsux G (1976) Am Rev Respir Dis 114:635-637
5. Klofkorn RW, Steigerwald JC (1976) Am J Med 60:583
6. Canton Sánchez JA, Mateo Bernardo I, Llorente de Jesús R (1979) Rev Clin Esp 155:189-194
7. Newton P, Sharp J, Barnes KL (1982) Ann Rheum Dis 41:1-6
8. Kuntz JL, Meyer R, Paille R, Bannwarth B, Asch L (1982) Rev Rheumatisme 49: 477-478
9. Martini M. Gottesman H, Martini-Benkeddache et Daoud A (1977) Rev Chir Orthop 63:539-55
10. Boulahbal F (1978) Ann Algerien Chirurg XII 1:25-28
11. Martini M (1978) Ann Algerien Chirurg XX 1:33-34
12. Davies PD, Humphries MJ, Byfield SP et al (1984) J Bone Joint Surg (Br) 66-B: 326-330
13. Halsey JP, Reeback JS, Barnes CG (1982) Ann Rheum Dis 41:7-10

Mycobacteria of Clinical Interest. M. Casal, editor

TUBERCULOUS AND NONTUBERCULOUS MYCOBACTERIAL INFECTION IN KOREA

SANGJAE KIM, YOUNG PYO HONG and BYOUNG WON JIN

Korean Institute of Tuberculosis, Korean National Tuberculosis Association, 121-150 Dangsandong, YoungdeungpoKu, Seoul-150 (Korea)

INTRODUCTION

The prevalence of tuberculous infection as demonstrated by tuberculin testing is an important epidemiological index most relevant to tuberculosis in Korea and to programme strategy (1). However, interpretation of the tuberculin reaction is complicated in some areas with a high prevalence of nonspecific cross-reactions induced apparently by environmental mycobacteria (1-2). Nontuberculous mycobacteria are commonly isolated from human sputum specimens; their pathogenic involvement has been found in many cases all over the world, suggesting that a considerable number of skin reactions are elicited by nontuberculous mycobacteria, even in nontropical countries, although prevalence may be lower than that of tropical countries. It is also conceivable that the increasing frequency of nontuberculous mycobacterial infection in early life may alter the epidemiology of subsequent tuberculous infection. This situation led us to study skin reactivity to nontuberculous mycobacterial sensitins in people in Korea where the annual risk of tuberculous infection is now sharply decreasing, while case reports of nontuberculous mycobacterial infection are increasing (3). The results obtained are analyzed and discussed in relation to frequency of clinical mycobacterial isolates from patients with pulmonary diseases or of isolates from the environment.

MATERIAL AND METHODS

Sensitins

Sensitins of *Mycobacterium tuberculosis* (RT23), *M. scrofulaceum* (RS95), *M. avium* (RS10), and *M. fortuitum* (RS20) were purchased from the Seruminstitut in Copenhagen, Denmark.

Skin testing

One TU of tuberculin RT23 was intracutaneously injected on the left forearm simultaneously with 0.1 μg of one of four nontuberculous mycobacterial sensitins by two nurses. Induration elicited was measured 72 hours after injection.

Subjects tested

In 1984 dual skin tests had been performed in 1356 6-year-old children who did not have a BCG scar and in 95 sputum-positive patients. In 1985 21,614 individuals under the age of 30 years sampled by a multi-stage stratified systemic method for the 5th nation-wide prevalence surgery were also tested;

their tuberculin reactions were compared with those of four previous surveys which have been carried out at five-year intervals since 1965 using exactly the same methods and sensitin.

Isolation of mycobacteria

Mycobacteria were isolated from 163,722 sputum specimens of patients with pulmonary diseases by the simple Ogawa method using acid-buffered medium during the period from January 1981 to May 1985; the species of isolates with definite or possible clinical significance were compared with isolates from soil samples.

Calculation of annual risk of tuberculosis infection

The annual risk of infection in the 5-9 year old age group were calculated on the basis of prevalences which had been serially investigated. An average value for risk of infection (qb) can be obtained with the aid of logarithms ($qb = (1-p)^{1/a}$, where q = average value for annual risk of escaping infection, b = born years of cohort, a = exact year of cohorts, p = prevalence). After transforming the annual risk of infection during the lifetime of the cohort, we can derive a formula for linear regression by risk of infection. We are then able to estimate the annual risk of infection by substituting the number in a year for X in formula $Y = a+bX$.

RESULTS

Of 1356 school children, 10.2% reacted with more than 9 mm induration to 1 TU of tuberculin RT23 which was significantly decreased compared with 26.1% in 1965 (Tables I and III). The number of 6 or 9 mm reactions to nontuberculous sensitins exceeded those to tuberculin; the percentage of children who reacted to RT23 with less than 10 mm induration but with a 3 mm or more reaction to nontuberculous sensitin and more than 5 mm reactions to RS95, RS20, RS10, and RS23, were 12.0%, 10.9%, 6.5%, and 3.4%, respectively (Table II). However, most of the 9 mm reactions to RT23 showed either nearly the same reaction to both RT23 and nontuberculous sensitins or a 3 mm or larger reaction to RT23, suggesting cross-reactions between tubercle bacilli induced hypersensitivity and nontuberculous sensitins. Almost all sputum-positive patients showed more than 9 mm reactions to both tuberculin and nontuberculous sensitins; the mean size of their induration (18.6 to 19.9 mm) was much greater than that of the more than 5 mm reactions in school children (11.1 to 14.9 mm). Only two patients who were both RS10 and RS95 reactors showed a more than 3 mm induration than that to RT23.

The prevalences of 9 mm reactions to RT23 in 6-year-old children and in study populations aged less than 20 years in past and present surveys, are summarized in Table III along with the annual risk of infections derived from prevalences in the 5-9 year old age group. A significant decrease in prevalence is noted in age groups under 15 years during the last two decades. Prevalence in children under 5 years of age, however, has remained unchanged since 1975. A steep decrease in the annual risk of infection has been observed in 5-9 year old children, showing a 6.5% annual reduction between 1965 and 1985.

Overall findings suggest that children in Korea are exposed to less and less tuberculosis in their environment, while a considerable number now appear to have been sensitized with nontuberculous mycobacteria, as found in 6 year old children. Of the sensitins tested, positive reactions to RS95 and RS20 were most common and this finding was fairly compatible with the frequency of clinical mycobacterial isolates (Table IV). *M. fortuitum* and *M. scrofulaceum* were commonly encountered in both soil and clinical specimens, but nonpathogenic mycobacteria prosperous in soils were often not isolated from clinical sources.

TABLE I

SKIN REACTIVITY TO TUBERCULOUS AND NONTUBERCULOUS MYCOBACTERIAL SENSITINS IN 6-YEAR OLD CHILDREN AND IN SPUTUM-POSITIVE PATIENTS

Sensitins	6-year old children			Sputum-positive patients		
	No. of testees	5 mm reactors (%)	9 mm reactors (%)	No. of testees	5 mm reactors (%)	9 mm reactors (%)
RT23	1356 (2.2±5.1)*	12.0 (14.9±4.8)	10.2	95	100.0(19.9±2.7)	100.0
RS95	334 (3.0±5.1)	21.6 (11.8±4.1)	13.2	48	97.9(19.9±2.5)	97.9
RS10	459 (2.4±4.6)	13.9 (12.8±4.1)	11.3	47	97.9(18.6±3.6)	97.9
RS23	234 (2.0±4.6)	13.0 (13.2±4.2)	10.5			
RS20	239 (2.2±4.4)	18.0 (11.1±3.0)	12.6			

*Numbers in parenthesis are mean induration ± SD

TABLE II

COMPARISON OF SKIN REACTIONS TO TUBERCULIN AND NONTUBERCULOUS SENSITINS IN 6-YEAR OLD CHILDREN

Sensitins	5 mm reactions to both sensitins	10 mm to RT23			9 mm to RT23		
		NSR*	LR**	SR***	NSR	LR	SR
RS95	76.9****	0.3	12.0	0.3	4.5	0.9	5.1
RS10	81.7	1.3	6.5	1.3	3.9	0.4	4.8
RS23	85.2	0.6	3.4	0.0	4.9	0.9	4.9
RS20	76.6	1.3	10.9	1.3	1.3	0.4	8.4
Total	80.5	0.9	7.9	0.7	3.8	0.7	5.5

*NSR = nearly same reactions within 3 mm differences; **LR & ***SR = 3 mm or more larger or smaller reactions to nontuberculous sensitins; ****Numbers in table indicate percent.

TABLE III

PREVALENCE OF 9 mm REACTIONS TO RT23 IN INDIVIDUALS WITHOUT BCG SCAR

Age groups	Prevalence (P) or risk of infection (RI) and its reduction (D)	Survey year						
		1965	1970	1975	1978	1980	1984	1985
5	P	10.1*	8.7	4.8		4.9		5.4
6	P	26.1	20.4	15.4	14.3	11.0	10.2	
5-9	P	34.7	28.9	15.9		12.6		9.0
	RI	5.73	3.87	2.61		1.76		1.19
	D	------------6.5------------						
10-14	P	70.0	57.1	49.6		32.1		28.6
15-19	P	71.0	73.7	69.6		71.1		64.2

*Numbers in table indicate per cent

TABLE IV

ISOLATION OF MYCOBACTERIA FROM SPUTUM SPECIMENS AND FROM SOIL SAMPLES

Species or groups	Clinically significant isolates	Clinical isolates with possible pathogenesis	Soil isolates
M. tuberculosis	32455	0	0
M. scrofulaceum	3	1	25
M. szulgai	1	1	0
M. gordonae	0	0	12
M. flavescens	0	0	3
M. avium-intracellulare complex	4	2	9
M. terrae complex	0	1	58
Other nonchromogens	0	0	24
M. chelonei subsp. *abscessus*	3	3	0
M. fortuitum	4	1	55
Other rapid growers	0	0	2
Total	32470	9	188

In summary we have found that children in Korea are now exposed to less tuberculosis as is apparent from the steeply decreasing annual risk of infection in the 5-9 year old age group. A considerable number of 6-year old children, however, showed a significantly larger reaction to nontuberculous sensitins than to tuberculin, indicating exposure to nontuberculous mycobacteria. Mycobacterial isolates from human sputum specimens and from soil samples seemed to be compatible with this finding.

REFERENCES

1. WHO (1974) WHO expert committee on tuberculosis - 9th report. Geneva

2. Palmer CE, Edwards LB (1968) Identifying the tuberculous infected, the dual test technique. Am J Med Assoc 205:167

3. Kim SJ, Bai GH, Yoo WH (1984) Skin test reactivity to mycobacterial antigens in patients with pulmonary tuberculosis and in normal population in Korea. Tuberc Resp Dis (Korean) 31:91-107

TUBERCULOUS INFECTION OF THE PLEURA: OUR EXPERIENCE

F. SANTOS-LUNA, L.M. ENTRENAS-COSTA, L. MUNOZ-CABRERA, A. SALVATIERRA-VELAZQUEZ, J. MUNOZ-ALGUACIL, F. SEBASTIAN-QUETGLAS, A. COSANO-POVEDANO and J. LOPEZ-PUJOL
Respiratory Division "Reina Sofia" Regional Hospital, 14004 Cordoba (Spain)

INTRODUCTION

The incidence of tuberculous pleural effusion still remains high within general pleural disorders. The diagnosis of this disorder has been greated aided by the introduction of pleural biopsy as part of the diagnostic protocol. We analyze here our experience in the management of these patients.

MATERIAL AND METHODS

From January 1982 to June 1985, 110 patients were admitted to our Respiratory Division for diagnostic evaluation of pleural effusion. In 42 patients (33.1%) pleural tuberculosis was the final diagnosis. There were 26 males and 16 females (age range 12-76 years). Age and sex distribution are shown in Table I.

Median age was 27.8 years and 75% of cases were in the 15 to 29 year age group.

Diagnostic methods included: (a) clinical features; (b) Mantoux skin test; (c) chest x-ray; (d) biochemical and cytological analysis of pleural fluid; (e) mycobacteriology and mycobacterial culture of sputum and/or bronchial aspirate, pleural fluid and pleural biopsy; (f) pleural optical findings by thoracoscopy; and (g) cytological and histological study of pleural biopsy.

TABLE I

AGE AND SEX DISTRIBUTION

Age (years)	Males	Females	Total
0 - 14	2	1	3
15 - 19	5	5	10
20 - 24	6	3	9
25 - 29	5	1	6
30 - 34	3	2	5
35 - 39	3	2	5
40 or more	2	2	4
ALL AGES	26	16	42

RESULTS

Fever, thoracic pain and cough were the most frequent clinical findings (Table II). Familial tuberculosis was present in 6 cases (14.2%) and smoking habits were documented in 13 cases (20.9%).

The Mantoux skin test was positive in 29 cases (66%).

The chest x-ray showed a right effusion in 23 patients (52.3%) and a left effusion in 19 cases (43.2%). In one patient with cardiac enlargement pericardial effusion was demonstrated by echocardiography.

Biochemical study of pleural fluid showed an exudate in all cases with total proteins greater than 30 g per liter. A high differential lymphocyte count was found in 39 cases (95.1%) with a mean percentage of 80.2%.

The proportion of bacteriologically proven cases is shown in Table III.

In 5 cases acid- and alcohol-fast bacilli were demonstrated in the pleural biopsy. As a rule, we do not perform routine culture of biopsy specimens.

TABLE II

CLINICAL FINDINGS

Sintoms	Number	(%)
Fever	40	95.2
Thoracic pain	35	83.3
Cough	32	75.7
Sweating	26	62.3
A.W.WL.*	23	54.7
Dyspnea	21	50.0
Others	4	9.5

* A.W.WL.=anorexia, weakness and weight loss.

TABLE III

MYCOBACTERIAL DIAGNOSIS

Material examined	Number	(%)
Ziehl-Neelsen staining of sputum	2	4.7
Ziehl-Neelsen staining of pleural fluid	2	4.7
Culture of pleural fluid	5*	11.9
TOTAL	7	16.6

* Including two cases with positive Ziehl-Neelsen staining of pleural fluid.

Thoracoscopy was performed in 41 patients (97.6%). The most frequent picture was reddened walls with small tubercles of a few millimeters in diameter or white plaques ranging from one to several millimeters in diameter on the pleural surfaces. Other less specific lesions were also found, particularly reddened pleura without tubercles or white fibrous pleura.

Histological findings were granulomatous pleuritis with or without central caseation necrosis in all cases. One patient showed granulomata without acid-fast bacilli on percutaneous needle biopsy of the liver.

DISCUSSION

The incidence of pleural tuberculosis varies from country to country and according to the social and economic conditions. It ranges from 0.9% (1) and 3.3% (2) in USA and 16% in France (3,4) to 44% in Tunisia (5). In Spain the frequency varies (Conde Yagüe et al 17.3%, Plans Bolibar et al. 45.6%, Varea Hernando et al 58.6%) (6-8).

In our study the incidence was 38.1%, showing a progressive increase since 1982 when 6 cases were found, 13 and 12 cases were found in 1983 and 1984 respectively, and 16 cases from January to June of this year. This is in contrast to other authors: Lamy in France found a decrease from 66.5% in 1960 to 26% in 1975 (4).

The presence of a febrile illness with pleural effusion in a young patient with a positive Mantoux skin test and a high differential lymphocyte count in pleural fluid is considered as an indication for thoracoscopy and pleural biopsy as elective diagnostic procedures. This agrees with several authors (3,4,9,10) due to the small number of cases with mycobacterial diagnosis (4.7% in our study, which is similar to or higher than that in other series) (10,11). If we include the positive cultures, our mycobacterial diagnosis increased to 11.9%. This is a lower percentage than that found in other series (4,7,10,12).

Because of the low positivity and the delay in culture we make the diagnosis from histological findings. This gives rapid results so that chemotherapy can be initiated immediately (4,9,10). In our experience this procedure allowed a diagnosis in all cases. When thoracoscopy is performed, pleural drainage accompanying this procedure accelerates the healing process since it ensures total evacuation of the pleural fluid, thus preventing pleural enlargement (9).

We have recently standardized our culture method of biopsy specimens so as to increase the percentage of bacteriological diagnoses. This has given encouraging results.

REFERENCES

1. Storey DD, Dines DE, Coles DT. Pleural effusion. A diagnostic dilemma. JAMA 1976; 236: 2183.

2. Rice R, van der Kuyp F. Clinical and epidemiological aspects of Childhood Tuberculosis. Am Rev Respir Dis (Annual Meeting Supplement) 1984; 129: A-200.

3. Wihlm JM, Roeslin N, Morand G, Pauli G, Witz JP. Résultats comparés de la ponction, de la biopsie à l´aiguille, de la pleuroscopie et de la thoracotomie dans le diagnostic des pleurésies chroniques. Poumon-Coeur 1981; 36: 83-94.

4. Lamy P, Canet B, Martinet Y, Lamaza R. Evaluation des moyens diagnostiques dans les épanchements plevraux. A propos de deux cent observations. Poumon-Coeur 1980; 36: 83-94.

5. Ayoub AK, Kerkeni AH. Le pH, la pCO2, la pO2 du liquide pleural. Variations et intéret diagnostique. Rev Pneumol Clin 1984; 40: 243-250.

6. Conde Yagüe R, Ledesma Castaño F, Prieto Ponga S, Carpintero Carcedo ME, Arias T, Fernández Pérez A, González Coterillo T, Zúñiga Pérez-Lemaur M. Células en el líquido pleural. Su valor en el diagnóstico diferencial de los derrames pleurales. Rev Clin Esp 1984; 173: 211-214.

7. Plans Bolibar C, Aranda Torres A, Roca Montanari A, Vidal Pla R, Sendra Salillas S, Vilaplana Soler M, Morell Brotad F, Morera Prat J. Pleuritis Tuberculosa; diagnóstico histológico y bacteriológico en 310 pacientes. Arch Bronconeumol 1980; 16: 106-110.

8. Varea Hernando HR, Yebra Pimentel MT, Martín Egaña MT, Masa Jiménez JF, Domínguez Juncal L, Fontán Bueso JM. Biopsia Pleural con aguja de Abrams. Análisis y rentabilidad en 207 casos. Arch Bronconeumol 1985; 21: 99-104.

9. Sebastián Quetglas F, Cosano Povedano J, Entrenas Costa LM, López Pujol J. Aspectos toracoscópicos de la Tuberculosis Pleural. Arch Bronconeumol 1985; 21: 14-18.

10. Huguenin-Dumittan S, Dottrens A. Résultats de la biopsie plevrale à l´auguille. Poumon-Coeur 1981; 37: 35-50.

11. Boutin C, Arnaud A, Uarette V, Cargnino P. La ponction-biopsie de plevre à l´aiguille d´Abrams. Médico-Rama 1971; 109: 2-10.

12. Hirsch A, Ruffie P, Nebut M, Bignon J, Chrétien J. Pleural Effusion: laboratory tests in 300 cases. Thorax 1979; 34: 106-112.

TREATMENT

NEW IN VITRO ANTIMICROBIAL POSSIBILITIES IN THE TREATMENT OF TUBERCULOSIS

M. CASAL

Department of Microbiology, School of Medicine, Cordoba University, 14004 Cordoba (Spain)

In the present worldwide struggle against tuberculosis one of the problems of physicians when treating patients is the small number of drugs that they can administer. Other problems are resistance to the drugs and possible toxicity. All this results in drugs often becoming useless and in certain cases the doctor just does not have any drugs to prescribe.

Since the introduction of rifampin and ethambutol to tuberculosis therapy, no new drugs have been added which could broaden the therapeutic arsenal against tuberculosis, contrary to what has been occurring in other infectious diseases.

This has interested investigators around the world, who have been trying to find some solutions to the problem.

TABLE I

SOME NEW THERAPEUTIC POSSIBILITIES IN TUBERCULOSIS I

1. New compounds (antituberculous group)
   - Amynoglucoside = Amikacine
   - Rifamycin = Ansamycin
2. Coenzyme Q and vitamin K analogues
   - 6- Cyclo-octilamino - 5 - 9 quinoline-quinone (CQQ)
3. Resensitizing strains compounds
   - Dimethyl sulphoxide (DMSO)
4. Immunotherapeutic agents

Apart from immunotherapy, various possibilities have emerged: (a) New antimicrobial drugs based on a similar action to those already in use, e.g. an aminoglucoside such as amikacin, which is undoubtedly the most active drug of this type. It has been tested *in vitro*, in animals and has also been used in human cases in the place of streptomycin.

(b) The use of analogue compounds of coenzyme Q and vitamin K, e.g. 6-cyclooctylamino-5,9-quinoline-quinone (CQQ) which has proven to inhibit the *in vitro* growth of *M. tuberculosis* or *M. avium*.

(c) It was observed that chemical compounds used *in vitro* and *in vivo* such as dimethylsulphoxide (DMSO) resensitize *M. tuberculosis* strains previously resistant to rifampin and isoniazid.

However, a new therapeutic model of interest which has not so far been used in tuberculosis has been used successfully in other infectious diseases.

The well-known resistance to penicillin by *Mycobacterium tuberculosis* and by mycobacteria in general is also found in the beta-lactam compounds. Most probably this is not due to a lack of action by the beta-lactam compounds at the site of infection, e.g. in membrane fragments of *M. smegmatis* D.D. carboxipeptidase activity has been found which is sensitive to the beta-lactam compounds, but rather to the failure of the antibiotic to reach the local site of action because of problems of permeability (1).

These drugs have therefore not been used in the treatment of tuberculosis and we think that the loss of interest in investigating new compounds of this type against mycobacteria is due to the fact that they are usually regarded as being resistant.

This is not totally true, since from time to time an isolated publication can be found giving interesting details which normally go unnoticed, as we will comment on later.

The beta-lactam compounds are known to be inactivated by three types of enzymes: (1) acetyl-esterase which hydrolates the acetyl group of the cephalosporins; (2) acylases which are usually penicillinase, but rarely cephalosporinase ; (3) the beta-lactamases which are generally included in the 3,5 subclasses of the hydrolases. The latter can break the beta-lactam ring of the penicillins, cephalosporins and monobactams since they play an important role in resistance to this type of anti-microbial compound by the micro-organisms which produce them.

The beta-lactamases can be classified in many ways. However, from a practical point of view 3 groups can be established: those with penicillinase activity, cephalosporinase activity, and broad spectrum activity. Differences are also established between the beta-lactamases of the gram-positive and the gram-negative bacteria and also between chromosomally controlled enzymes or the plasmid-mediated enzymes.

With regard to the beta-lactamases of the mycobacteria, we know less about them than about other types of bacteria and the facts we do know do not allow us as yet to categorise them.

If the beta-lactamases of mycobacteria are to be categorised, we might do so as follows: intracellular, constitutive, non-inducible, broad spectrum, more like gram-negative than gram-positive bacteria, with penicillinase and cephalosporinase activity (2).

Beta-lactamases have been found in many species of mycobacteria, although in some of them in such small quantities that a special technique such as isotechnique focussing is required.

It has been known for a long time that *Mycobacterium tuberculosis* has a beta-lactamase that inactivates substrates such as benzyl-penicillin.

The role of these beta-lactamases in mycobacteria is not thought to be related to the resistance of the antimicrobial agents since mycobacteria of *M. avium* species, which are highly resistant to antimicrobial agents and beta-lactamases, have not normally been found in them.

However, certain facts make us think that these beta-lactamases may play a part in resistance of microbial agents in certain species. If this role is well understood, it could be used in our favour.

It has been known for a long time that methicillin remains stable in *M. tuberculosis* cultures which inactivate benzylpenicillin. It is also known that mycobacteria show major sensitivity to beta-lactamase resistant penicillin, and it has been noted that methicillin and other beta-lactamase resistant penicillins inhibit the inactivation of benzylpenicillin in mycobacterial cultures, e.g. bicloxacillin which has been shown to act synergistically with benzylpenicillin against tuberculosis.

M.avium, which does not seem to have beta-lactamases (so far, there is only one published description) or at least not in such great quantities as in other mycobacteria, has been shown to be sensitive *in vitro* to 10 g of benzylpenicillin, in contrast to *M. tuberculosis* which has beta-lactamase activity that shows resistance. If *M. avium* is usually resistant to a large number of beta-lactam compounds, it could be due to problems of permeability (4).

Recently, *M. fortuitum* complex species have been shown *in vitro* and *in vivo* sensitivity to cephalosporins of the cephamycin type, e.g. cephoxitin and cephmetazol, the chemical structure of which gives them high resistance to beta-lactamases (5,6).

Using the term 'beta-lactamase inhibitor' we can group a number of substances of varying origin, chemical composition, mechanism of action, inhibition spectrum and ability to increase the antibacterial activity of the beta-lactam compounds.

Within the enzymatic inactivator groups in which the compound reacts with the enzyme at its catalytic centre, we find the better known compounds derived from penicillin such as meticillin, cloxacillin, cephalotin, cephoxitin, etc. or natural products such as the olivianic acids (i.e. MM4550) which have a broad spectrum against penicillinases and cephalosporinases, and clavulanic acid, perhaps the best known of this group.

Clavulanic acid is a potent inhibitor of beta-lactamases, particularly the penicillinase type, and is produced by *Streptomyces clavuligerus* (7). It has been proven that the addition of clavulanic acid increases the inhibitory action of amoxicillin against beta-lactam resistant bacteria, by inhibiting the destruction of amoxicillin both within and without the bacterial cell wall *in vitro* and *in vivo* (8). This has been applied clinically so far in treatment of urinary or other infections caused by amoxicillin-resistant bacteria, using the two compounds in combination. In these cases, the pharmacokinetics are fairly compatible, and a synergistic effect and a good therapeutic response have been observed.

This combination has not so far been used in the treatment of human mycobacteriosis. We also know that beta-lactamases exist in a great number of species of mycobacteria and the action which these beta-lactamases seem to have causes the resistance so as to determine the beta-lactam compounds. It would seem logical to try potent beta-lactamase inhibitors such as clavulanic acid against mycobacteria in association with beta-lactam compounds to observe whether they have an effect on the resistance of the mycobacteria to the beta-lactam compounds.

We think therefore that the possible mechanism of antimicrobial action of clavulanic acid against mycobacteria is based on the following: In mycobacteria, there could be building enzymes in cell walls, as has been demonstrated in *M. smegmatis* where D.D. carboxypeptidase has been found. In mycobacteria, beta-lactamases could exist which should protect the mycobacteria by destroying the beta-lactam compounds, preventing the accumulation of these in concentrations high enough to form inhibition of the peptidase membrane. It is not known where these beta-lactamases are located.

If the problem of resistance of mycobacteria to beta-lactam compounds were due solely to the failure in permeability of the antibiotic, the inhibition of beta-lactamase would not have any advantage. If, on the other hand, all species of mycobacteria had beta-lactamases, and these beta-lactamases played an essential role in the resistance of these species to the beta-lactam compounds, its inhibition would give us the possibility of destroying mycobacteria with beta-lactam compounds. The beta-lactam compounds would accumulate, if there were no failure of permeability, in sufficient quantities to have an effective action on the building pectidases of cell walls.

In this way it is known how clavulanic acid inhibits and inactivates beta-lactamases in the cell wall of bacteria, allowing the antibiotic molecules to reach the membrane where the antibiotic would then act on the pectidases thus interfering with the formation of the bacterial cell wall. This theoretical model has been developed in our laboratory and has given encouraging results:

(a) Detection of beta-lactamase activity in all *Mycobacterium tuberculosis* strains but not in the *Mycobacterium avium* strains tested.

(b) The MIC values of different beta-lactam compounds such as amoxicillin, cephtazidime and thienamycin against *M. tuberculosis*, *M. bovis* and *M. africanum* are reduced by ten times when clavulanic acid is used in combination with these compounds.

(c) The MIC values of a cephamicin such as cephoxitin against *M. tuberculosis*, *M. bovis* and *M. africanum* are reduced by five times when clavulanic acid is added.

(d) MIC values obtained which were below the serum levels obtained with therapeutic doses of a carboxypenicillin such as ticarcillin and a cephalosporin such as cephotaxime when clavulanic acid is added against *M. tuberculosis, M. bovis* and *M. africanum*.

(e) There is no variation in the MIC, which remains high, thus indicating resistance to some beta-lactam compounds such as ureidopenicillin and piperacillin, even when clavulanic acid is used in combination against *M. tuberculosis, M. bovis* and *M. africanum*.

(f) There is no variation in the MIC values for any of the beta-lactam compounds when clavulanic acid is added against *M. avium*.

These results allow us to conclude that:

(1) Not all beta-lactam inhibitors are active against mycobacteria. Hence we know that cephoxitin alone is inactive against *M. tuberculosis* and that clavulanic acid favours the action of these compounds. The effectiveness of beta-lactamase inhibitors against mycobacteria will depend on the type of beta-lactamase inhibited or on the different mechanism of action which it has against the same type of beta-lactamase.

(2) With the use of clavulanic acid not all beta-lactam compounds used in association with it are active against mycobacteria. For example, amoxicillin is active but not piperacillin. This different effect will also depend on other factors such as the permeability or the site of action of the antibiotic.

(3) All these facts will differ depending on the mycobacteria species and on the beta-lactam compounds: for example, *M. avium*, which is sensitive to penicillin, is resistant to a great number of other beta-lactam compounds, alone or in combination with clavulanic acid, and the opposite occurs with *M. tuberculosis*.

In view of all the above and the differing behaviour of species such as *M. tuberculosis* and *M. avium*, we think that these facts must be important in the resistance of mycobacteria to the beta-lactam compounds. Hence, *M. tuberculosis* usually has beta-lactamase activity which is easily detectable, in contrast to *M. avium* which either does not have any, or if it does, in such minute quantities that detection is extremely difficult. These two mycobacteria species respond different to clavulanic acid: *M. tuberculosis* is re-sensitized and *M. avium* remains resistant, in most cases, to high concentrations of beta-lactam compounds.

It seems that there are species such as *M. tuberculosis* in which one type of resistance mechanism predominates (e.g. beta-lactamase production), and other species such as *M. avium* in which permeability would be important.

The practical results could therefore be the following:

(1) In species such as *M. avium* more investigations should be made.

(2) In species such as *M. tuberculosis*, *M. bovis* or *M. africanum*, the therapeutic possibilities already proven *in vitro* with the beta-lactam compounds, need to be

TABLE II

MYCOBACTERIUM TUBERCULOSIS, M. BOVIS, M. AFRICANUM

β-Lactams	μg/ml	Serum levels	Strains inhibited
ceftazidime	8 + CA	20-80	100%
cefotaxime	8 + CA	18-70	100%
thienamycin	16 + CA		100%
ticarcillin	32 + CA	250	100%
amoxicillin	5 + CA	4-8	100%

CA = clavulanic acid (5 mcg/ml)

developed *in vivo*.

The time has come to investigate by experimental studies in animals and in human controlled therapeutic studies the true role of beta-lactam compounds and the new beta-lactamase inhibitors such as clavulanic acid in combination with rifampin, ethambutol, isoniacid, streptomycin and pyrazinamide, as another weapon in the modern therapeutic arsenal against tuberculosis.

There is another possibility different from all those mentioned above, namely the use of new molecules of antimicrobial agents different from the classical antituberculous compounds. We believe that new possibilities are presented by the use of the pyridone-carboxylic acid derivatives of the DNA gyrase inhibitor group.

This type of molecule has a chemical structure formed by condensed hexagonal heterocycles, diversely substituted, which retain half of the pyridone-carboxylic acid of the first compound of this group which is nalidixic acid. The last compounds of this group have many advantages over the old ones which would allow much wider use in infectious diseases and not only for urinary infections.

These have a broad spectrum of action and much greater activity, in some compounds up to 100% more than nalidixic acid (9,10).

Moreover, there is good intestinal absorption, which allows oral intake, good serum levels and half-life, and also low toxicity (11,12). Another interesting quality is their bactericidal activity, since they act on the sub-unit A of DNA gyrase, the enzyme that acts on the formation of the chains constituting the DNA helix. Inhibition of RNA synthesis reduces the bactericidal effect of quinolones. However, addition of rifampicin (RNA inhibition) is less antagonistic on ciprofloxacin or ofloxacin (13,14).

With regard to bacterial resistance to these compounds, mechanisms of resistance related to plasmid or transposons are not known. Resistance seems to be due to mutation in the bacterial chromosome. The frequency of changes in relation to resistance of the new compounds of this group such as norfloxacin, enoxacin, ofloxacin and ciprofloxacin seems to be much less than with older compounds such as nalidixic acid.

The clinical indications of these compounds so far seem to be, apart from urinary infections, venereal diseases, gastrointestinal osteomyelitis, respiratory infections, etc. They may also be useful in systemic infections.

The toxicity of these compounds seems to affect only about 5% of patients treated (skin changes, nausea, vertigo). Laboratory parameters were not altered.

Given the limited clinical experience with these compounds, it would seem advisable to assess their effectiveness carefully when administering them for longer periods, or when microorganisms with more restricted margins of security are involved.

According to our *in vitro* investigations (*M. tuberculosis* strains), in Middlebrook 7H12 liquid medium with the Bactec TB 460 system and also studying MICs in Middlebrook 7H10 solid medium, 95% of *M. tuberculosis* strains tested against ofloxacin and ciprofloxacin were inhibited by concentrations attainable in human serum or sputum after administering therapeutic doses.

TABLE III

Sputum levels (μg/ml)	Quinolones	Serum levels (μg/ml)	Oral doses	*Mycobacterium tuberculosis* MIC (μg/ml)
1.5	Ciprofloxacin	1.5	500 mg	1 (95%)
3.1	Ofloxacin	2	300 mg	0.5 (80%)
		5	400 mg	2 (95%)
				3 (97%)

Results with ofloxacin of animal experiments and tests carried out in human subjects affected by tuberculosis were favourable.

Finally, and along the same line of utilization of new molecules of antibiotics from different groups of classical antituberculosis drugs, we have a new macrolide - RU28965 - which according to our tests *in vitro* seems to have definite antimycobacterial activity. RU28965 has a spectrum of activity similar to that of erythromycin, since it is effective against both aerobic and anaerobic gram-positive bacteria, certain gram-negative bacteria such as Neisseria, Haemophilus and Legionella, and against Chlamydia and Mycoplasma (15,16).

It is effective against pulmonary infections. *In vivo*, it is twice to 14 times more active than erythromycin because of its better biodegradability and slow elimination which allow higher concentrations of antibiotics to be maintained 4 to 6 hours after intake. *In vitro* tests against *M. tuberculosis* showed that more than 90 or 95% of the strains are inhibited at lower concentrations than the serum peak found in human serum after an oral dose of 400 mg and that 80% of the

TABLE IV

ANTIMYCOBACTERIAL ACTIVITY OF THE NEW MACROLIDE, RU28965

Mycobacteria	MIC/μg/ml	Serum levels	Dose
M. tuberculosis	<5	5-11 μg/ml	400 mg oral
M. avium	16 1*		

*in combination with sulphonamides

strains continued to be inhibited by the concentration determined in serum three hours after intake.

The combination of RU28965 with other drugs against mycobacteria proved to be extra effective and synergistic, but never antagonistic, thus managing to lower the MIC necessary to inhibit mycobacteria. Combination with other drugs such as rifampin, INH, EB, PZA etc. should also be tested.

To conclude, there are now 3 new groups of antimicrobial agents to be tested in infections caused by mycobacteria. Although we have referred almost exclusively to *M. tuberculosis*, it is possible that these drugs will also be found to be effective against other typical mycobacteria such as *M. avium* using the new macrolide in combination or *M. leprae* which is being investigated with the new quinolones.

The three new groups of antimicrobials to be tested are: (1) the beta-lactamase inhibitors in combination with beta-lactam;(2) the new quinolone derivatives; and (3) the new macrolide RU28965, alone or in combination.

TABLE V

SOME NEW THERAPEUTIC POSSIBILITIES IN TUBERCULOSIS. II. (CASAL, 1985)

New antimicrobial compounds (non-antituberculous family)

a) Beta-lactamase inhibitors
   clavulanic acid + beta-lactam compounds
b) Pyridone-carboxylic acid derivatives of the DNA gyrase inhibitors
   ciprofloxacin and ofloxacin
c) New macrolide (ether oxime derivative of erythromycin)
   RU28965

All these agents will have to be tested individually and also in combination with other effective antituberculous drugs such as rifampin, pirazinamide, INH, ethambutol and streptomycin to see whether there are any synergistic effects. This would be a good base for animal experiments and for human clinical trials.

TABLE VI

NEW POSSIBILITIES TO INVESTIGATE IN THE THERAPY OF TUBERCULOSIS

Conventional antituberculous drugs	New antimicrobial compounds
Rifampin	β-Lactamase inhibitors + β-Lactam compounds
INH	New quinolones
Pirazinamide	RU28965
Ethambutol	
Streptomycin etc.	
possible synergistic effect	

REFERENCES

1. Eun HM, Yapo A. Petit FJ. (1978) DD-Carboxypeptidase activity of membrane fragments of Mycobacterium smegmatis, Europ. J. Brochem 86:97-103.

2. Kasik JE, Pearcham L. (1968) Properties of B-Lactamases produced by three species of Mycobacteria. Brochem, J. 107:675-682.

3. Sparks J, Ross GW. (1981) Isoelectric foensing studies on Mycobacterium chelonei. Tubercle 62:189-193.

4. Kasik JE, Monick M, Schwarz B. (1980) B-Lactamase activity in slow growing nonpigmented mycobacteria and their sensitivity to certain B-Lactam antibiotics. Tubercle 61:213-219.

5. Casal M, Rodriguez F. (1982) In vitro susceptibility of Mycobacterium fortuitum and Mycobacterium chelonei to cefoxitin. Tubercle 63:125-127.

6. Casal M, Rodriguez F. (1985) In vitro susceptibility of Mycobacterium fortuitum and Mycobacterium chelonei to cefmetazole. Antimicrob. Agents Chemother 27:282-283.

7. Reading C, Cole M. (1977) Clavulanic acid: a Beta-Lactamase inhibiting Beta-Lactam from Streptomices clavuligerus. Antimicrob. Agents Chemother 11:852-857.

8. Cole M. (1982) Biochemistry and action of clavulanic acid. Scott. Med J. 27:510-516.

9. Sato K, Matsuura F. (1982) In vitro and in vivo activity of Ofloxacin a new oxazine derivate. Antimicrob. Agents Chemother 22:4:548-553.

10. Daschner F, Just H. (1985) In vitro activity of five quinoline derivatives against nosocomial isolates of staphylococcus and pseudomonas species. Eur. J. Clin Microbiol 4: 72-73.

11. Crump B, Wise R, Deut J. (1983) Pharmacokinetics and tissue penetration of Ciprofloxacin. Antimicrob. Agents and Chemotherapy 24:784-786.

12. Wise R, Andrews J M, Eduards L J. (1983) In vitro activity of Bay 09867 a new quinoline derivate, compared with those of other antimicrobial agents. Antimicrob. Agents and Chemotherapy 23:559-564.

13. Smith JT (1984) Pharm J 1:229.

14. Smith JT, Ratcliffe NT. (1984) 4th Mediterranean Congress Chemother (Rhodos) Abstract 732.

15. Iones RN, Barry A, and Thornsberry C. (1983) In vitro Evaluation of three new nacrolide. Antimicrobial agents, RU 28965, RU 29065 and RU 29702 and comparisons with other orally Administered Drugs. Antimicrob. Agents. Chemother 24:209-215.

16. Barlam T, New HC. (1984) In vitro comparison of the Activity of RU 28965 a new Macrolide with that of Erytromycin against aerobic and anaerobic bacteria. Antimicrob. Agents Chemother 25:529-531.

Mycobacteria of Clinical Interest. M. Casal, editor

A MULTICOMPARTMENTAL MODEL FOR THE STUDY OF THE KINETICS OF THE TUBERCULOUS POPULATION IN RELATION TO VARIOUS CONDITIONS OF TREATMENT AND WITH SPECIAL REFERENCE TO THE DEVELOPING COUNTRIES

GIANNI ACOCELLA, WALTER POLLINI* and LAURA PELATI
Centre of Reference for the Chemotherapy of mycobacterial infections, Institute of Phthysiology and Diseases of the Chest, University of Pavia and *Lepetit Research Laboratories, Milan (Italy)

Epidemiological data indicate that the number of new cases of tuberculosis in the world is increasing (1). This is happening in spite of the existence of highly effective chemotherapeutic regimens which have been shown to cure close to 100% of treated patients.

The hypothesis has been put forward that the main cause of the phenomenon is the administration of regimens characterized by a low to very-low therapeutic efficacy (2).

In an attempt to identify the cause of the phenomenon, a multicompartmental kinetic model has been developed to simulate some epidemiological and clinical parameters related to tuberculosis over time.

The model has been called 'ESKIMO' (Epidemiological Simulation KInetic MOdel) and will be referred to as such in the text.

RATIONALE

It has been assumed that the size of a given population, in the absence of a tuberculosis reservoir (patients with active disease), is governed by the rate of entry (birth rate) and the rate of exit (death rate due to causes other than tuberculosis). If, however, such a reservoir exists, a certain number of individuals will interact at any given age with patients capable of transmitting the disease (sputum + patients) and, of these, a certain number will show the clinical and bacteriological signs of the disease.

The number of such new cases over time (incidence) will depend on the number of patients transmitting the disease and on the probability of contact between "healthy" and "sputum positive" individuals.

One of the assumptions of the model is that the incidence is proportional to the size of the sputum + population. It is in fact obvious that if the number of such patients is high and the chances of their coming into close contact with healthy individuals is also high, the number of new cases of tuberculosis will be correspondingly high.

Poor sanitary and housing conditions together with overcrowding have been long recognized as factors of relevance in the perpetuation of the disease through the transfer of mycobacteria from one host to another.

Breaking this biological, but mainly socio-economic cycle, through the separation of patients with active disease from the rest of the population was certainly one of the main reasons for the decrease of tuberculosis in Europe before the advent of successful chemotherapy.

The aim of any antituberculosis program is the reduction of the sputum positive sub-population and eventually to reduce the number to zero.

This can be achieved in several ways, namely:

1) Making individuals of the "healthy" sub-population insensitive to the infecting effect of their interaction with the "sputum positive" patients. This can be achieved through vaccination. This is a medium to long-term measure as it does not directly affect the patients in the "sputum positive" sub-population.
In the model the effect of vaccination has been disregarded on the basis of the most recent results of the large scale vaccination program with BCG in India.
2) Increasing the rate of transfer of patients from the "sputum positive" sub-population to the "healthy" sub-population. This can be achieved through chemotherapy and to a very minor extent through spontaneous healing.
3) Increasing the tuberculosis death rate. This is an hypothesis which, for obvious reasons, is reported here for the sake of completeness only.

In the model, the transfer rates of subjects between the different sub-populations tested ("healthy"; "untreated sputum positive"; "untreated sputum negative"; "treated sputum positive B.K.S. (Mycobacteria in the sputum sensitive to the drugs used)" and "treated sputum positive B.K.R. (Mycobacteria in the sputum resistant to the drugs used") have been derived from the fate of patients in the corresponding conditions as described in the literature (3).

The census data and the parameters characterizing the various treatments utilized in ESKIMO are: population; birth rate; non-tb death rate; incidence; prevalence; % of patients with sputum positivity for mycobacteria (of total patients identified); coverage (% of patients treated of total identified); % of sputum positive patients treated (of total patients identified); % of sputum conversion at 2 months of treatment; cure rate (% of patients cured of total treated); length of treatment in months; % of patients with mycobacteria resistant to the drugs used in the regimen, of the total number of treated patients.

A hypothetical country has been defined introducing data collected from different sources, all characterizing a developing country.

The validity of the simulations of ESKIMO has been controlled, introducing "time 0" parameters for a given country as listed above and having the model projecting the simulated values, which were in fact already known (prevalence and tb-deaths in Japan 1953-1957; tb-deaths in Thailand 1979-1983).

The results were highly satisfactory as a very good correspondence between the measured and the simulated values was found.

Of the several parameters which can, and have been, simulated by ESKIMO we will confine ourselves here to the number of sputum positive cases over a period of 5 years in different conditions of treatment.

For a country with a population of 88 millions and a birth rate of 3.5% and non-tb death rate of 1.4%, values of incidence of new cases of 100/100,000 and of prevalence of 0.3% were set.

Assuming a percentage of sputum positive patients of 50, a coverage of 30% and in cases where 100% sputum positive patients are treated, three different situations were simulated.

The first was that of no treatment. The second was that of administrating a regimen lasting 12 months and producing a sputum conversion rate of 20% at 2 months and an overall cure rate of 40%. The third was that of a regimen lasting 6 months and characterized by 95% sputum conversion at 2 months and the same percentage of overall cure rate.

The results of the simulation have indicated that the number of sputum positive cases (135,000 at the beginning of simulation) increases to 469,561 at the 5th year in the case of no treatment.

In the case of administration of the first regimen, the number of sputum positive cases decreases from 135,000 to 63.258 after 5 years.

In the case of administration of the second regimen, the number of sputum positive cases decreases from 135,000 to 20,544 after 5 years.

At this point we addressed the question of how many patients needed to be treated with regimens of different efficacy if one wanted to be left with 20,000 sputum positive cases after 5 years of administration of the various regimens.

The regimens were characterized by two values as follows (the first value refers to the % of sputum conversion at 2 months and the second to the overall cure rate): 20/40; 30/60; 40/80; 50/95; 95/95.

As can be noted, the last two regimens have the same overall cure rate of 95% but differ in that, with the first, 50% of sputum conversion occurs at 2 months whereas with the second 95% sputum conversion is achieved.

The results have indicated that to remain with 20,000 cases of sputum positive patients after 5 years, one needs to treat 476,000 patients with the first regimen (20/40); 427,000 with the second (30/60); 359,000 with the third (40/80); 295,000 with the fourth (50/95) and 218,000 with the fifth (95/95).

Attributing to the various regimens a cost based on recent estimates of drug cost, and assuming that this cost represents approximately 10% of the overall cost of treating a patient, it has been possible to demonstrate that even if the cost of the most effective regimen is more than 10 times that of the least effective ones, the overall expenditure is lower with the former, than with the latter.

The ESKIMO analysis seems to indicate that the most important single factor in characterizing the epidemiological value of a given regimen is the percent sputum conversion at 2 months, a fact which is not surprising from the clinical viewpoint.

This conclusion has relevant economic implications since it clearly indicates that it is far less expensive, both in purely financial and medical terms, to allocate the available funds to regimens of high efficiency.

ESKIMO is extremely simple to use and requires at present, the availability of a personal computer (Hewlett-Packard HP 150).

REFERENCES

1. Rouillon A. (1982) Bull. U.I.C.T. 57: 200 (No. 3-4)
2. Grzybowki S., Enarson D.A. (1978) 53: 70
3. Grzybowki S. (1985) Bull. U.I.C.T. 66: 69

EXPERIMENTAL CHEMOTHERAPY OF TUBERCULOSIS AND LEPROSY

JACQUES H.GROSSET, CHANTAL TRUFFOT, HERVE LECOEUR, CLAIRE-CECILE GUELPA-LAURAS and JEAN-LOUIS CARTEL
Department of Bacteriology-Virology, Faculte de Medecine Pitie-Salpetriere, 75634 Paris cedex 13 (France)

INTRODUCTION

In the last two decades, tremendous progress has been made in the field of chemotherapy for mycobacterial diseases. Tuberculosis and leprosy have benefited much more than other mycobacterioses from the progress which was due not only to the discovery of new antibacterial drugs, but also to a better use of pre-existing drugs and a better knowledge of the mechanisms of drug activity against the microbial subpopulations inside the lesion. The aim of this presentation is to recall the main contributions of the experimental work, first to the chemotherapy of tuberculosis, and second, to the chemotherapy of leprosy.

CHEMOTHERAPY OF TUBERCULOSIS

In 1960 chemotherapy of clinical tuberculosis had a mean duration of 18 months to 2 years, was performed mainly in hospitals or sanatoria, gave side-effects in about 15% of the patients and was followed at best by at least 5% failures during treatment and 10% relapses after stopping treatment. In 1985, the optimal duration of chemotherapy was 6 months, the treatment was mainly ambulatory, the incidence of side-effects less than 5% and the rate of failures-relapses was also less than 5%. As everyone knows, such a dramatic change results from the discovery of rifampicin in 1965 and from the standard use of combinations of the three key-drugs isoniazid, rifampicin and pyrazinamide which act upon all microbial subpopulations (1) present in the lesions.

The role of isoniazid

Although streptomycin (1944) and para-amino-salycilic acid (1949) were the first available antituberculous drugs, chemotherapy of tuberculosis began in fact with the discovery of the antituberculosis activity of isoniazid (INH) in 1952.

Before the era of INH, patients with tuberculosis as well as experimental animals (2) treated with streptomycin (SM) alone experienced delayed, but regular therapeutic failures, the mechanism of which was clearly understood. As the proportion of SM-resistant mutants in a wild strain of *M. tuberculosis* is approximately 10^{-6}, it is easy to calculate that in a tuberculous cavity containing 10^{8} multiplying sensitive organisms, the number of SM-resistant mutants is of the order of 10^{2}. Under treatment with SM alone, the sensitive organisms are killed whereas the resistant mutants are progressively selected in such a way that the short initial phase of apparent improvement is followed by a secondary phase of deterioration, the two phases corresponding to the classical "fall and rise" phenomenon.

To observe in the experimental animal the same phenomenon, one condition is necessary: the microbial population in the lesions should be more or less the same size as the microbial population in the lung cavity of human tuberculosis, that is 10^8 organisms, because only large microbial populations contain drug-resistant mutants. Such a condition is obtained when mice, 28 days old, are infected by the intravenous route with 0.1 mg *M. tuberculosis* from a strain highly virulent for the mouse, cultivated on tween 80 broth, and are kept without treatment until the 14th day. On that day, the microbial population in the lungs as measured by quantitative cultures has reached 10^8 organisms and treatment with SM alone is followed in three months time by the selection of resistant mutants (Grumbach, 1953; Bartmann, 1960) as in man (2).

The crucial interest of the combination INH-SM was its ability, not only to kill sensitive organisms (streptomycin alone was able to do so), but also to prevent the selection of both streptomycin and isoniazid-resistant mutants, and thus to cure tuberculosis. However, in the mouse, the cure of tuberculosis was never perfect. As shown by Grumbach (2) during treatment some rare failures (5%) were regularly observed with selection of isoniazid-resistant mutants and, after 12 or even 18 months of daily treatment, half of the mice still harbored few colonies of drug-sensitive organisms in their lungs.

The role of rifampicin and other ansamycins

As soon as rifampicin (RMP) was introduced (1967), Grumbach and Rist demonstrated the apparent cure of experimental tuberculosis of the mouse after 4 months of treatment with the combination of INH plus RMP. This demonstration initiated a series of experiments (3) performed by many workers, mainly in mice, but also in guinea-pigs, which confirmed the initial findings that, after 4 to 6 months of treatment with the combination 25 mg/kg INH + 25 mg/kg RMP, the lung and spleen cultures were negative in 100% of the animals. In no cases was selection of isoniazid or rifampicin-resistant mutants observed, a result never achieved with the combination INH + SM.

Because the combination INH + RMP was so effective, the next step was to assess the relapse rates after stopping treatment. As shown in Table 1, the relapse rate after 6 months treatment of mice with INH + RMP was 20%, a fairly high rate when considered alone, but quite low in comparison with the relapse rate of 75% after 18 months treatment with INH + SM. When the treatment with INH + RMP was extended until the 9th month, no relapses at all occurred on the condition that RMP had been given all the way through. This condition emphasized the special sterilizing effect of RMP and, in comparison, the relatively poor sterilizing effect of INH, at least in the mouse. From these data it could be predicted as early as in 1969 that, giving INH + RMP in man, for 9 months might be more effective than standard INH + SM chemotherapy for 18 months which was regarded, at this time, as very effective.

TABLE I

RELAPSES IN MICE TREATED DAILY WITH EITHER ISONIAZID+STREPTOMYCIN OR ISONIAZID+RIFAMPICIN.

Drug regimen	Duration of treatment (months)	Percentage of negative animal at the end of treatment	after follow-up of 3-month	6-month
INH + SM	6	0	0	
INH + SM	18	65	25	
INH + RMP	6	100	-	80
INH + RMP	9	100	100	100
6-month INH + RMP followed by 3-month INH alone	9	100	-	80

INH : 25 mg/kg isoniazid SM : 200 mg/kg streptomycin RMP : 25 mg/kg rifampicin.

In many countries giving daily rifampicin-containing regimens is difficult for financial and/or compliance reasons. Therefore, once and twice weekly regimens were tested. Mice treated with the combination 25 mg/kg RMP and 25 mg/kg INH once or twice weekly gave significantly worse results than mice treated with the same regimens given daily, as shown in Table II.

Because the only resistant organisms that were selected with the intermittent regimens were INH-resistant mutants, it might be suspected that intermittency decreased more the effectiveness of RMP than that of INH. Actually, this hypothesis was confirmed by two subsequent experiments (4) performed in mice with limited bacillary populations containing no drug-resistant mutants. In this first experiment mice were treated with 10 mg/kg RMP given either once, twice, three times or six times a week. The antimicrobial effectiveness was assessed by quantitative cultures of the spleen, done before the beginning of treatment, and after 14, 28 and 56 days of treatment. As shown in Table III, the mean number of colony-forming units (c.f.u.) in the mice treated three times a week was significantly higher than in those treated six times a week ($p < 0.001$) but was significantly lower than in those treated twice a week ($p < 0.05$). There were no significant differences in the number of c.f.u. from control mice and mice treated once or twice weekly, the difference being at the limit of statistical significance ($p=0.05$) for twice-weekly treated mice. In the second experiment, mice were treated twice a week with 25 mg/kg INH and 10 mg/kg RMP given alone or in combination. Assessments of effectiveness were also made by quantitative spleen cultures before the beginning of treatment and after 14, 28 and 56 days of treatment. Because the combination INH + RMP did not give better results than INH alone, it can be concluded (Table IV) that RMP when given twice a week, is poorly effective in the mouse, although its pharmacokinetics are much more favourable in the mouse than in man.

The overall conclusion of these experiments is that rifampicin activity decreases when the interval between each dose increases. Several experiments performed in the mouse or in the guinea-pig have shown that increasing the daily dosage of rifampicin can compensate the decreased effectiveness, due to intermittency (3), whereas increased dosages of RMP given intermittently are badly tolerated in human beings, and therefore are of no practical value. Although intermittent administration of RMP would facilitate the organization and supervision of ambulatory chemotherapy, it cannot be recommended as a routine procedure. The best solution would be to have a slow-release rifampicin that, given once or twice a week, would be as active as daily rifampicin. The cyclopentyl-rifamycin, called DL 473, could be an example of such a slow-release rifampicin. Its minimal inhibitory concentration (MIC) is 0.12 to 0.25 mg/l. As there is cross-resistance between DL 473 and RMP, the interest of DL 473 lies mainly in its pharmacokinetics. Its serum half-life in the mouse is about 24 h, whereas the half-life of RMP is only 8 h. When tested in the mouse in the initial phase of chemotherapy, as well as in the continuation phase, 10 mg/kg DL 473 given once a week appears as effective as 10 mg/kg RMP given daily (4). Thus it is easy to speculate on the operational advantages of such a drug. For example, after two months of initial intensive daily multiple-drug chemotherapy, it might be possible to treat patients with DL 473 alone, given once a week, and to insure a fully supervised and extremely effective chemotherapy.

The role of pyrazinamide

Pyrazinamide (PZA) is a very special drug that is active only at an acid pH and, therefore, not active against all microbial subpopulations, as is rifampicin. It is the reason for which its real contribution in the chemotherapy of tuberculosis was difficult to establish, in man as well as in the mouse, although as early as in 1956 McCune, Tompsett and McDermott were able to render the organs of mice culture-negative after only 3 months of treatment with the combination of INH plus PZA. In their experiment these authors used a daily dosage of 4000 mg/kg PZA which gave blood levels in the mouse of about 400 to 600 mg/l, not comparable with the 40-80 mg/l obtained in man with the usual daily dosage of 30-35 mg/kg. Actually, as demonstrated later (3), the daily dosage to be used in the mouse to obtain blood levels comparable to those obtained in man is 150 mg/kg. With this dosage a series of experiments were performed in mice that gave an overall assessment of the value of pyrazinamide in the short-course chemotherapy of tuberculosis (4). First of all as shown in Table V, 150 mg/kg PZA given daily for six months, in combination with daily 25 mg/kg INH and 10 mg/kg RMP decreases significantly the relapse rate after stopping chemotherapy. When given only in the continuation phase of chemotherapy, it decreases also the relapse rate, but the decrease is not statistically significant (Table VI). Therefore, it can be concluded that PZA exhibits a sterilizing activity particularly when given in the initial phase of chemotherapy in the mouse, as in man (5). Due to this activity short-course chemotherapy with INH and RMP can be reduced to six months when PZA is added. The question of the precise length of time PZA should be given during the initial phase had also to be answered. As shown in Table VII, from more recent experiments in the mouse, the optimal length of time appears to be 2 months, a duration that is currently in use in clinical chemotherapy without definite evidence! Last but not least, it should be emphasized that PZA can play also a major role in the prevention of rifampicin resistance, PZA being active only at an acid pH.

It was supposed that the main bacillary subpopulation located in the cavity at a neutral pH escaped the killing action of PZA. In the case of chemotherapy, with the three-drug regimen INH + RMP + PZA, patients with primary INH resistance were considered at high risk to acquire rifampicin resistance, because their main microbial subpopulation located in the cavity would be submitted only to the action of rifampicin. To prevent this risk it was, and still is, common to give four drugs, namely INH, RMP, PZA and ethambutol (EMB), the role of the latter being to prevent RMP resistance in the case of primary INH resistance. However, in some countries, short-course chemotherapy relied only upon the triple combination INH, RMP and PZA and did not give rise to numerous cases of acquired RMP resistance in patients with primary INH resistance, though PZA had actually the ability to prevent the selection of RMP-resistant mutants. To test this ability, mice were inoculated with 2×10^6 c.f.u. of *M. tuberculosis* and two weeks later, when the microbial population in their lungs had reached approximately 10^8 organisms, were treated daily with 10 mg/kg RMP given either alone or together with 150 mg/kg PZA. After 3 months of treatment, lungs and spleen were

TABLE 2

COMPARATIVE EFFECTIVENESS OF DAILY AND INTERMITTENT 25 MG/KG ISONIAZID PLUS 25 MG/KG RIFAMPICIN COMBINATIONS GIVEN FOR SIX MONTHS.

rhythm of administration	% negative cultures at 6 months
Daily	100
Twice weekly	8*
Daily one month followed with	
either twice weekly	60**
or once weekly	0**

* 25 % of the positive cultures yielded organisms resistant to INH but sensitive to RMP.

** All of the positive cultures yielded organisms sensitive to INH and RMP.

TABLE 3

ROLE OF INCREASING THE INTERVALS BETWEEN 10 MG/KG DOSES OF RIFAMPICIN ON THE EFFECTIVENESS OF THE DRUG AGAINST M. TUBERCULOSIS IN THE MOUSE.

rhythm of rifampicin administration per week	Log_{10} c.f.u.* in the spleen of mice at day			
	0	14th	28th	56th
Control	5.65±0.24	4.61±0.56	5.05±0.30	4.85±0.37**
6 times		3.45±0.91	2.90±0.34	0.65±0.62
3 times		4.65±0.38	3.91±0.35	3.42±0.86
2 times		4.39±0.75	4.22±0.53	4.43±0.53**
Once		4.62±0.49	4.81±0.22	4.86±0.30**

* Mean number from ten mice ± standard deviation. ** Not significantly different.

TABLE 4

COMPARATIVE EFFECTIVENESS AGAINST M. TUBERCULOSIS OF ISONIAZID AND RIFAMPICIN GIVEN TWICE WEEKLY EITHER ALONE OR IN COMBINATION TO THE MOUSE.

Twice weekly regimen*	Log_{10} c.f.u.** in the spleen of mice at day 0	14th	28th	56th
Control	5.96±0.39	5.13±0.25	5.38±0.46	5.01±0.52****
INH alone		4.03±0.25	3.30±0.33	2.24±0.22***
RMP alone		5.28±0.30	4.42±0.80	3.65±0.36****
INH + RMP		4.22±0.28	3.89±0.25	1.98±0.41***

* 25 mg/kg INH and 10 mg/kg RMP given each week on days 1st and 4th

** Mean number from ten mice ± standard deviation

*** $p = 0.05$

**** $p < 0.001$

TABLE 5

ROLE OF PYRAZINAMIDE IN THE INITIAL PHASE OF 6-MONTH CHEMOTHERAPY AS MEASURED BY THE RELAPSE RATE AFTER 6-MONTH FOLLOW-UP.

Drugs given daily from: 0 to 2nd month	2nd to 6th month	Follow-up after 6 months: Total mice	Lung and/or spleen positive mice N	%
INH + RMP	INH + RMP	52	19*	36.5
INH + RMP + PZA	INH + RMP	47	5*	10.7

* $p < 0.05$

INH : 25 mg/kg isoniazid RMP : 10 mg/kg rifampicin PZA : 150 mg/kg pyrazinamide

TABLE 6

ROLE OF PYRAZINAMIDE IN THE CONTINUATION PHASE OF 6-MONTH CHEMOTHERAPY AS MEASURED BY THE RELAPSE RATE AFTER 6-MONTH FOLLOW-UP.

Drugs given daily from		After 6-month follow-up		
0 to 3d month	3d to 6th month	total mice	lung and/or spleen positive mice N	%
INH + RMP	RMP	53	35	66
INH + RMP	INH + RMP	52	31	60
INH + RMP	INH + RMP + PZA	55	32	58

TABLE 7

`THE OPTIMAL DURATION OF THE INITIAL SUPPLEMENT OF PYRAZINAMIDE IN THE MOUSE TREATED WITH ISONIAZID AND RIFAMPICIN.

6-month daily INH+RMP with an initial supplement of daily PZA for	After 6-month follow-up Total mice	lung and/or spleen positive mice N	%
0-month	51	35	68.6
1-month	39	18	46.1*
2-month	51	6	11.8*
3-month	54	13	24.1**
4-month	48	11	22.9**

* p = 0.001   ** p = 0.44 with 2-month PZA

INH = 25 mg/kg isoniazid   RMP = 10 mg/kg rifampicin   PZA = 150 mg/kg pyrazinamide.

TABLE 8

THE CAPABILITY OF PYRAZINAMIDE TO PREVENT THE SELECTION OF RIFAMPICIN RESISTANT MUTANTS IN THE MOUSE.

Daily treatment	Total mice	Mice with lungs culture positive			
		Total		RMP resistant	
		N	%	N	%
RMP alone	40	38	95	33	87
RMP + PZA	108	40	37	0	-

RMP : daily 10 mg/kg rifampicin   PZA : daily 150 mg/kg pyrazinamide

respectively culture positive in 95% and 90% of mice treated with RMP alone and in 37% and 24% of mice treated with RMP + PZA ($p < 10^{-5}$). The strains isolated from lungs and spleen were resistant to RMP in, respectively, 87% and 75% of mice treated with RMP alone, and in none of those treated with RMP + PZA (Table VIII). From the results of this experiment it can therefore be concluded that PZA has the ability to prevent the selection of RMP-resistant mutants.
As sputum conversion occurs more rapidly in patients treated with PZA, than in patients without PZA (5), and more favourable results are obtained in patients with primary INH resistance when they are treated with INH + RMP + PZA than when they are treated with the same combination without PZA, one is let to infer that PZA is active against the main microbial population located in the tuberculous cavity. As it is active only at an acid pH, that would mean that a large part of the microbial population of the tuberculous cavity is in an acid, and not in a neutral environment, as was thought until now (6). In practice one may consider it ethical to treat patients only with the three-drug combination INH, RMP and PZA, which is operationally much more easy to handle than the four-drug combination INH, RMP, PZA and EMB.

Future prospects in experimental chemotherapy for tuberculosis

Although antituberculosis chemotherapy has now reached a point of extreme activity, there are still problems that need to be solved. The first concerns the long-lasting rifamycin-derivatives, which could permit a fully-supervised highly-effective continuation phase of chemotherapy.

Other semi-synthetic rifamycin-derivatives (ansamycins) than cyclopentul rifamycin (DL 473) may be interesting, like LM 427 or CGP 29861.

The second problem is to find new drugs active against tubercle bacilli, resistant to INH and RMP, that are isolated in case of failures of badly implemented modern chemotherapy. To retreat the patients, new drugs are needed. Among all of the antibacterial drugs that have been recently developed, the fluoroquinolones are the most promising (7, 8).

The third problem is the treatment of mycobacterial infections, due to the so-called atypical mycobacteria or mycobacteria other than the tubercle bacillus, and leprosy bacillus. The atypical mycobacteria are, with rare exceptions, resistant to the majority of existing drugs. Their role as opportunistic organisms is increasing, particularly among patients with A.I.D.S., and drugs active against atypical mycobacteria are urgently needed, as is all available modern biotechnology.

CHEMOTHERAPY OF LEPROSY

Experimental chemotherapy of leprosy was born in 1960 when C.C. Shepard demonstrated the multiplication of *Mycobacterium leprae* in the foot-pad of the mouse. Before that time, *M. leprae*, perhaps the only one Mycobacterium with obligatory intracellular multiplication, had only been studied in man. Until the discovery of C.C. Shepard, experimental chemotherapy of leprosy was experimental chemotherapy in man. Since 1960, it has been possible to use the mouse to assess the antibacterial activity of the available antileprosy drugs (dapsone, clofazimine, rifampicin and the thioamides) against reference strains of *M. leprae* as well as to test the drug susceptibility of various strains of *M. leprae*. As the only way to measure the multiplication of *M. leprae* in the foot-pad of the mouse is to count the acid-fast bacilli (AFB) in foot-pad suspensions stained by the Ziehl-Neelsen method, and because the multiplication of *M. leprae* is limited and extremely slow, studies based on experimental chemotherapy of leprosy are difficult to perform and their results not always easy to interpret. At present, three main methods (9) are currently in use in experimental chemotherapy of leprosy: (i) the "continuous" method devised by Shepard to determine whether a drug has activity against *M. leprae*, (ii) the "kinetic" method, also devised by Shepard to determine whether a drug or a combination of drugs could have bactericidal activity against *M. leprae*, and (iii) the "proportional bactericidal test" devised by Hilson and Banerjee to study the bactericidal activity of drugs against *M. lepraemurium* and adapted by Colston et al for the assessment of bactericidal activity against *M. leprae*. Using these methods, it has been demonstrated that rifampicin and the thioamides had a bactericidal activity against *M. leprae*, whereas dapsone and clofazimine had an activity of the bacteriostatic type. One important point of discussion with rifampicin, is its apparently more powerful activity against *M. leprae* in man, than in the mouse, despite its more favourable pharmacokinetics in the mouse than in man. This point raises important questions on the relations between the viability of *M. leprae* in the tissues and the response to rifampicin. A number of studies in man and in the mouse (conventional and athymic) should certainly be performed to answer these questions, which are important to determine the optimal length of time the dosage, and the rhythm of administration that rifampicin has to be given to the patients.

At present, another objective of experimental chemotherapy is to test the value of the new fluoroquinolones (7,8) against *M. leprae*. It is hoped that new drugs will become available in the near future to treat patients resistant to the existing drugs, especially dapsone and RMP.

REFERENCES

1. Grosset J. (1980) : Bacteriologic basis of short-course chemotherapy for tuberculosis. Clinics in Chest Medicine 1 : 231-241.

2. Grumbach F. (1965) : Etudes chimiothérapiques sur la tuberculose avancée de la souris. Adv. Tuberc. Res. 14 : 31-96.

3. Grosset J. (1978) : The sterilizing value of rifampicin and pyrazinamide in experimental short-course chemotherapy. Tubercle 59 : 287-297.

4. Grosset J., Truffot-Pernot Ch., Bismuth R., Lecoeur H. (1983) : Recent results of chemotherapy in experimental tuberculosis of the mouse. Bull. Int. Union Against Tub. 58 : 90-96.

5. East African/British Medical Research Study (1981). Controlled clinical trial of five short-course (4 months) chemotherapy regimens in pulmonary tuberculosis. Second Report of the 4th study. Amer. Rev. Resp. Diseases 123 : 165-170.

6. Grosset J., Truffot-Pernot Ch., Poggi S., Lecoeur H., Chastang Cl. (1985) : prevention de la resistance à la rifampicine par le pyrazinamide dans la tuberculose experimentale de la souris. Rev. Fr. Mal. Resp., in press.

7. Montay G, Goeffon Y, Roquet F. (1984) : Absorption, distribution, metabolic fate and elimination of pefloxacin mesylate in mice, rats, dogs, monkeys and humans. Antimicrob. Agents and Chemoth. 25 : 463-472.

8. Tsukamura M., Nakamura E., Yoshii S., Amano H. (1985). Therapeutic effect of a new antibacterial substance ofloxacin (DL 8280) on pulmonary tuberculosis. Amer. Rev. Resp. Diseases 131 : 352-356.

9. Colston M.J., Hilson G.R.F., Banerjee D.K. (1978) : The "Proportional Bactericidal Test" : a method for assessing bactericidal activity of drugs against Mycobacterium leprae in mice. Lepr. Rev. 49 : 7-15.

Mycobacteria of Clinical Interest. M. Casal, editor

BACTERIOSTATIC AND BACTERICIDAL EFFECTS OF RIFABUTINE[R] (ANSAMYCIN LM427) ON *MYCOBACTERIUM AVIUM* CLINICAL ISOLATES

LEONID B. HEIFETS, MICHAEL D. ISEMAN, PAMELA J. LINDHOLM-LEVY
National Jewish Center for Immunology and Respiratory Medicine,
Denver, CO 80206, U.S.A.

INTRODUCTION

Rifabutine[R] (ansamycin, LM427, spiro-piperidyl rifamycin), developed by Farmitalia Carlo Erba (Milan, Italy), is widely used in the U.S.A. as an experimental drug for the treatment of disseminated *M. avium* - *M. intracellulare* infection in AIDS patients. The clinical effect of this treatment is so far unknown. To establish valid criteria of susceptibility to the drug in future clinical trials, it is essential to decide what kind of *in vitro* test results should be compared with clinical response to Rifabutine[R]. In our approach to this problem, we determined the range of Minimal Inhibitory Concentrations (MIC) of Rifabutine[R] in 7H11 agar plates and in 7H12 broth, evaluated the ratio of MICs to Minimal Bactericidal Concentrations (MBC), and compared the range of MICs with ascertained serum levels of Rifabutine[R] in the patients. We also compared the *in vitro* susceptibility to this drug of *M. avium* isolates from AIDS and non-AIDS patients.

MATERIALS AND METHODS

Strains. Clinical isolates from 100 patients with disseminated infection (from blood) and from 111 patients with localized pulmonary infection (from sputum) were the subject of this study. Subcultures from transparent (SmT) type colonies were used and preserved as frozen aliquots in broth at -70°C.

MIC determination. Eight concentrations in a range of 0.015 µg/ml to 2.0 µg/ml were used to determine MIC in both solid (7H11 agar plates) and liquid (7H12 broth) media.

The conventional proportion method in 7H11 agar plates and the radiometric proportion method in 7H12 broth (1,2) employing the BACTEC 460-TB system (Johnston Laboratories, Towson, MD) were used for testing all 211 strains. The conventional tube dilution method was used to determine the effect of the drug on the growth curve (CFU/ml determination by sampling from 7H12 broth and plating) of six strains. More details about these three methods of MIC determination were given in our previous communications (3,4).

MBC determination. 7H12 broth cultures containing different concentrations of the drug were sampled and plated daily to determine the killing curve. The lowest concentration that killed at least 99.9% of the bacterial population was considered as the MBC. Six strains were tested, and their MBCs were compared with their MICs.

RESULTS

In the agar plate proportion method, 82% of the strains were susceptible to 1.0 μg/ml of Rifabutine[R] (Table I), a finding which is in agreement with previous studies (3,5). Testing with concentrations lower than 1.0 μg/ml showed MICs of 1.0 μg/ml for 23.2%, 0.5 μg/ml for 31.3%, 0.25 μg/ml for 17.5%, and 0.125 μg/ml and lower for 10.0% of strains. These data indicate that the *in vitro* susceptibility testing by this method should include not only 2.0 and 1.0 μg/ml, but one or two lower concentrations as well.

TABLE I. THE MICs OF RIFABUTINE[R] FOR *M. avium* ISOLATES DETERMINED BY CONVENTIONAL AGAR PLATE METHOD

Rifabutine Concentrations (μg/ml)	AIDS		Non-AIDS	
	No. of Cultures	%	No. of Cultures	%
0.015	0	--	0	--
0.031	1	1.0	3	2.7
0.062	2	2.0	1	0.9
0.125	6	6.0	8	7.2
0.25	14	14.0	23	20.7
0.5	23	23.0	43	38.8
1.0	25	25.0	24	21.6
2.0	29	29.0	9	8.1

TABLE II. THE MICs OF RIFABUTINE[R] FOR *M. avium* ISOLATES DETERMINED IN LIQUID MEDIUM (BACTEC METHOD)

Rifabutine Concentrations (μg/ml)	AIDS		Non-AIDS	
	No. of Cultures	%	No. of Cultures	%
0.015	2	2.0	4	3.6
0.031	1	1.0	4	3.6
0.062	16	16.0	16	14.4
0.125	39	39.0	26	23.4
0.25	25	25.0	31	27.9
0.5	13	13.0	24	21.7
1.0	2	2.0	6	5.4
2.0	2	2.0	0	--

TABLE III. COMPARISON OF MIC AND MBC FOR SIX *M. avium* STRAINS

Strain No.	MIC (μg/ml)	Bactericidal Effect		Ratio of MIC/MBC
		CFU/ml When Drug Added	MBC (μg/ml)	
800-4	0.031	185,000	1.0	1/32
242-5	0.125	169,000	8.0	1/64
1317-2	0.25	464,000	32.0	1/128
1408-2	0.25	163,000	16.0	1/64
168-4	0.5	123,000	16.0	1/32
3009-4	1.0	237,000	32.0	1/32

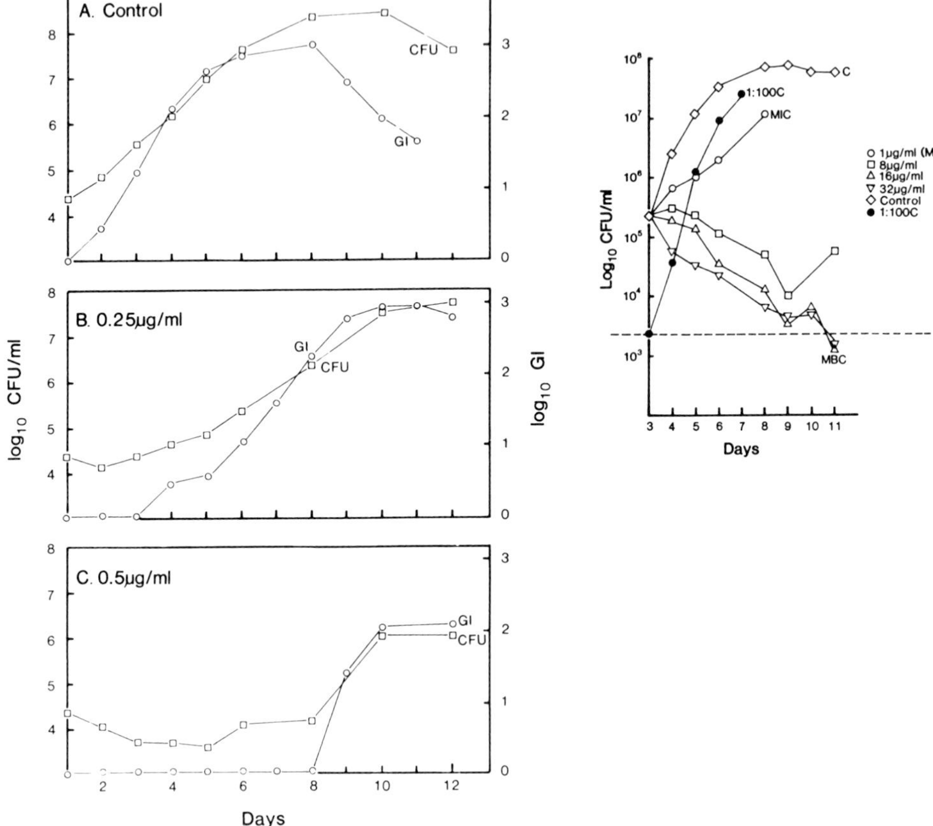

Fig. 1. Correlation between the growth curve (CFU/ml) and the daily GI readings. (A) Control; (B) with Rifabutine in subinhibitory concentration; (C) with Rifabutine in a concentration considered as the MIC for this strain.

Fig. 2. An example of MBC and MIC determinations

The validity of determining the MIC in 7H12 broth by the radiometric method was established in our previous report (4). It was based on correlations found between growth curves (CFU/ml) and daily Growth Index (GI) curves. An example of such correlations is presented in Fig. 1. Comparison of the MIC values determined by the two methods indicated that most of the isolates appeared more susceptible in 7H12 broth (Table II) than in agar plates (Table I): the broth-determined MICs were 0.25 µg/ml or lower for 77.7% of the strains, while only 27.5% of the same strains were in this range when tested in agar plates. This difference is a result of partial degradation of the drug during the longer period of exposure to 37°C in agar plates. The determination of MIC in broth by the radiometric method requires only a few days, therefore this method provides more precise information about MIC values.

Whether determined by agar plate or broth methods, no significant differences were found in MICs between the strains from AIDS and non-AIDS patients.

The bactericidal effect of Rifabutine[R] was determined as 99.9% killing effect, which is a decrease in the number of CFU/ml below the dotted line in the example presented in Fig. 2. The radiometric GI readings were used to determine the most appropriate time to add the drug during the exponential phase of growth. The actual number of CFU/ml at the moment of drug addition (48 to 72 hrs after beginning of cultivation) appeared to be between 1×10^5 to 5×10^5 (Table III). The MIC/MBC ratio was 1/32 for three strains, 1/64 for two strains, and 1/128 for one strain (Table III).

The achievable serum peak level of Rifabutine[R] in patients is reported to be 0.38 µg/ml (4) and 0.49 µg/ml (6), and in tissues 10 to 20 times higher (6). Comparison of these data with the results of MBC determination (Table III) suggests that the bactericidal effect might appear in tissues, but should not be expected in the serum. If the levels of susceptibility in agar plates (Table I) were taken as true MIC, the data would indicate that the susceptibility of most isolates is 0.5 µg/ml or higher, a level which cannot be translated into attainable serum concentrations, and from which a suboptimal clinical response might be expected. If the levels of susceptibility in broth (Table II) were taken as the MIC, then 70% or more of the patients could achieve a blood level equal to one or more MICs. Considering the wide range of MICs and the low blood concentrations of the drug, the determination of the MIC (in broth by the radiometric method) of an isolate ought to have a predictive value in chemotherapy.

REFERENCES

1. Siddiqi SH, Libonati JP, Middlebrook G (1981) Evaluation of a rapid radiometric method for drug susceptibility testing of *Mycobacterium tuberculosis* J Clin Microbiol 13: 908-912
2. Tarrand JJ and Groschel DHM (1985) Evaluation of the BACTEC radiometric method for detection of 1% resistant population of *Mycobacterium tuberculosis* J Clin Microbiol 21: 941-946
3. Heifets LB and Iseman MD (1985) Determination of *in vitro* susceptibility of mycobacteria to ansamycin Am Rev Respir Dis 132 (3): 710-711
4. Heifets LB, Iseman MD, Lindholm-Levy PJ, Kanes W (1985) Determination of ansamycin Minimal Inhibitory Concentrations for *M. avium* in liquid medium by radiometric and conventional methods Antimicrob Agents Chemother 28: Oct issue
5. Woodley CL, Kilburn JO (1982) *In vitro* susceptibility of *M. avium* comp. and *M. tuberculosis* strains to a spiro-piperidyl rifamycin Am Rev Respir Dis 126: 586-587
6. Della Bruna C, Schioppacassi G, Ungheri D, Jables D, Morvillo E, Sanfilippo A (1983) LM427, a new spiropiperidylrifamycin: *in vitro* and *in vivo* studies J Antibiot (Tokyo) 36: 1502-1506

Mycobacteria of Clinical Interest. M. Casal, editor

RIFABUTINE (LM 427): *IN VITRO* and *IN VIVO* ACTIVITY AGAINST RIFAMPICIN-RESISTANT AND ATYPICAL MYCOBACTERIA

DOMENICO UNGHERI*, VALERIA PENATI** and ALDO GIOBBI**

*Research & Development, Farmitalia Carlo Erba S.p.A. - Nerviano (Italy)

**Consorzio Provinciale Antitubercolare (C.P.A.), V.le Zara, 81 - Milano, (Italy)

INTRODUCTION

Rifabutine, a member of the class of the spiropiperidyl rifamycin S derivatives, has been studied in our laboratories and found to be a potent antibiotic *in vitro* and *in vivo*[1] against *M.tuberculosis*, including a certain percentage of rifampicin-resistant strains [2-3].

Other studies have demonstrated that rifabutine is active against *M.avium intracellulare*[3] and many atypical Mycobacteria of possible clinical interest[4]. Recently, the high activity of rifabutine against *M.leprae* both sensitive[5] and resistant to rifampicin[6] has been reported.

The aim of this study was to extend our knowledge on the susceptibility to rifabutine of rifampicin-resistant and atypical clinical isolates.

MATERIAL AND METHODS

Strains. During an epidemiological study carried out in 1984 at the Consorzio Provinciale Antitubercolare (C.P.A.) in Milano, 655 strains of *M.tuberculosis* were isolated; 112 (17.1%) were rifampicin-resistant, and 32 of them were tested for sensitivity to rifabutine in comparison with rifampicin.

In the course of the same study 10 strains of *M.kansasii* and 17 strains of *M.fortuitum* were isolated and tested as above.

In vitro activity on typical and atypical Mycobacteria. The MICs were performed in Dubos Albumin Broth (Difco) by the serial twofold dilution technique. The inoculum was a 2-7 day old culture in the same medium diluted 1:100 in the test tubes. The solutions of the antibiotic were prepared in dimethylformamide (1 mg/ml) and subsequently diluted in phosphate buffer 1/15 M (Sorensen) pH 6.8 (rifabutine) and pH 7 (rifampicin). MICs were read after 2-10 days of incubation at 37°C.

Therapeutic activity. Groups of 8-10 female CD1 albino mice (COBS) weighing 21±2 g were infected by the i.v. route with 3 LD_{50} of 2 selected strains of *M.tuberculosis*. The animals were treated orally by administration of antibiotic solutions, 0.1 ml/10 g body weight. The solutions were prepared in phosphate buffer 1/15 M (Sorensen), pH 6.8 (rifabutine) and pH 7 (rifampicin) + 5% dimethylacetamide. Treatment started 3 days after infection and continued for 5 weeks, 5 days/week.

Deaths were recorded daily; ED_{50} was determined on the basis of the survival rate at the end of the experiment.

RESULTS AND CONCLUSIONS

The *in vitro* inhibitory activity of rifabutine (LM 427) on 32 clinical isolates of *M.tuberculosis* selected for rifampicin resistance is reported in Table I.

Rifabutine inhibited 40% of the tested strains at 1.56 mcg/ml and 90% at 10 mcg/ml, while rifampicin was active at 20 mcg/ml and >80 mcg/ml, respectively.

10 strains of *M.kansasii* (Table II) appeared sensitive to 0.075 mcg/ml of rifabutine in comparison with 1.25 mcg/ml of rifampicin.

Table III shows the MICs of the two antibiotics on 17 strains of *M.fortuitum* complex. 50% of them were inhibited by 1.25 mcg/ml of rifabutine and 90% by 2.5 mcg/ml in comparison with 5 and >20 mcg/ml of rifampicin.

Two rifampicin-resistant strains of *M.tuberculosis* susceptible to rifabutine at 1.8 and 2.5 mcg/ml were tested on mouse experimental tuberculosis.

The ED_{50}s were 3.95 and 15 mg/kg for rifabutine and >25 mg/kg and >50 mg/kg for rifampicin (Table IV).

The data presented in this paper confirm the profile of rifabutine as a broad spectrum antimycobacterial agent.

The degree of *in vitro* susceptibility observed in 40% of rifampicin-resistant *M.tuberculosis* strains and the therapeutic activity displayed on mouse infection do not exclude the possible clinical efficacy of rifabutine, at least in polychemotherapy, in some cases of rifampicin resistance.

An analogous perspective can be suggested for *M.fortuitum* complex; it is important to remember that the infections determined by these pathogens are known, so far, as untreatable diseases. The high degree of susceptibility of *M.kansasii* to rifabutine suggests possible therapeutic use.

TABLE I

SUSCEPTIBILITY TO RIFABUTINE AND RIFAMPICIN OF 32 CLINICAL ISOLATES OF M. TUBERCULOSIS

	MIC (mcg/ml)		
	Geometrical mean	$MIC_{40\%}$	$MIC_{90\%}$
Rifabutine	3,2	1,56	10
Rifampicin	26,0	20,0	>80

TABLE II

SUSCEPTIBILITY TO RIFABUTINE AND RIFAMPICIN OF 10 CLINICAL ISOLATES OF *M. KANSASII*

	MIC (mcg/ml)		
	Geometrical mean	$MIC_{50\%}$	$MIC_{90\%}$
Rifabutine	0.05	0.075	0.075
Rifampicin	0.7	1.25	1.25

TABLE III

SUSCEPTIBILITY TO RIFABUTINE AND RIFAMPICIN OF 17 CLINICAL ISOLATES OF *M. FORTUITUM* COMPLEX

	MIC (mcg/ml)		
	Geometrical mean	$MIC_{50\%}$	$MIC_{90\%}$
Rifabutine	1.2	1.25	2.5
Rifampicin	7.3	5.0	$>$ 20

TABLE IV

IN VITRO AND *IN VIVO* ACTIVITY OF RIFABUTINE AND RIFAMPICIN ON 2 STRAINS OF RIFAMPICIN-RESISTANT *M. TUBERCULOSIS*

	M. tuberculosis N. 2		*M. tuberculosis* DU2	
	Rifabutine	Rifampicin	Rifabutine	Rifampicin
MIC (mcg/ml)	1.8	$>$ 40	2.5	$>$ 40
DT_{50} (mg/Kg)	3.95	$>$ 25	15	$>$ 50

REFERENCES

1. Della Bruna C. et al. (1983) In: LM 427, a new spiropiperidyl rifamycin: in vitro and in vivo studies. J. Antibiotic, 36, 1502.

2. Ungheri D., Della Bruna C., Sanfilippo A. (1984) In: Studies on the mechanism of action of the spiropiperidyl rifamicyn LM 427 on rifampicin-resistant M. tuberculosis. Drugs Exptl. Clin. Res., X (10), 681.

3. Woodley C.L., Kilburn J.O. (1982) In: In vitro susceptibility of Mycobacterium avium complex and Mycobacterium tuberculosis strains to a spiropiperidyl rifamycin. Am. Rev. Respir. Dis., 126, 586.

4. Ungheri D., Morvillo E., Sanfilippo A. (1983) In: Attività delle ansamicine LM 427 e FCE 22250 su micobatteri atipici in vitro. G. Ital. Chemioter., 30, 97.

5. Hastings R.C., Jacobson R.R. (1983) In: Activity of ansamycin against Mycobacterium leprae in mice. Lancet, II, 1079.

6. Hastings R.C., Richard V.R., Jacobson R.R. (1984) In: Ansamycin activity against rifampicin-resistant Mycobacterium leprae. Lancet, I, 1130.

Mycobacteria of Clinical Interest. M. Casal, editor

SUSCEPTIBILITY OF RAPIDLY GROWING MYCOBACTERIA TO 21 ANTIMICROBIAL AGENTS

M. LUQUIN, B. MIRELIS, V. AUSINA, MaJ. CONDOM and L. MATAS

Departamento de Microbiologia, Hospital de la Sta. Creu i Sant Pau, Facultad de Medicina de la Universidad Autónoma de Barcelona (Spain)

INTRODUCTION

Rapidly growing group IV mycobacteria are now more frequently recognized as a cause of infections than in previous years (1). Treatment of infections caused by these mycobacteria has been complicated by their resistance to antituberculous and other antimicrobial agents (2). Because of the recent outbreaks of nosocomial infections due to highly resistant strains (3), every effort should be made to evaluate new antimicrobial agents with activity against these organisms (4,5,6).

The present study was undertaken to determine the *in vitro* susceptibility of rapidly growing mycobacteria to 21 antibacterial agents.

MATERIAL AND METHODS

A total of 28 strains of *M. fortuitum*, 18 strains of *M. chelonae*, 10 strains of *M. fallax* and 24 strains of Runyon group IV nonpigmented mycobacteria that could not be identified as to species were used in this study. Organisms were isolated from clinical and environmental sources and identified by conventional tests (7, 8) and thin-layer chromatographic (TLC) analysis of their respective mycolic acids (9).

Agar dilution testing was performed with a Steers replicator and Mueller-Hinton agar (pH 7.2 to 7.4) to which OADC (oleic acid, albumin, dextrose and catalase) had been added in a ratio by volume of 10:1 (10). Antibiotics were prepared by serial two-fold dilutions of a concentrate prepared from standard diagnostic powders of 21 antibacterial agents: kanamycin, tobramycin, amikacin, erythromycin, tetracycline, doxycycline, minocycline, ampicillin, cefoxitin, aztreonam, N-formimidoyl-thienamycin, nalidixic acid, norfloxacin, ciprofloxacin, rifampin, ansamycin, clofazimine, vancomycin, sulfapyridine, sulfamethoxazole and sulfamethoxazole-trimethoprim (SXT). The organisms were maintained on Löwenstein-Jensen slants at room temperature until testing was performed. The inoculum was prepared by growing the organisms for 24-48 h in Mueller-Hinton broth supplemented with 0.02% Tween 80 and adjusting the turbidity to equal 0.5 McFarland standard. To determine the number of colony-forming units (CFU) in a suspension equivalent to a 0.5 McFarland standard, counts were done on two strains of each species tested. All counts ranged from $1x10^7$ to $2x10^8$ CFU/ml. The final inoculum was between 10^3 and 10^4 CFU per spot. All test plates were incubated at 35^oC and read at 2, 3 and 4 days; MICs were read as the lowest concentrations that completely inhibited growth or that allowed no more than one colony to grow. For sulfonamides MICs were read as the concentrations inhibiting 80 to 90% of the control growth.

RESULTS

Table I summarizes the ranges of MICs of 15 antimicrobial agents that were

TABLE I

MICs OF 15 ANTIMICROBIAL AGENTS FOR 28 STRAINS OF *M. FORTUITUM*

Agents	MIC_{50}	MIC_{90}	RANGE
Kanamycin	8	16	2-64
Tobramycin	8	16	2-64
Amikacin	1	2	<0.25-8
Tetracycline	16	128	<0.25->128
Doxycycline	64	>128	<0.25->128
Minocycline	16	>128	<0.25->128
Cefoxitin	16	32	8-128
N-formimidoyl thienamycin	2	4	0.5-8
Norfloxacin	0.5	1	<0.25-1
Ciprofloxacin	<0.25	<0.25	<0.25
Ansamycin	1	8	<0.25-8
Clofazimine	0.5	0.5	<0.25-1
Sulfapyridine	>128	>128	4->128
Sulfamethoxazole	64	>128	1->128
SXT (a)	2	>8	<0.25->8

(a) Sulfamethoxazole/trimethoprim ratio: 19:1. MIC expressed only as trimethoprim concentration.

TABLE II

MICs OF 10 ANTIMICROBIAL AGENTS FOR 18 STRAINS OF *M. CHELONAE*

Agents	MIC_{50}	MIC_{90}	RANGE
Tobramycin	8	16	2-32
Amikacin	32	128	8-128
Erythromycin	64	>128	1->128
N-formimidoyl thienamycin	64	128	0.5->128
Norfloxacin	32	64	0.5->128
Ciprofloxacin	4	8	<0.25->128
Ansamycin	32	64	2-64
Clofazimine	1	2	<0.25-8
Sulfamethoxazole	>128	>128	32->128
SXT (a)	>8	>8	0.5->8

active for at least some strains of *M. fortuitum*. The most active antimicrobics against *M. fortuitum* based on achievable blood levels were: amikacin, N-formimidoyl thienamycin, norfloxacin, ciprofloxacin, ansamycin, clofazimine and SXT. Drugs with no useful activity against *M. fortuitum* were as follows: erythromycin, ampicillin, aztreonam, nalidixic acid, rifampin and vancomycin.

The most active antimicrobial agents against *M. chelonae* strains are listed in Table II. Amikacin and tobramycin were the most active aminoglycosides. More than 75% of the *M. chelonae* strains had clofazimine MICs ≤1 μg/ml. The only other antimicrobics that were active for at least some strains of *M. chelonae* were: erythromycin, N-formimidoyl thienamycin, norfloxacin, ciprofloxacin, ansamycin, sulfamethoxazole and SXT. None of the remaining drugs tested appeared to be specially active against *M. chelonae*.

The *M. fallax* and other group IV strains were the most susceptible of the groups tested (Table III). Seventy-five percent or more of *M. fallax* and other

TABLE III

MICs OF 21 ANTIMICROBIAL AGENTS FOR 10 STRAINS OF M. FALLAX AND 24 OTHER STRAINS OF NONPIGMENTED RAPIDLY GROWING MYCOBACTERIA.

	M. fallax			Other group IV strains		
Agents	MIC_{50}	MIC_{90}	Range	MIC_{50}	MIC_{90}	Range
Kanamycin	0.5	0.5	0.5	0.5	1	≤0.25-1
Tobramycin	≤0.25	2	≤0.25-2	4	4	2-16
Amikacin	0.5	0.5	≤0.25-1	1	2	≤0.25-2
Erythromycin	≤0.25	≤0.25	≤0.25	2	4	≤0.25-16
Tetracycline	≤0.25	≤0.25	≤0.25	1	32	≤0.25-32
Doxycycline	≤0.25	≤0.25	≤0.25	≤0.25	>128	≤0.25->128
Minocycline	≤0.25	≤0.25	≤0.25	≤0.25	32	≤0.25-32
Ampicillin	2	4	1-4	4	8	1-16
Cefoxitin	4	8	2-8	8	8	4-16
Aztreonam	>128	>128	>128	>128	>128	>128
N-formimidyol thienamycin	≤0.25	0.5	≤0.25-1	0.5	2	0.5-2
Nalidixic acid	16	32	8-64	32	32	16-64
Norfloxacin	≤0.25	0.5	≤0.25-0.5	2	4	0.5-8
Ciprofloxacin	≤0.25	≤0.25	≤0.25	≤0.25	0.5	≤0.25-0.5
Rifampin	≤0.25	4	≤0.25-16	>128	>128	4->128
Ansamycin	≤0.25	1	≤0.25-1	8	16	1-16
Clofazimine	≤0.25	0.5	≤0.25-0.5	0.5	0.5	≤0.25-0.5
Vancomycin	1	8	1-8	32	128	2->128
Sulfapyridine	64	>128	8->128	>128	>128	32->128
Sulfamethoxazole	64	>128	2->128	>128	>128	16->128
SXT	≤0.25	8	≤0.25-8	≤0.25	1	≤0.25-1

group IV strains were susceptible to clinically attainable levels of aminoglycosides, erythromycin, tetracyclines, ampicillin, cefoxitin, N-formimidoyl thienamycin, norfloxacin, ciprofloxacin, ansamycin, clofazimine and SXT.

In this study, the MICs inhibiting 50 and 90% of isolates (MIC_{50} and MIC_{90}, respectively) were significantly lower for ciprofloxacin than for norfloxacin with each species tested, as has been shown for gram-positive and gram-negative bacteria in another *in vitro* study (11).

Several antimicrobials tested in this study (N-formimidoyl thienamycin, norfloxacin, ciprofloxacin, ansamycin and clofazimine) have had little or no previous evaluation regarding their in vitro activity against rapidly growing mycobacteria. Additional studies are required on *in vitro* animal and clinical grounds, to assess the efficacy of these drugs, particularly clofazimine and ciprofloxacin, in the treatment of rapidly growing mycobacterial infections.

REFERENCES

1. Wallace, R.J.Jr., Swenson, J.M. (1983). Rev. Infect. Dis. 5:657-679.
2. Sanders, W.E.Jr., Eldert, C.H., Schneider, N.J., Cacciatore, R., Valdez, H. (1977). Antimicrob. Agents Chemother. 12: 295-297.
3. Hines, M.P.R., Waggoner, P.R., Vernon, T.M.Jr. (1976). Morbid. Mort. Weekly Rep. 25: 238-239.
4. Welch, D.F., Kelly, M.T. (1979). Antimicrob. Agents Chemother. 15: 754-757.
5. Swenson, J.M., Thornsberry, C., Silcox, V.A. (1982). Antimicrob. Agents Chemother. 22: 86-192.
6. Casal, M.J., Rodríguez, F.C. (1983). Ann. Microbiol. 134A: 73-78.
7. Silcox, V.A., Good, R.C., Floyd, M.M. (1981). J. Clin. Microbiol. 14: 686-691.
8. Lévy-Frébault, V., Daffé, M., Goh, K.S., Lanéelle, M.A., Asselineau, C., David, H.L. (1983). J. Clin. Microbiol. 17: 744-752.
9. Minnikin, D.E., Hutchinson, I.G., Caldicott, A.B., Goodfellow,M. (1980). J. Chromatogr. 188: 221-233.
10. Wallace, R.J., Dalovisio, J.R., Pankey, G. (1979). Antimicrob. Agents Chemother. 16: 611-614.
11. Gay, J.D., De Young, D.R., Roberts, G.D. (1984). Antimicrob. Agents Chemother. 26: 94-96.

Mycobacteria of Clinical Interest. M. Casal, editor

IN VITRO ACTIVITY OF DIFFERENT ANTIMICROBIAL COMBINATIONS AGAINST *MYCOBACTERIUM FORTUITUM, M. CHELONEI* AND *M. AVIUM*

M. CASAL, F. RODRIGUEZ, M.C. BENAVENTE and R. VILLALBA
Department of Microbiology, School of Medicine, University of Cordoba (Spain)

INTRODUCTION

M. fortuitum, M. chelonei and *M. avium* are multiresistant to classical tuberculostatic and non-tuberculostatic drugs such as antifungals and β-lactamase and other groups of antimicrobials. We have investigated a series of combinations of several drugs for their action against these microorganisms to assess their possible use in clinical human tuberculosis.

MATERIAL AND METHODS

A total of 33 *M. fortuitum* strains, 15 *M. chelonei* strains and 13 *M. avium* strains were studied against the following antimicrobial combinations: cephapirin + cephaloridin + entamycin; cephapirin + gentamycin + clavulanic acid; cephaloridin + gentamycin + clavulanic acid; RU 28965 + sulphamethoxipiridazin; RU 28965 + ciprofloxacin; and RU 28965 + ciprofloxacin + cephaloridin.

The determination of the Minimum Inhibitory Concentration (MIC) was found by the dilution method in Mueller-Hinton agar for *M. fortuitum*, Mueller-Hinton agar with OADC (Difco) at a ratio of 1/75 for *M. chelonei* and Middlebrook 7H10 with OADC (Difco) at a ratio of 1/75 for *M. avium*.

To combine the different antibacterials we used the levels reached by each one in a medium dose, decreasing and increasing the concentration following a logarithmic scale to the base 2, with the exception of two combinations which included clavulanic acid, where they were administered at 5 µg/ml of the latter for each of the different concentrations of assayed antibiotic.

The microorganisms were suspended in sterile distilled water reaching an opacity corresponding to scale 1 of McFarland. The plates were inoculated with Steer replicator with a dilution 1/1000 of the suspension. Incubation was at 37°C for *M. fortuitum* and 28°C for *M. chelonei*, and the readings were made on the 5th and 7th days. *M. avium* plates were inoculated from a dilution of 1/10 of the scale 1 of McFarland. The incubation was kept at 37°C and readings were made on the 20th and 40th days.

M. fortuitum control strains were ATCC 6841 and *E. coli* ATCC 25922.

RESULTS

M. fortuitum was generally more sensitive and showed the following percentages of inhibited strains in relation to the usual serum concentrations at medium doses of the following combinations: RU 28965 + ciprofloxacin, 100% for *M. fortuitum*, 92% for *M. chelonei* and 0% for *M. avium*; RU 28965 + ciprofloxacin + cephaloridin, 100% for *M. fortuitum* and 46% for *M. chelonei*; cephapirine + gentamycin 77% for *M. fortuitum* and 0% for *M. chelonei* and *M. avium*; cephapirin + gentamycin + clavulanic acid gave a result of 70% for *M. fortuitum* and 0% for *M. chelonei*; cephaloridin + gentamycin 74% for *M. fortuitum* and 0% for *M. chelonei* and *M. avium*; cephaloridin + gentamycin + clavulanic acid, 77% for *M. fortuitum* and 7% for *M. chelonei*; RU 28965 + sulphametoxipiridazin, 77% for *M. fortuitum* and 100% for *M. chelonei* and *M. avium*.

In general the MICs obtained with the combinations were inferior to those reached in previous studies of the same antibacterial agents assayed individually.

In Tables I, II and III the results obtained with these combinations are listed.

TABLE I

ACTIVITY OF DIFFERENT ANTIMICROBIAL COMBINATIONS AGAINST *M. FORTUITUM*, *M. CHELONEI* AND *M. AVIUM*

Species studied	No. of souches		MIC µg/ml										
		CEP	0.12	0.25	0.5	1	2	4	8	16	32	64	64
		+											
		GEN	0.6	0.12	0.25	0.5	1	2	4	8	16	32	32
M. fortuitum	27							9(33)	21(77)	27(100)			
M. chelonei	14										4(28)	14(100)	
M. avium	10									10(100)			
		CEP	0.12	0.25	0.5	1	2	4	8	16	32	64	64
		+											
		GEN	0.6	0.12	0.25	0.5	1	2	4	8	16	32	32
		+											
		AC	5	5	5	5	5	5	5	5	5	5	5
M. fortuitum	27						1(4)	7(26)	19(70)	27(100)			
M. chelonei	14									1(7)	5(36)	14(100)	

CEP - cephapirin GEN - gentamicin AC - clavulanic acid

TABLE II

ACTIVITY OF DIFFERENT ANTIMICROBIAL COMBINATIONS AGAINST *M. FORTUITUM*, *M. CHELONEI* AND *M. AVIUM*

Species studies	No of souches	MIC µg/ml											
		CEL + GEN	0.12 0.6	0.25 0.12	0.5 0.25	1 0.5	2 1	4 2	8 4	16 8	32 16	64 32	>64 >32
M. fortuitum	27						1(4)	8(30)	20(74)	27(100)			
M. chelonei	14									1(4)	6(43)	14(100)	
M. avium	10									10(100)			
		CEL + GEN + AC	0.12 0.6 5	0.25 0.12 5	0.5 0.25 5	1 0.5 5	2 1 5	4 2 5	8 4 5	16 8 5	32 16 5	64 32 5	>64 >32 5
M. fortuitum	27						2(7)	13(48)	21(77)	27(100)			
M. chelonei	14								1(7)	3(21)	11(79)	14(100)	

CEL = cephaloridin GEN = gentamicin AC = clavulanic acid

TABLE III

ACTIVITY OF DIFFERENT ANTIMICROBIAL COMBINATIONS AGAINST *M. FORTUITUM*, *M. CHELONEI* AND *M. AVIUM*

Species studied	No. of souches	MIC µg/ml											
		RU + SUL	0.05 0.25	0.1 0.5	0.3 1	0.6 2	0.12 4	0.25 8	0.5 16	1 32	2 64	4 128	4 128
M. fortuitum	30						7(23)	9(30)	15(50)	17(57)	30(100)		
M. chelonei	14							2(14)	8(57)	14(100)			
M. avium	13									13(100)			
		RU + CI	0.03 0.06	0.03 0.06	0.06 0.12	0.12 0.25	0.25 0.5	0.5 1	1 2	2 4	4 8	8 16	8 16
M. fortuitum	21		12(57)		18(86)		(100)						
M. chelonei	13								12(92)	13(100)			
M. avium	11											11(100)	
		RU + CI + CEL	0.03 0.06 0.5	0.03 0.06 0.5	0.06 0.12 1	0.12 0.25 2	0.25 0.5 4	0.5 1 8	1 2 16	2 4 32	4 8 64	8 16 128	8 16 128
M. fortuitum	21		1(5)		15(71)	18(86)	21(100)						
M. chelonei	13							2(15)	6(46)	13(100)			

RU = RU 28965 SUL = sulfametoxipiradizin CI = ciprofloxacin CEL = cephaloridin

SUSCEPTIBILITY OF *MYCOBACTERIUM FORTUITUM* COMPLEX TO 35 ANTIBACTERIAL AGENTS

J. GIL, C. RUBIO-CLAVO, A. REZUSTA, M.A. VITORIA and M.S. SALVO

Department of Microbiology, School of Medicine, University of Zaragoza, 50009 Zaragoza (Spain)

INTRODUCTION

Organisms belonging to the *Mycobacterium fortuitum* complex have been implicated in several forms of clinical disease (1,2).

These organisms are known to be resistant to most of the classical antimycobacterial agents (1). Within the past few years, several investigators have reported that other antibacterial agents are active against these organisms (3,4,5).

The present study was undertaken to determine the susceptibility of the *Mycobacterium fortuitum* complex (*M. fortuitum*, *M. chelonei* subp. *abscessus*, *M. chelonei* subp. *chelonei* and *M. chelonei*-like) to 35 antibacterial agents.

MATERIAL AND METHODS

Organisms. A total of 26 *M. fortuitum* complex were obtained from patients with subcutaneous abscesses (2 cases) and several respiratory diseases (24 cases).

The strains were isolated in BACTEC and Lowenstein-Jensen (LJ) medium and LJ with pyruvate.

Identification procedures. All strains were identified as *Mycobacterium fortuitum* complex by standard procedures (6,7). The organisms were subsequently identified as to species and subspecies by methods previously described by Silcox (8).

Susceptibility test. Disk diffusion susceptibility studies were performed by the Wallace method (5) on Mueller-Hinton agar plates (MHA) and MHA supplemented with 10% OADC. We used 35 available commercial antimicrobial disks (Difco and Institute Pasteur): penicillin, cloxacillin, ampicillin, carbenicillin, ticarcillin, piperacillin, azlocillin, mezlocillin, cephalothin, cefamandole, cefoxitin, cefotaxime, moxalactam, cefoperazone, gentamicin, tobramycin, dibekacin, amikacin, netilmicin, erythromycin, lincomycin, clindamycin, rifampin, tetracycline, minocycline, colistin, chloramphenicol, fosfomycin, trimethoprim-sulfamethoxazole, nalidixic acid, pipemidic acid, nitrofurantoin, vancomycin, pinmecillinam and fusidic acid.

RESULTS AND DISCUSSION

The 26 strains of *M. fortuitum* complex were identified: 4 as *M. fortuitum*, 2 as *M. chelonei* subp. *abscessus* and 1 as *M. chelonei* subp. *chelonei*. Nineteen strains showed a low semiquantitative catalase test result, were negative in the

TABLE I. ANTIMICROBIAL SUSCEPTIBILITY OF 26 STRAINS OF *M. FORTUITUM* COMPLEX TO 35 ANTIMICROBIAL AGENTS.

Antimicrobial agent	M.fortuitum(4)		M.chelonei subp. abscessus(2)		M.chelonei subp. chelonei(1)		M.chelonei-like(19)	
	N°. of strain	%	N°. of strain	%	N°. of strain	%	N°. of strain	%
Penicillin	0	0	0	0	0	0	0	0
Cloxacillin	0	0	0	0	0	0	0	0
Ampicillin	0	0	0	0	0	0	14	73.6
Carbenicillin	0	0	0	0	0	0	0	0
Ticarcillin	0	0	0	0	0	0	0	0
Piperacillin	0	0	0	0	0	0	2	10.5
Azlocillin	0	0	0	0	0	0	4	21.0
Mezlocillin	0	0	0	0	0	0	4	21.0
Cephalothin	0	0	0	0	0	0	17	89.4
Cefamandole	0	0	0	0	0	0	19	100
Cefoxitin	2	50.0	0	0	0	0	18	94.7
Cefotaxime	0	0	1	50.0	0	0	18	94.7
Moxalactam	0	0	0	0	0	0	0	0
Cefoperazone	0	0	0	0	0	0	0	0
Gentamicin	2	50.0	0	0	0	0	18	94.7
Tobramycin	2	50.0	0	0	0	0	17	89.4
Dibekacin	2	50.0	1	50.0	1	100	15	78.9
Amikacin	4	100	2	100	1	100	18	94.7
Netilmicin	3	75.0	1	50.0	1	100	18	94.7
Erythromycin	1	25.0	1	50.0	1	100	19	100
Lincomycin	0	0	0	0	0	0	0	0
Clindamycin	0	0	0	0	0	0	0	0
Rifampin	0	0	1	50.0	0	0	0	0
Tetracycline	0	0	0	0	0	0	16	84.2
Minocycline	0	0	0	0	0	0	15	78.9
Colistin	1	25.0	0	0	0	0	0	0
Chloramphenicol	1	25.0	0	0	0	0	18	94.7
Fosfomycin	0	0	0	0	0	0	0	0
Trimethoprim/ sulfametoxazole	0	0	0	0	0	0	17	89.4
Nalidixic Acid	0	0	0	0	0	0	0	0
Pipemidic Acid	4	100	0	0	0	0	0	0
Nitrofurantoin	1	25.0	0	0	0	0	6	31.5
Vancomycin	3	75.0	0	0	0	0	14	73.6
Pinmecillinam	0	0	0	0	0	0	0	0
Fusidic Acid	2	50.0	0	0	0	0	14	73.6

catalase test at 68°C and pH 7, and utilized mannitol for growth. According to Silcox (8) they belong to the *Mycobacterium chelonei*-like group. In addition, the organisms identified as *M. chelonei*-like showed a higher sensitivity to the antimicrobial agents tested (9).

The sensitivity of *Mycobacterium fortuitum* complex to the 35 antimicrobial drugs tested are shown in Table I.

All *Mycobacterium chelonei* strains were resistant to pipemidic acid. In contrast, all strains of *M. fortuitum* were found susceptible. This result is in agreement with those obtained by other authors (10,11).

M. fortuitum strains tested were 100% sensitive to amikacin and only 50% to cefoxitin, according to Swenson (9), Casal (12), Cynamon (3), Wallace (5,13), Dalovisio (4) and Welch (14).

Amikacin, erythromycin and netilmicin were the most effective antibiotics against *M. fortuitum* complex.

REFERENCES

1. Wolinsky, E. 1979. Non tuberculous mycobacteria and associated diseases. Am. Rev. Respir. Dis. 119: 107-159.
2. Finegold, S.M., W.T. Martin, and E.G. Scott. 1978. Bailey and Scott's diagnostic microbiology, 5th ed. Mosby, St. Louis.
3. Cynamon, M.H., and A. Patapow. 1981. In vitro susceptibility of Mycobacterium fortuitum to Cefoxitin. Antimicrob. Agents Chemother. 19: 205-207.
4. Dalovisio, J.R., and G.A. Pankey. 1978. In vitro susceptibility of Mycobacterium fortuitum and Mycobacterium chelonei to Amikacin. J. Infect. Dis. 137: 318-321.
5. Wallace, R.J., Jr., J.R. Dalovisio, and G.A. Pankey. 1979. Disk diffusion testing of susceptibility of Mycobacterium fortuitum and Mycobacterium chelonei to antibacterial agents. Antimicrob. Agents Chemother. 15: 754-757.
6. Runyon, E.H., A.G. Karlsol, G.P. Kubika, and L.G. Wayne. 1980. In Lennette, E.H., A. Balows, W.J. Hausler, Jr, and J.P Truant (ed). Manual of Clinical Microbiology. 3rd. American Society for Microbiology. Washington D.C.
7. Vestal, A.L. 1975. Procedures for the isolation and identification of mycobacteria. U.S. Public Health Serv. Publ. 75-8230. Center for Disease Control, Atlanta, Ga.
8. Silcox, V.A., R.C. Good, and M.M. Floyd. 1981. Identification of clinically significant Mycobacterium fortuitum complex isolates. J. Clin. Microbiol. 14: 686-691.
9. Swenson, J.M., C. Thornsberry, and V.A. Silcox. 1982. Rapidly growing mycobacteria: testing of susceptibility to 34 antimicrobial agents by broth microdilution. Antimicrob. Agents Chemother. 22: 186-192
10. Casal, M.J., and F.C. Rodriguez. 1981. Simple new test for rapid differentation of the Mycobacterium fortuitum complex. J. Clin. Microbiol. 13: 989-990.

11. Sommers,H.M., and R.C. Good. 1985. Lennette,E.H., A. Balows, W.J. Hausler, Jr., H.J. Shadomy (ed) In Manual of Clinical Microbiology. 4th. American Society for Microbiology. Washington D.C.

12. Casal, M., F. Rodriguez y M.C. Benavente. 1985. Investigación preliminar de la acción in vitro de nuevos antibioticos beta-lactámicos sobre _Mycobacterium fortuitum_ y _Mycobacterium chelonei_. Infectologika 2/85: 49-51.

13. Wallace, R.J.,Jr., J.M. Swenson, V.A. Silcox, and R.C. Good. 1982. Disk diffusion testing with Polymyxin and Amikacin for differentiation of _Mycobacterium fortuitum_ and _Mycobacterium chelonei_. 16: 1003-1006

14. Welch, D.F., and T.K. Michael. 1979. Antimicrobial susceptibility testing of _Mycobacterium fortuitum_ complex. Antimicrob. Agents Chemother. 15: 754-757.

Mycobacteria of Clinical Interest. M. Casal, editor

IN VITRO ACTIVITY OF CLAVULANIC ACID IN COMBINATION WITH DIFFERENT β-LACTAMINES AGAINST *MYCOBACTERIUM FORTUITUM*, *MYCOBACTERIUM CHELONEI*, *MYCOBACTERIUM AVIUM* AND *MYCOBACTERIUM TUBERCULOSIS*

M. CASAL, F. RODRIGUEZ, M.C. BENAVENTE, M.D. LUNA, R. VILLALBA and M. RUBIANO
Department of Microbiology, School of Medicine, University of Cordoba (Spain)

INTRODUCTION

Given the usual multiresistance of *M. fortuitum* and *M. chelonei* to the classical antituberculosis agents (1), and in the search for new antimicrobial agents not tuberculostatic which could be active against these mycobacteria (2,3,4,5,6), a comparative study was made of the *in vitro* activity of clavulanic acid in combination with amoxicillin (augmentin), ticarcillin, cephapirin and cephaloridin. At the same time a study was made of these antimicrobial agents against *M. tuberculosis* and *M. avium* to assess their possible clinical use against these mycobacteria.

MATERIAL AND METHODS

Thirty-one strains of *M. fortuitum*, 17 strains of *M. chelonei*, 30 strains of *M. tuberculosis* and 13 strains of *M. avium* were used.

The MIC was determined by the Mueller-Hinton agar dilution method for *M. fortuitum* and Mueller-Hinton agar with Dubos Medium Albumina (Difco) at a ratio of 1/75 for *M. chelonei*. The antimicrobial concentration used was of 0.25 μg/ml to 128 μg/ml for ticarcillin, cephapirin and cephaloridin adding 5 μg/ml of clavulanic acid to each different antibiotic tested. The combination of amoxicillin and clavulanic acid (augmentin) was 2 parts amoxicillin and 1 clavulanic acid in the scale mentioned above. The bacteria were suspended in sterile distilled water reaching an opacity corresponding to scale 1 of MacFarland. The plates were inoculated with a Steers replicator at a dilution of 1/1000 of the suspension mentioned above. Incubation was at 37°C for *M. fortuitum* and 28°C for *M. chelonei*. Readings were made on the 5th and 7th days. Middlebrook 7H10 agar with OADC (Difco) medium was used with the 30 strains of *M. tuberculosis* and 13 strains of *M. avium* studied. They were inoculated at a dilution of 1/100 and 1/10 respectively from an original concentration of MacFarland's 1 and incubated at 37°C. Readings were made after 20 and 40 days.

M. fortuitum ATCC 6841 and *E. coli* ATCC 25922 and *M. tuberculosis* H37Rv strains were used as controls.

TABLE I

AMOXICILLIN AND TICARCILLIN ACTIVITY IN COMBINATION WITH CLAVULANIC ACID AGAINST *M. FORTUITUM*, *M. CHELONEI* AND *M. TUBERCULOSIS*

Species studied	No. of souches		MIC µg/ml										
		AMOX + AC	0.16 0.08	0.32 0.16	0.66 0.33	1.33 0.66	2.66 1.33	5.33 2.66	10.66 5.33	21.33 10.66	42.6 21.3	85.3 42.6	>85.3 >42.6
M. fortuitum	31							10(32)	25(81)	31(100)			
M. chelonei	17								1(6)		2(12)	5(29)	17(100)
M. tuberculosis	30					15(50)	18(60)	25(83)	30(100)				
		TI + AC	0.25 5	0.5 5	1 5	2 5	4 5	8 5	16 5	32 5	64 5	128 5	>128 5
M. fortuitum	31									1(3)	4(13)	10(32)	31(100)
M. chelonei	17												17(100)
M. tuberculosis	16				1(6)	3(19)	9(56)	13(81)	16(100)				

AMOX = amoxicillin TI = ticarcillin AC = clavulanic

TABLE II

CEPHAPIRIN, CEPHALORIDIN IN COMBINATION WITH CLAVULANIC ACID AGAINST *M. FORTUITUM* AND *M. AVIUM*

Species studied	No. of souches	MIC µg/ml										
		CEP	0.25	0.5	1	2	4	8	16	32	64	128	>128
		+ AC	5	5	5	5	5	5	5	5	5	5	5
M. fortuitum	27							2(7)	14(51)	27(100)			
M. chelonei	14									1(7)		8(57)	14(100)
M. avium	13											13(100)	
		CEL	0.25	0.5	1	2	4	8	16	32	64	128	>128
		+ AC	5	5	5	5	5	5	5	5	5	5	5
M. fortuitum	27						4(14)	6(22)	16(58)	24(89)	27(100)		
M. chelonei	14								1(7)			5(36)	14(100)
M. avium	13											13(100)	

CEP = cephapirin   CEL = cephaloridin   AC = clavulanic acid

RESULTS

Amoxicillin at a concentration of 10.66 μg/ml in combination with clavulanic acid at a concentration of 5.33 μg/ml proved to be very active against *M. tuberculosis* and *M. fortuitum*; it inhibited 100% and 81% of the strains studied respectively, but *M. chelonei* proved to be resistant.

Cephaloridin in combination with clavulanic acid was not totally active against *M. fortuitum*, inhibiting 58% of the strains tested.

The other combinations tested were hardly or not at all active against the microorganisms studied.

The results are shown in Tables I and II.

REFERENCES

1. CASAL,M. (1978) Estudio de las resistencias a las drogas de las mycobacterias atipicas patogenas para el hombre.Rev. Clin. Esp. 148:343-346
2. Casal MJ and FC RODRIGUEZ.(1981) Etude de l'activite in vitro des beta-lactamines et des aminosides sur M.fortuitum. Ann. Microbiol (Paris). 132 B : 51-56
3. CASAL MJ and FC RODRIGUEZ (1982) In vitro susceptibility of Mycobacterium fortuitum and Mycobacterium chelonei to cefoxitin. Tubercle. 63:125-127
4. CASAL MJ and FC RODRIGUEZ (1983) In vitro susceptibility of Mycobacterium fortuitum to non-antituberculous antibacterial agents. Ann. Microbiol. (Paris). 132 A : 73-78
5. CYNAMON MH and GS PALMER (1982) In vitro susceptibility of Mycobacterium fortuitum to N-formidoyl thienamycin and several cephamycins. Antimicrob. Agents. Chemother. 22 : 1079-1081
6. CYNAMON MH and A PATAPOW (1981) In vitro susceptibility of Mycobacterium fortuitum to cefoxitin. Antimicrob. Agents. Chemother. 19 : 205-207

PRELIMINARY INVESTIGATION OF IN VITRO ACTIVITY OF NEW ANTIMICROBIAL AGENTS AGAINST MYCOBACTERIUM TUBERCULOSIS BY THE BACTEC 460 TB SYSTEM

M. CASAL, J. GUTIERREZ and P. RUIZ

Department of Microbiology, School of Medicine, University of Cordoba, Cordoba (Spain)

INTRODUCTION

One of the major problems we have been confronted with for a long time in the treatment of tuberculosis is the great number of *M. tuberculosis* strains resistant to the normal drugs. This is what motivates investigators specialised in this field to find new antimicrobial agents which can totally inhibit the mycobacteria.

Another great problem we have had is the length of time which *M. tuberculosis* needs for growth and therefore the time needed to perform a sensitivity test to the drugs. The new systems such as Bactec, which is rapid in detection of growth, identification and sensitivity to the drugs is a great advantage for obtaining results rapidly (1,2,3).

The aim of our work, therefore, was to evaluate the activity of a series of antimicrobial agents and of combinations of antimicrobial agents in relation to possible use in future investigations of antituberculosis therapy with the Bactec 460 (TB) system.

MATERIAL AND METHODS

We have studied the *in vitro* activity against *M. tuberculosis* of the following antimicrobial agents: benzyl-penicillin (10 µg/ml), ofloxacin (1 µg/ml), ciprofloxacin (1 µg/ml), norfloxacin (1 µg/ml), RU 28965 (5 µg/ml), ticarcillin (16 µg/ml), thienamycin (16 µg/ml), cefotaxime (8 µg/ml), and ceftazidime (8 µg/ml). The following combinations were also studied: clavulanic acid (3 µg/ml) plus amoxicillin (6 µg/ml), clavulanic acid (5 µg/ml) plus ticarcillin (16 µg/ml), clavulanic acid (5 µg/ml) plus cefotaxime (8 µg/ml) and clavulanic acid (5 µg/ml) plus ceftazidime (8 µg/ml).

All the concentrations of antimicrobial agents used were equal or inferior to the serum levels reached by these antimicrobial agents.

The method used was the same as that used for the study of sensitivity to the usual drugs (streptomycin, rifampin, ethambutol and isoniazid) (3).

RESULTS

The results are reported in Table I.

TABLE I

Antimicrobial agents	Concentrations (µg/ml)	Strains inhibited (%)
Benzyl-penicillin	10	36.6
Ofloxacin	1	63.6
Ciprofloxacin	1	80.6
Norfloxacin	1	81.8
RU 28965	5	77.2
Ticarcillin	16	0
Thienamycin	16	92.3
Cefotaxime	8	42.5
Ceftazidime	8	0
Clavulanic acid + amoxicillin	3 + 6	80.6
Clavulanic acid + ticarcillin	5 + 16	77.4
Clavulanic acid + thienamycin	5 + 16	93.5
Clavulanic acid + cefotaxime	5 + 8	42.5
Clavulanic acid + ceftazidime	5 + 8	46.6

CONCLUSIONS

In view of the results and given the activity of some of these antimicrobial agents against *M. tuberculosis*, it would be interesting to perform more explicit studies, such as an investigation to find a possible therapeutic role in animals and clinical assays in humans.

REFERENCES

1. Middlebrooc G, Reggiardo Z, Tigertt WD (1977). Automatable radiometric detection of growth of Mycobacterium tuberculosis in selective media. Am. Rev. Respir. Dis. 115: 1067-1069.

2. Laszlo A, Siddiqi SH (1984). Evaluation of a rapid radiometric differentiation test for the Mycobacterium tuberculosis complex by selective inhibition with p-nitro- α -acetylamino- β -Hydroxypropiophenone. J. Clin. Microbiol. 19: 694-698.

3. Siddiqi SH, Libonati JP, Middlebrook G (1981). Evaluation of a rapid radiometric method for drug susceptibility testing of Mycobacterium tuberculosis. J. Clin. Microbiol. 13: 908-912.

CORRELATION BETWEEN THE DISC-PLATE AND MIC METHODS OF CHLORHEXIDINE GLUCONATE SOLUTION AGAINST 95 STRAINS OF *FORTUITUM-CHELONEI* COMPLEX

M. CASAL, R. FDZ-CREHUET and A. SERRANO
Department of Microbiology, School of Medicine, Cordoba (Spain)

INTRODUCTION

The evaluation of antibacterial agents in a solid medium requires a methodical study of all the factors which can influence either the diffusion in the medium or those which can inactivate this substance. This method was systematized for antibiotics and chemotherapeutics (1).

El Nakeeb (2) and Alvarez Alcántara (3) have created a diffusion method in solid medium to evaluate bacterocides derived from quaternary ammonium and chlorhexidine gluconate by studying the medium of culture, pH and quantity of agent and its action on the diffusion of these antiseptics.

Delmotte et al (4) have studied the sensitivity of *Ps. aeruginosa* against different disinfectants and concluded that it was necessary in order to use the antiseptogram to adequately select some characteristics of medium of culture to facilitate the diffusion of the antiseptic and the growth of the bacterial strains to be studied.

Alvarez Alcántara et al (3) find that with chlorhexidine the diffusion is greater in Mueller-Hinton pH 8 medium.

The aim of our study was to find the correlation between the disc-plate method by diffusion in agar and the MIC method, both tested on 96 strains of Mycobacterium fortuitum-chelonei.

MATERIAL AND METHODS

Material

Bacterial strains: 68 *M. fortuitum* strains and 33 *M. chelonei* were used from the collection of mycobacteria at the School of Medicine of Cordoba University.

Antiseptics: Chlorhexidine gluconate was used at 20% commercialised by ICIFARMA as habitane 20%.

Medium of culture: Mueller-Hinton agar (DIFCO) graded at pH 8 using phosphate tampon.

Method

MIC: To determine the MIC chlorhexidine gluconate solution was added to Mueller-Hinton medium at final concentrations of 0.25 mgr/l, 0.5 mgr/l, 1 mgr/l, 2 mgr/l, 4 mgr/l, 8 mgr/l, 16 mgr/l, 32 mgr/l, 64 mgr/l, and 128 mgr/l. Each one of the plates was inoculated by Stern replicator the 95 strains of *M. fortuitum-chelonei* in suspensions equivalent to the dilution 1/1000 from 1 of MacFarland scale. In each of the tests, the *E. coli* strain ATCC25922 was used as control.

Disc plate: For the disc-plate technique, sterile discs of 6 mm diameter (BIO-MERIEUX) were impregnated with a disinfectant solution at concentrations of 50 and 100 μg/d using for each strain two discs of each of the concentrations according to the Kirby-Bauer technique (5). The inoculation of mycobacteria was the same as that used to determine MIC.

The inoculated plates for both methods were incubated at 37°C and read after 3 and 5 days.

RESULTS

The susceptibility of *M. fortuitum* and *M. chelonei* showing the MIC at which 50, 75 and 90% of strains were inhibited and the maximum MIC found are summarized in Table I.

TABLE I

		MIC (mg/l)			
	No	50	75	90	Max
M. fortuitum	66	2	2	8	64
M. chelonei	33	2	4	16	64

The mean of the halos obtained with 100 μg/l was of 37.57 ± 7.73 and with 50 μg/d of 34.7 ± 7.7.

The mean of the halos obtained with 100 μg/d for the 62 strains of *M. fortuitum* was 39.3 ± 6.03 and for the 33 strains of *M. chelonei* 34.2 ± 9.4.

Comparing the mean obtained with 100 μg/d for *M. fortuitum* and *M. chelonei* a significant difference can be found with SD = 1.8 ($p < 0.01$).

The coefficient of correlation obtained between MIC and the halos obtained with 50 μg/d for the total of the strains was $r = -0.59$ with confidence limits of -0.58-0.82 ($p < 0.05$).

The coefficient of correlation obtained between MIC and the halos obtained with 100 μg/d was $r = -0.64$ with confidence limits of -0.61-0.89 ($p < 0.05$).

REFERENCES

1. Ericsson HM, Sherris JC (1971). Antibiotic sensitivity testing. Report of an International collaborative study. Acta Path. et Microbiol. Scand. Sect. B. Suppl. 217.

2. El-Nakeeb (1976). Development of an Agar-Diffusion Method for the Assay of Quaternary Ammonium Germicides. Arzneim-Forsch (Drug Res), 26: 14-20.

3. Alvarez Alcántara A, Cueto Espinar A, Gálvez Vargas R (1982). Estudio del pH y medio de cultivo en la difusión y actividad de la clorhexidina en medio sólido. Laboratorio 434: 117-143.

4. Delmotte A, Beumer J, Cotton E, Dekeysser-Delmotte N (1971). Etude sur la sensibilité du bacille pyocyanique (Ps. aeruginosa) aux antiseptiques et aux antibiotiques II. De la notion statistique de la sensibilité du bacille pyocyanique aux antiseptiques a la conception de l'antiseptogramme. Thérapie 26: 645-654.

5. Bauer AW, Kirby WMM, Sherris JC, Turck M (1966). Antibiotic susceptibility testing by a standardized single disk method. Am. J. Clin. Pathol. 36: 493-496.

Mycobacteria of Clinical Interest. M. Casal, editor

EFFECTS OF RIFAMPICIN ON THE STRUCTURE OF *M. TUBERCULOSIS*

LJUBISA MARKOVIC and MILOS JOVANOVIC

Institute of Microbiology and Immunology, School of Medicine, University of Belgrade, P O Box 497, 11000 Belgrade (Yugoslavia)

INTRODUCTION

Rifampicin rapidly inhibits synthesis of all classes of RNA in susceptible bacteria. This inhibition of RNA synthesis is due to a specific inhibition of bacterial DNA-dependent RNA polymerase. Mycobacteria are, in general, susceptible to inhibition by rifampicin. The action of rifampicin on mycobacteria is similar to its action on *E. coli* (1).

Konno et al (2) reported their observations on the fine structural changes produced in *M. tuberculosis* H37Ra by rifampicin. In the present investigation, the fine structural changes occurring in a virulent strain of *M. tuberculosis* treated with rifampicin were studied.

MATERIAL AND METHODS

A fresh clinical isolate of *M. tuberculosis* was used as the test strain. A seven-day culture of this virulent strain in Youman's medium was used for investigation. Rifampicin was added to the medium at a concentration of 10 µg/ml, and the cells were exposed to the drug for 12, 24 and 48 hours, respectively.

The cells prepared for electron microscopy were fixed in 1% solution of osmic acid for 72 hours at 4°C according to the method of Ryter and Kellenberger (3), then dehydrated with graded alcohol and embedded in epoxy resin. Thin sections were cut on a LKB Ultrotome III, stained with uranyl acetate and lead citrate, and examined under a Philips EM 300 electron microscope.

RESULTS AND DISCUSSION

Fine structure of untreated *M. tuberculosis*: As can be seen on the electron micrographs of the sectioned cells of untreated *M. tuberculosis* (Figs. 1 and 2), the cells have a three layer envelope: external, which is amorphous (capsule ?), cell wall and plasma membrane. The cell wall consists of two electron-dense layers separated by a less dense layer. The plasma membrane of the mycobacterial cell appears in sections as the usual triple-layered "unit membrane". The cytoplasm of mycobacteria often contains polyphosphate granules and lipid droplets. Mycobacterial ribosomes seem to be typical bacterial ribosomes. Vesicular membranous structures may sometimes be found in the cytoplasm and are described as mesosomes (Fig. 2). Mesosome-like structures may be lamellar or vesicular and their membranes are apparently continuous with the plasma membrane.

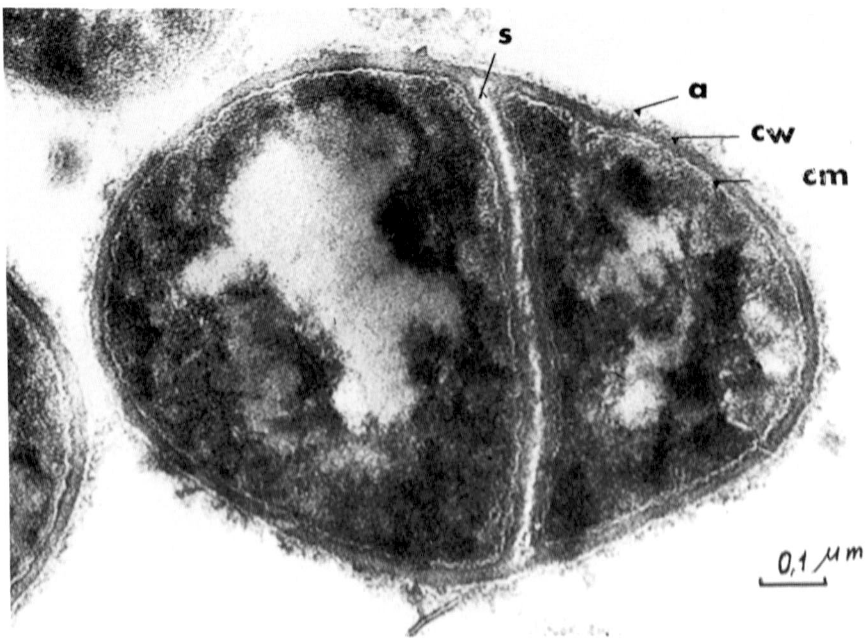

Fig. 1. Fine structure of untreated M. tuberculosis: cw - cell wall; cm - cytoplasmic membrane; s - septum; a - amorphous material (capsule ?).

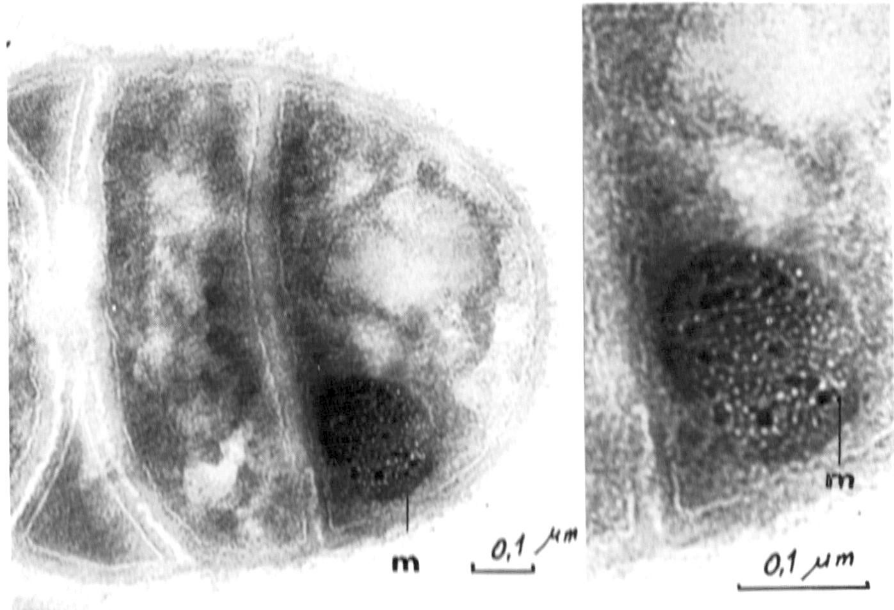

Fig. 2. Fine structure of untreated M. tuberculosis; m - mesosome.

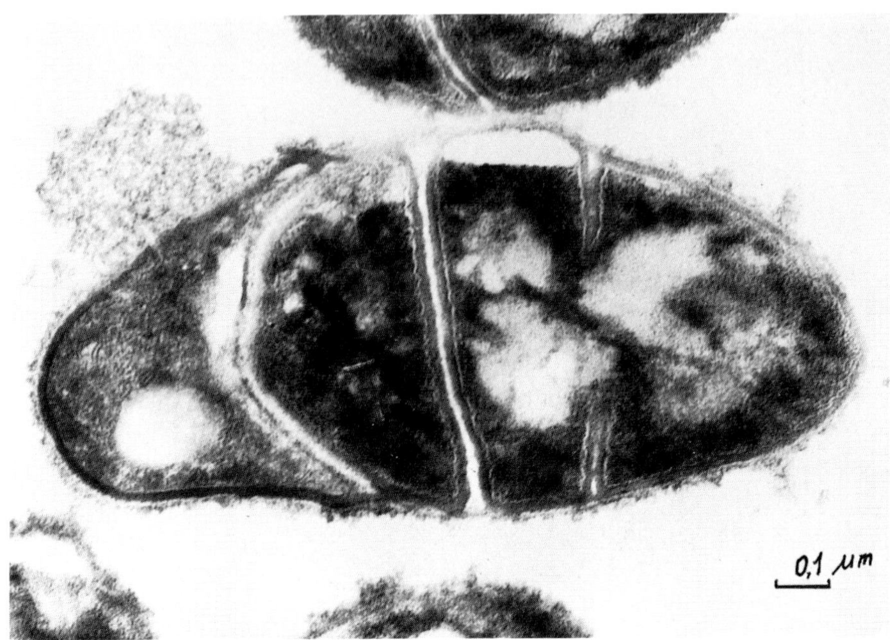

Fig. 3. Fine structure of *M. tuberculosis* treated with rifampicin for 24 hours.

The process of cell division in most mycobacterial cells resembles that of most Gram-positive bacteria. Centripetal invagination of the plasma membrane is followed by formation of a cross wall that splits in order to form the poles of the two new cells (Fig. 1). However, it has also been noticed that, at the same time, the bacilli divide as well into more unequal parts when the cell wall and the cytoplasmic membrane gradually advance from outside inwards between the two dividing fragments (Figs. 2 and 3).

Fine structure of *M. tuberculosis* treated with rifampicin. The cells treated with rifampicin for 12 hours showed few structural changes.

A thin section of a cell treated with rifampicin for 24 hours is shown in Figure 3. The most remarkable changes were seen in the cytoplasm which lost its compact structure, ribosomes disappeared and defects such as vacuoles appeared in the cytoplasm. The vesicular structure of the mesosome was also lost. The fine structure of the cytoplasmic membrane and the cell wall were well preserved.

More marked changes were noticed in the cells treated with rifampicin for 48 hours (Fig. 4). The cytoplasm was almost completely destroyed and replaced by vacuoles. This was followed by the separation of the cytoplasm from the cell wall and the disorganization of the mesosomes.

These findings are in keeping with the fact that the action of rifampicin is initially reversible but later becomes irreversible. The changes in the cytoplasm and the disappearance of ribosomes from the cell suggest that ribonucleic acid polymerase and protein synthesis of the *M. tuberculosis* are inhibited by rifampicin.

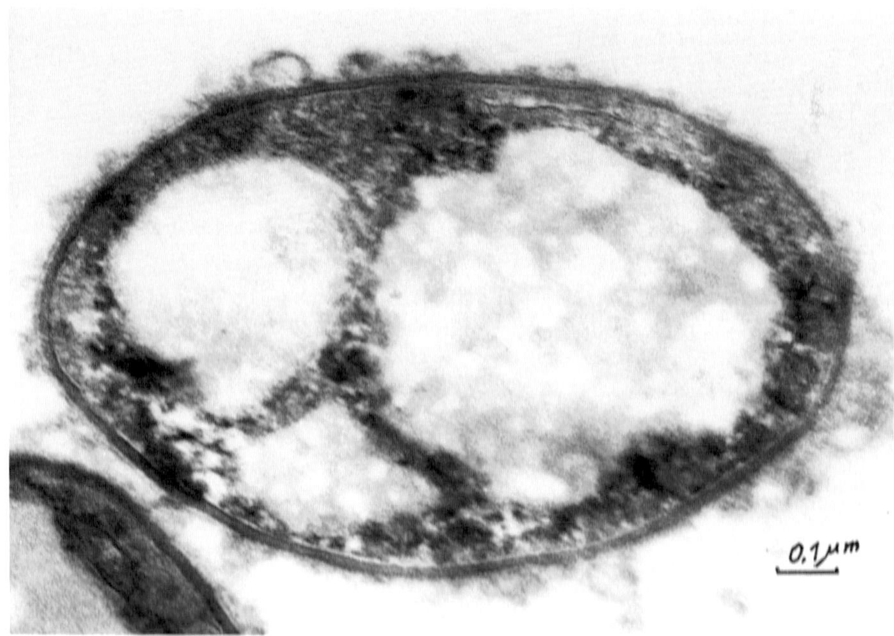

Fig. 4. Fine structure of *M. tuberculosis* treated with rifampicin for 48 hours.

REFERENCES

1. Ratledge C, Stanford J (1982) The Biology of the Mycobacteria, Volume 1, Academic Press, London

2. Konno K, Oizumi K, Araiji F, Yamagushi J, Oka S (1973) Mode of Action of Rifampicin on Mycobacteria. American Review of Respiratory Disease 107 : 1002 - 1005

3. Kellenberger E, Ryter A, Sechand J (1958) J Biophys Biochem Cytol 4 : 671

Mycobacteria of Clinical Interest. M. Casal, editor

MYCOBACTERIUM TUBERCULOSIS DRUG RESISTANCE DURING A PERIOD OF 9 YEARS

T. GONZALEZ FUENTE, N. MARTIN CASABONA, T. BASTIDA VILA, L. VINAS DOMENECH and E. FITE REIG

Service of Microbiology, Service of Pneumonology, Ciudad Sanitaria "Vall d'Hebron" Barcelona, Generalitat de Catalunya, Departament de Sanitat i Seguretat Social (Spain)

INTRODUCTION

In this study we present the resistance statistics for *Mycobacterium tuberculosis* to the most commonly used tuberculostatics. We have accumulated data of resistance over the past 9 years and the changes observed in resistance to various drugs, as well as statistical evaluation. Most of the patients included in the study were from our hospital, a general hospital, were non-tuberculous (TB) and were treated for 9 months with 3 tuberculostatics and controlled as out-patients. However, some came from other hospitals and TB clinics.

MATERIAL AND METHODS

From 1976-1984 inclusive, 6700 strains of mycobacteria were isolated in our department which correspond to 5664 *M. tuberculosis* isolated from 3670 patients. Antibiotic susceptibilities were checked on 1754 strains of which 1405 were pleuropulmonary samples, 151 urogenital and 195 from other sources (CSF, osteo-articular, adenitis, etc.). The antibiotic susceptibility testing was carried out according to Canetti's method (1) against Isoniazide (INH), p-aminosalicylic acid (PAS), Streptomycin (SM), Rifampicin (RF) and Ethambutol (EB).

RESULTS

Of the 1754 strains studied, 340 (19.38%) (290 pleuropulmonary, 25 urogenital and 25 from other sources) showed total resistance (TR) to one or more tuberculostatics. There was no significant difference between the percentage resistance of strains obtained from pleuropulmonary specimens (20.64%) and those of urogenital origin (16.56%) ($p > 0.05$). However, there was a difference between pleuropulmonary specimens and those from other sources (12.6%) ($p = 0.04$). According to the drug used, the greatest number of strains were resistant to SM followed by INH (Table I).

Nearly 40% of resistant strains (128 strains) were resistant to more than one drug and most often INH and SM resistance were found associated (Table II). A further 15 strains showed other associated resistance, but in each case this was never more than 3 strains. The 6 strains resistant to 5 tuberculostatics were obtained from patients from a TB clinic who had undergone various treatments.

TABLE I

MT STRAINS WITH TR TO EACH DRUG

Drug	Number of strains	% of 1.754 strains
SM	254	14.48
INH	151	8.61
EB	59	3.36
RF	58	3.31
PAS	18	1.03

TABLE 2

MT STRAINS RESISTANT TO VARIOUS DRUGS

Drug association	Number of strains	% of 1.754 strains
INH + SM	49	2.79
INH + SM + RF + EB	17	0.97
RF + SM	13	0.74
INH + SM + RF	9	0.51
SM + EB	7	0.40
INH + RF	6	0.34
INH + SM + EB	6	0.34
INH + SM + PAS + RF + EB	6	0.34

Determination of changes in resistance statistically over the 9 years to each drug showed a decrease for PAS ($p = 0.006$, correlation coefficient -0.77) and for INH ($p = 0.006$, correlation coefficient -0.77). SM, on the contrary, showed an increase ($p = 0.03$, correlation coefficient +0.62). No statistical changes in resistance were observed for RF or EB or in total resistance (TR) over the 9 years.

Leaving aside secondary or acquired resistance due to previous indiscriminately treated infection or recently treated with one of the drugs tested, 6.28% of cases studied (110 strains) showed primary or transmitted resistance (i.e. non-selective chromosomic resistance). SM was the drug with which most cases of primary resistance (PR) were seen (Table III).

Looking at PR and specimens, we find that nearly all strains with PR are of pleuropulmonary origin. Urogenital specimens have no strains with PR and other specimens have 7 strains with PR (CSF, 4; osteoarticular, 2; pericardium, 1). As with TR, no statistical change can be seen in PR over the 9 years.

TABLE 3
MT STRAINS WITH PR TO RESPECTIVE DRUG

Drug	Number of strains	% of 1.754 strains
INH	23	1.31
PAS	1	0.05
SM	74	4.23
EB	3	0.17
INH + SM	8	0.46
SM + RF	1	0.05

COMMENTS

Studies of *M. tuberculosis* resistance to tuberculostatics indirectly reflect the stage of treatment programmes within a country or geographic region. As indicated by Grzybowski (2), PR shows what happens with the impact of epidemiology on treatment programmes, whereas TR gives a more direct idea of the quality of the programme. For this reason, strains of *M. tuberculosis* with TR vary greatly from one country to another (2,3,4,5) or even within a country from one community to another (6,7,8).

For these 9 years in our hospital, no significant changes in % PR or TR have been observed in the strains, which indicates that the control and treatment of tuberculosis in our hospital and its catchment area are acceptable.

Statistically, for the change of resistance for these drugs, we find that PAS resistance has fallen off, a drug hardly ever used for treatment over the 9 years. Similarly, INH resistance has dropped, partly due to the use of EB and RF at the same time and partly due to the fact that INH resistant strains are less contagious (2,9) and have biochemical deficiencies (10,11). SM resistance, on the other hand, is increasing as it is still used for tuberculosis and in non-specific processes. The virulence and infectivity of SM resistant strains has not decreased. There is no statistical evidence for change in TR for EB and RF. EB resistance is almost always associated with resistance to another drug and RF resistance has only been observed in conjunction with resistance to other drugs.

The fact that there is no difference in TR for strains isolated from pleuropulmonary or urogenital specimens can be explained by the fact that urogenital cases are almost always secondary infections following a previously treated pulmonary infection. This is confirmed by the fact that none of the RT strains from urogenital specimens show PR. On the contrary, there was a statistical difference in TR between pleuropulmonary and other specimens, since in some of the latter cases (CSF, Lymphatic ganglions, osteoarticular), pulmonary tuberculosis had not been present before and thus not treated. This is verified by the presence of PR in certain strains (7/25).

In terms of PR, SM is the tuberculostatic for which most PR was observed, followed by INH with a marked difference. This is in agreement with other authors (5,6,7).

PR to EB is hardly mentioned (6,8); we came across 3 strains isolated from a patient with meningitis, one with pulmonary tuberculosis and one with pleuritis, all less than 18 years. They had never had prior treatment, nor were there any family antecedents.

In agreement with all other authors (6,8), we found no PR to RF. One case has been cited (12) but, like ours, the patient had pulmonary tuberculosis with SM resistance as well, which would suggest that the patient's history was not complete.

With the PR of the 5 drugs tested mentioned above, we feel optimistic that the given resistances should not change the epidemiological treatment of tuberculosis, as SM, for which resistance is greatest, is used less and less in triple treatment.

RP to INH (mostly at low level) is eliminated by combination with EB and RF or, according to the new 6 month guidelines which have started to be followed in our hospital, Pirazinamide should be used in combination with INH, EB and RF.

REFERENCES

1. Canetti G, Rist M, Grosset J (1963) Rev Tub Pneum 27:217-272
2. Grzybowski S (1985) Tubercle 66:69-72
3. Canetti G, Gay PH, Le-lirzin M (1972) Tubercle 53:57-65
4. Valenzuela MT and cols. (1984) Boletin de UICT 59:191-192
5. Tuberculosis Reseach Committee (Ryoken), Japan (1970) Tubercle 51:152-171
6. Ortega A (1980) Boletin Epidemiológico Semanal del Ministerio de Sanidad y Seguridad Social 1430:129-130
7. March Ayuela P (1978) Terapeútica Moderna en Tuberculosis. Avances en Terapeútica 9. Salvat Editores S.A. Barcelona
8. Ausina V and cols. (1984) Med Clin (Barc) 82:741-744
9. March Ayuela P (1977) Chest 72:683-684
10. Tsukamura M (1974) Am Rev Respir Dis 110:101-103
11. Youmans GP (1979) Tuberculosis. Edit WB Saunders Company, London
12. Stottmeimer KD (1976) J of Inf Dis 133:88-90

Mycobacteria of Clinical Interest. M. Casal, editor

INITIAL AND PRIMARY DRUG RESISTANCE OF TUBERCLE BACILLI IN FRANCE DURING 1978-1984

HUGO L. DAVID, ANTOINETTE FEUILLET and KYE S. GOH
Unité de la Tuberculose et des Mycobacteries, Institut Pasteur, Paris (France)

INTRODUCTION

From 1962 to 1972 Canetti and his associates published data on a series of surveys on primary drug resistance of tubercle bacilli in France (3-7). Since then primary drug resistance has been regularly surveyed in the Pasteur Institute (Paris), and the results collected during 1978-1984 are reported here.

Primary drug resistance (PDR) is an epidemiological concept that refers to patients initially infected by *M. tuberculosis* resistant to one or more of the drugs used in chemotherapy. Judging from worldwide data (15,24) PDR has remained relatively stable; however, it has been well established that it may vary within a country according to the characteristics of the population under observation (for a recent review, see ref. 15). The stability of PDR may be attributed to the mechanism of drug resistance in tubercle bacilli and to the mechanism of transmission of tuberculosis infection. Because drug resistance is the result of spontaneous and rare mutations (12-14) followed by the selective pressures of the drugs (15,17,19,22) when adequate treatment is given, the development of acquired resistance is avoided. Consequently, the potential sources of drug-resistant tubercle bacilli are rare and PDR is also infrequent (1,6,18). Although these notions have been well confirmed throughout the years, it is necessary to continue surveillance of PDR for reasons summarized below.

Within a country like France where the diagnosis, treatment and follow-up of patients is satisfactory, PDR has remained at a low level, but a significant number of persons immigrated from countries where tuberculosis control may not be satisfactory. In practice it is often difficult to ascertain whether these individuals received treatment before entering the country, and it is also difficult to evaluate them as potential sources of transmission of tubercle bacilli. Recently, phage typing of tubercle bacilli showed that in these patients the phage types encountered approached that found in their countries of origin (8) but the data available is still too limited to conclude about the usefulness of phage typing in epidemiological surveys. Thus, the study of PDR remains a useful epidemiological tool.

Although all evidence concerning drug resistance in tubercle indicated that it is caused only by sponanteous mutations the finding of plasmids in various mycobacteria, but not tubercle bacilli (9,10,11,16,20,21), and the recent finding of plasmid-mediated drug resistance in bacterial genera in which this kind of drug resistance was not known (14,15) raises the possibility that plasmids may be transferred from non-tuberculous mycobacteria into tubercle bacilli. The continuing survey of PDR is thus also necessary to identify the eventuality of plasmids being established in tubercle bacilli. Other aspects of drug resistance in mycobacteria were excellently reviewed by Gangadharam (15).

MATERIAL AND METHODS

During 1978-1984 a total of 3891 strains of tubercle bacilli were tested. When more than one strain was received for the same patient, only the first was included in the statistical analysis. One thousand and two strains were from patients in whom there was definitive evidence that they had not received any treatment before the strain was isolated.

According to the policy of this laboratory, strains of tubercle bacilli referred for drug studies must be cultures isolated at the time a diagnosis of tuberculosis was made. For the purpose of this report data on these strains are designated the Initial Drug Resistance survey.

Because disease caused by *M. bovis* and *M. africanum* is relatively frequent in France, drug resistance studies on these organisms are also reported.

Drug susceptibility testing was performed using a simplified version of the proportion method of Canetti and associates (2,6). The cultures were incubated at 37°C, and the proportion of resistant colonies was checked after 28 days of incubation. From previous studies (unpublished data), it was found that the 42 day reading recommended by Canetti and associates (2,5) was not necessary.

RESULTS AND DISCUSSION

The types and frequencies of drug resistance are depicted in Table I. The analysis of the data indicated that the initial drug resistance did not differ significantly from previous data on PDR as summarized in Table II. Furthermore, the distribution of PDR estimated from 1002 strains received from patients who did not receive treatment, did not differ significantly from the distribution of drug resistance estimated from the total of 3891 strains examined. Therefore it was not necessary to present the data separately.

In the French nationals, resistance to streptomycin alone was the most frequent type (3.48%); in the foreign population resistance to isoniazid alone (5.30%) and to streptomycin alone (4.68%) were the most frequent types. The average initial drug resistance was higher in the foreign population (14.51%) than in French nationals (7.75%).

During 1978-1984, 5171 strains of tubercle bacilli were referred to the Pasteur Institute among which 4954 were *M. tuberculosis* (95.80%), 158 were *M. bovis* (3.06%), and 59 were *M. africanum* (1.14%). Considering that disease caused

Table 1. Frequency of Initial Drug-resistance in France (1978-1984).

Types of resistance	French population No.	%	Foreign population[1] No.	%	Algerians No.	%	Portuguese No.	%	Moroccans No.	%	Black Africans No.	%	South-east Asians No.	%
INH alone	49	1.51	34	5.30	10	4.33	5	5.32	4	4.44	6	5.71	5	4.13
SM alone	113	3.48	30	4.68	9	3.90	2	2.13	0	-	3	2.86	10	8.26
PAS alone	7	0.22	2	0.31	0	-	0	-	0	-	0	-	2	1.65
RIF alone	2	0.06	0	-	0	-	0	-	0	-	0	-	0	-
EMB alone	2	0.06	1	0.16	0	-	1	1.06	0	-	0	-	0	-
ETH alone	11	0.34	0	-	0	-	0	-	0	-	0	-	0	-
INH+														
SM	31	0.95	16	2.50	4	1.73	2	2.13	0	-	1	0.95	7	5.79
PAS	6	0.18	2	0.31	1	0.43	0	-	0	-	1	0.95	0	-
RIF	15	0.46	2	0.31	0	-	0	-	1	1.11	0	-	0	-
EMB	3	0.09	0	-	0	-	0	-	0	-	0	-	0	-
ETH	5	0.15	4	0.63	0	-	0	-	0	-	2	1.40	1	0.83
SM+														
PAS	6	0.18	2	0.31	0	-	0	-	0	-	0	-	0	-
RIF	1	0.03	0	-	0	-	0	-	0	-	0	-	0	-
EMB	0	-	0	-	0	-	0	-	0	-	0	-	0	-
ETH	1	0.03	0	-	0	-	0	-	0	-	0	-	0	-
Multiple[2]	64	1.87	46	7.18	10	4.33	2	2.13	4	4.44	8	7.62	10	8.26
Total R	316	9.72	139	21.68	34	14.7	12	12.7	8	10.0	21	20.0	36	29.7
Total less multiple R	252	7.75	93	14.51	24	10.3	10	10.6	5	5.56	13	12.3	26	21.4
Total strains tested	3250	-	641	-	231	-	94	-	90	-	105	-	121	-

1) Including patients from Turkey,Spain,Italy,UK,India,Lebanon,Tunisia, and Yugoslavia.
2)Resistance to more than 2 drugs.

Table 2. Evolution of primary drug-resistance in France from 1962 to 1984.

Survey period	No. of strains tested	Frequency of resistance		References
		French	Foreigners	
1962-1964	2144	9.2	13.1	3,7
1965-1966	3004	9.7	12.8	4,7
1966-1970	6495	8.9	12.8	6,7
1975-1976	1907	8.9	12.8	24
1978-1984	3891	7.7	14.5	This survey
TOTAL	17441	8.8	13.2	-

Table 3. Frequency and types of drug-resistance in *M. africanum* and *M. bovis*.

Types of Resistance	*M. africanum*[1] No.	%	*M. bovis*[2] No.	%
INH-alone	3	5.36	3	5.17
SM-alone	1	1.79	1	1.72
PAS-alone	0	-	7	12.07
RIF-alone	1	1.79	0	-
EMB-alone	0	-	0	-
ETH-alone	0	-	0	-
INH+PAS	0	-	6	11.37
INH+ETH	0	-	1	1.72
SM+PAS	0	-	1	1.72
SM+RIF	1	1.79	0	-
Multiple	7	12.50	6	11.37
Total	13	23.21	25	43.10
Total less multiple	6	10.71	19	32.76
Total tested	56	-	58	-

1) All strains of M. africanum were from black africans, except one from Saoudi Arabia.
2) All strains of M. bovis were from French nationals, except one from a patient from North Africa.

by *M. bovis* and *M. africanum* occurred in significant frequencies it was also of interest to report the types of drug resistance found in these bacteria. As shown in Table III drug resistance in *M. bovis* was more frequent than in *M. tuberculosis* and *M. africanum*. Although not shown, it should be indicated that all strains of *M. bovis* were resistant to pyrazinamide.

As shown in this report, PDR remained stable in France for a period of 20 years. We think that in countries like France the main purpose of PDR studies may not be so much concerned about the transmission of drug resistant bacteria in the general population as identifying subpopulations at high risk. In these countries a detailed investigation to ascertain treatment before isolation of the cultures may not be necessary unless the frequency of drug resistance observed is significantly above the background data on individual drugs as known from previous PDR studies. We also think that when the frequency of drug resistance is high, and an investigation of previous treatment and follow-up is difficult to apply, drug resistance studies are recommended for every newly diagnosed patient.

REFERENCES.

1. Canetti, G. (1962) Tubercle 43,301.
2. Canetti, G.,Rist,N.,Grosset,J. (1963) Rev. Tuberc.(Paris) 27,237-272.
3. Canetti,G.,Kreis,B.,Thibier,R.,Grosset,J.,Gluszky,J.(1964) Rev. Tuberc. Pneumol.28,1115-1158.
4. Canetti,G.,Kreis,B.,Thibier,R.,Gay,Ph.,Le Lirzin,M. (1967) Rev. Tuberc. Pneumol. 31:433-474.
5. Canetti,G.,Fox,W.,Khomenko,A.,Mahler,H.T.,Menon,M.K.,Mitchison,D.A.,Rist,N.,Semelev,N.A.(1969)Bull. World Health Org. 41, 21-43.
6. Canetti,G.,Le Lirzin,M.,Gay,Ph.,Thibier,R.,Kreis,B.,Grosset,J. (1972) Rev. Tuberc. Pneumol. 36,337-356.
7. Canetti,G.,Gay,Ph.,Le Lirzin,M. (1972) Tubercle 53,57-83.
8. Clavel-Sérès,S.,Clément,F. (1984)Ann. Microbiol.(Institut Pasteur) 135B,35-44.
9. Crawford,J.T.,Bates,J.H. (1979) Infect. Immun. 24,979-981.
10. Crawford,J.T.,Cave,M.D.,Bates,J.H. (1981)Rev. Infect. Dis. 3,949-952.
11. Crawford,J.T.,Cave,M.D.,Bates,J.H. (1981) J. Gen. Microbiol. 127,333-338.
12. David,H.L. (1970) Appl. Microbiol.20,810-814.
13. David,H.L. (1980) Clinics in Chest Medicine 1,227-230.
14. David,H.L. (1981) Rev. Infect. Dis. 3,878-884.
15. Gangadharam,P.R.J. (1984). "Drug resistance in Mycobacteria", CRC Press Inc., Florida, USA.
16. Labidi,A.,Dauguet,C.,Goh,K.S.,David,H.L. (1984) Curr. Microbiol. 11,235-240.
17. McClatchy,J.K. (1980) in"Antibiotics in Laboratory Medicine",V. Lorian editor, Williams and Wilkins,Baltimore.
18. Mitchison,D.A. (1968) in "Recent Advances in Respiratory Tuberculosis", F. Heaf and N.L. Rusby editors, J&A. Churchill, London, UK.

19. Mitchison,D.A. (1961) Bull. Int. Union Ag. Tuberc.32,81.
20. Mizuguchi,Y.,Fukunaga,M.,Taniguchi,H. (1981) J. Bacteriol. 146,656-659.
21. Mizuguchi,Y.,Udon,T.,Yamada,T. (1983) Microb. Imm.27,425.
22. Pyle,M.M. (1947) Proc. Mayo Clinic 22,465.
23. Scott Meissner,P.,FalkinhamIII,J.O. (1984) J. Bacteriol. 157,669-672.
24. World Atlas of initial drug-resistance, H.H. Kleeberg and M.S. Boshoff, editors, Int. Un. Against Tuberc.,1078.

Mycobacteria of Clinical Interest. M. Casal, editor

THE INCIDENCE AND CLINICAL RELEVANCE OF PRIMARY DRUG RESISTANCE IN PERUVIAN ISOLATES OF MYCOBACTERIUM TUBERCULOSIS

W. A. BLACK, B. GANTER, S. GRZYBOWSKI, P. HOPEWELL, J. L. ISAAC-RENTON, A. LASZLO, M. SANCHEZ HERNANDEZ

Provincial Public Health Laboratory and University of British Columbia, Vancouver, Canada; National Reference Centre for Tuberculosis, Ottawa, Canada; University of California, San Francisco, U.S.A; Pan American Health Organization and Ministry of Health, Lima, Peru

INTRODUCTION

Chemotherapy for pulmonary tuberculosis can be nearly 100% successful provided patients adhere to the prescribed drug regimen and the tubercle bacilli are susceptible to the drugs used. Prior evaluation studies of two different treatment regimens in two large cohorts of patients in Peru showed that the major factor limiting the success of treatment was a very high rate of failure to complete the treatment (1). A possible second factor could have been a high prevalence of resistance of the tubercle bacilli to one or more of the antituberculous agents used, even in patients who had not been treated previously (primary resistance). In order to determine if this was the case, and to provide information to guide the selection of antituberculous drug regimens in the future, we conducted a study of bacillary resistance to antituberculous agents in patients residing in Lima, Peru, who had not been treated previously.

MATERIAL AND METHODS

Sputum specimens were obtained from 89 young, smear-positive, previously untreated patients living in Peru. At the time of specimen collection each patient completed a special questionnaire relating to personal history such as place of birth and length of stay in Lima, and clinical history such as type and duration of symptoms, previous BCG vaccination, etc.

The 89 specimens were sent by air to the Provincial Public Health Laboratory in Vancouver, B.C., where culture and sensitivity tests were performed; part of each specimen was also sent to the Canadian National Reference Laboratory in Ottawa for sensitivity testing by the Bactec radiometric method (2).

Laboratory Methods

The Vancouver laboratory processed the sputum specimens using the following procedures: decontamination with an equal volume of 3% sodium hydroxide for 40 minutes at 37°C, neutralization with 2.5 N hydrochloric acid, centrifugation and decanting. The resulting sediment from each specimen was used to make a smear, which was stained by the Ziehl-Neelsen technique, and to inoculate one slope each of three different culture media, viz., Lowenstein-Jensen with pyruvate, Lowenstein-Jensen with nalidixic acid and Tarshis blood agar. The inoculated media were incubated aerobically at 37°C and examined for growth at weekly intervals for up to 8 weeks if necessary.

Direct and indirect susceptibility testing was done by the proportion method, using techniques employed at the Centres for Disease Control, Atlanta (3). The antimicrobial agents and concentrations used were streptomycin (SM) at 2.0 μg/ml, isoniazid (INH) at 0.2 μg/ml, rifampin (RIF) at 1.0 μg/ml and ethambutol (EMB) at 7.5 μg/ml.

RESULTS

Six patients of the original 89 patients whose sputum specimens were submitted had to be excluded; in two the specimen failed to grow tubercle bacilli and in four the questionnaire revealed a record of previous treatment. Thus, 83 patients were left for final analysis. Table I shows their ages.

TABLE I
AGE OF THE PATIENTS

Age group	14-19	20-24	25-29	30-34	Total
Number	39	25	16	3	83

As can be seen, 47% of the patients were under the age of 20 and 77% were under the age of 25.

Sixty-two isolates (74%) were sensitive to all four antimicrobial agents and 12 (26%) showed resistance. No strain was resistant to all four drugs; one strain was resistant to three drugs (INH, SM, RIF); six strains were resistant to two drugs (INH, SM), and 14 strains were resistant to one drug (six to SM and eight to INH). Resistance to SM, INH, RIF and EMB was shown at 13, 15, 1 and 0 strains respectively.

In all but two instances there was agreement on the results between the Vancouver laboratory and the Ottawa laboratory. In these two instances testing in the Vancouver laboratory showed resistance to INH whereas the Ottawa laboratory found the organisms to be susceptible.

DISCUSSION

This pilot study confirms the necessity of basing drug regimens for tuberculosis upon the anticipated response of the organism, based in turn upon in vitro susceptibility testing. The importance of this approach, at least on a periodic basis, has previously been expressed by other workers (4, 5).

The present study shows that the problem of drug resistance of tubercle bacilli in Peru is likely to be quite significant with about a quarter of all the younger cases of pulmonary tuberculosis showing resistance to one or more drugs. The high prevalence of isoniazid resistance, in particular, indicates that some modifications of the existing treatment regimen is indicated. The best solution would seem to be a regimen of six to eight months that uses rifampin throughout. Unfortunately, economic factors in Peru preclude the use of rifampin or ethambutol in every case. The second solution would be the provision of such a regimen to cases proven to be resistant by susceptibility studies; however, existing laboratory facilities are inadequate for this purpose on a routine basis. Thus, the best practical approach may be to employ in vivo susceptibility testing, i.e. cases which are still smear-positive after five or six months of standard chemotherapy would receive a full course of a rifampin regimen. If possible, in vitro susceptibility studies should be done on these cases prior to changing therapy. Such a modified approach to therapy of tuberculosis appears essential for Peru and for other countries with limited financial resources and a sizeable problem of drug resistance of tubercle bacilli.

REFERENCES

1. Hopewell P. Personal communication.
2. Laszlo A, Gill P, Handzel V et al (1983) Conventional and radiometric drug susceptibility testing of _Mycobacterium tuberculosis complex_. J Clin Microbiol 18:1335-1339.

3. Vestal AL (1975) Procedures for the isolation and identification of mycobacteria. U.S. HEW Publication (CDC) No.76-8230.

4. Canetti G, Fox W, Kohmenko A et al (1969) Advances in techniques of testing mycobacterial drug sensitivity, and the use of sensitivity tests in tuberculosis control programmes. Bull WHO 41:21-43.

5. Grzybowski S (1985) The impact of treatment programmes on the epidemiology of tuberculosis. Tubercle 66:69-72.

Mycobacteria of Clinical Interest. M. Casal, editor

TOXICITY OF TWO ANTITUBERCULOUS THERAPEUTIC SCHEDULES OF 6 AND 9 MONTHS. THE ROLE OF PYRAZINAMIDE IN IATROGENESIS

ENRIQUETA FITE, RAFAEL VIDAL, JUAN RUIZ-MANZANO, JAVIER DE GRACIA, Mª LUISA MARTINEZ and TEOFILO GONZALEZ*

Pneumology Section, *Service of Bacteriology, Hospital General Vall d'Hebrón, Barcelona (Spain)

INTRODUCTION

It is well known that all antitubercular drugs administered individually or in combination can induce side effects (1). Generally these adverse reactions are not very important, but in some cases they can be potentially grave if they are not recognised in time (2). This is one of the reasons why it is advisable to carry out periodic controls during treatment. Special attention has been given in recent years to the hepatotoxicity developing in short-course regimens that include isoniazid (H), rifampicin (R), and pyrazinamide (Z) (2-5). However, the studies carried out show that hepatitis is not more frequent in this type of treatment than in those of standard duration which include H and other associated drugs, during one year or more (6).

We have compared the toxicity of a therapeutic schedule of 9 months which also includes Z.

MATERIAL AND METHODS

Patients

In this study we only analyze patients who correctly took the medication during the time prescribed and attended the established controls. From 1979 until February 1985 they fulfilled the following conditions: 429 patients had been treated with a 9-month regimen and 97 with one of 6 months.

The controls were carried out after 21 days, 2, 4, and 6 or 9 months. They consisted of a questionnaire aimed at detecting side effects and blood tests (bilirubin, SGOT, SGPT, alkaline phosphatase, hemogram and uric acid). In those cases in which hepatitis was detected, Australia antigen was found.

Treatment

The combinations of drugs, duration and dosing are detailed in the following Tables (I and II).

STATISTICAL ANALYSIS

Statistical comparisons were made by chi-square analysis with continuity correction. A p value of 0.05 or less was considered significant.

TABLE I

NINE-MONTH AND SIX-MONTH REGIMENS.DRUG REGIMENS

	HRE 9	HREZ 6
First 2 months	Isoniazid Rifampicin Ethambutol	Isoniazid Rimfapicin Ethambutol Pyrazinamide
Continuation Therapy	Isoniazid-Rifampicin For next 7 months	Idem for next 4 months

TABLE II

NINE-MONTH AND SIX-MONTH REGIMENS.DOSAGE OF DRUGS

Isoniazid	300 mg. daily, in one dose.
Rifampicin	600 mg. " " " "
....	450 mg. " for patients below 40 Kg.
Ethambutol	25 mg./Kg. daily, in one dose
Pyrazinamide	30 mg./Kg. daily, in one dose

RESULTS

The nature of the adverse reactions, their incidence in the two regimens and the statistical comparison between them (according to the chi-square method) are shown in Table III. The withdrawals due to drug toxicity are shown in Table IV.

CONCLUSIONS

1) A change of therapy because of adverse reactions was necessary in 2.09% of the 429 patients receiving treatment lacking pyrazinamide and in 5.15% on the treatment schedule including pyrazinamide.

2) There was no significant difference as regards serious hepatotoxicity between the usual schedule of 9 months and that of 6 months with pyrazinamide.

3) Asymptomatic alteration of liver function found in the 6-month schedule (17.5%) was superior to the 9-month (8.85%), but this level is similar to that found in schedules of isoniazid chemoprophylaxis.

4) Joint disorders only appeared in the group of patients taking Z. They were not important except in the case of a patient who suffered from gout. We believe that patients with a history of gout should be excluded from regimens with Z.

TABLE III

DRUG TOXICITY. NINE-MONTH AND SIX-MONTH REGIMENS.

ADVERSE REACTION	HRE 9(429 patients) Patients Nº	%	HREZ 6(97) Nº	%	Signification
Hepatitis	8	1.8	2	2.06	NS
Symptomless adnormal Liver function test	38	8.85	17	17.05	0.001
Gastrointestinal upset	13	3.08	7	7.2	0.025
Peripheral neuropathy	2	0.46	0	0	NS
Optical neuritis	4	0.93	1	1.03	NS
Rash	2	0.46	3	3.09	0.025
Leukopenia	4	0.93	2	2.06	NS
Thrombocytopenia	1	0.24	0		NS
"Flu" syndorme	0		1	1.03	NS
Arthralgias	0		2	4	
Gout	0		1	1.03	NS
TOTAL	72	16.7	36	37	

TABLE IV
WITHDRAWALS DUE TO DRUG TOXICITY

Adverse Reaction	9 HRE/429 patients Nº	%	6 HREZ/97 patients Nº	%	
Hepatitis	8	1.8	2	2.06	NS
Gastrointestinal Upset	1	0.23	1	1.03	NS
Rash	0	0	1	1.03	NS
Gout	0	0	1	1.03	NS
TOTAL	9	2.09	5	5.15	

CONCLUSIONS

1) Therapy change because of adverse reactions was necessary in 2,09% of the 429 patients with treatment lacking Pyrazinamide and in 5,15% of the treatment schedule including Pyrazinamide.

2) There was no significant difference in the serious hepatotoxicity between the habitual schedule of 9 moths and that of 6 months with Pyrazinamide.

3) Asymptomatic alteration of liver function found in the 6 month schedule (17,5%) was superior to the 9 month (8,85%), but this level is similar to that found in schedules of isoniazid chemoprophylaxis.

4) Joint disorders only appeared in the group of patients which took Z. They were not important except in the case of the patient who suffered from gout. We beleive that the patients with a history of gout should be excluded from the regimens with Z.

REFERENCES

1. GIRLING, D. J. (1984). Efectos adversos de los medicamentos antituberculosos. Boletín de la Unión Internacional contra la Tuberculosis, 59: 153-164.

2. GIRLING, D. J. (1978). The hepatic toxicity of antituberculosis regimens containing isoniazid. Rifampicin and Pyrazinamide. TUBERCLE 59: 13-32.

3. ANGEL, J. H. (1979). TOXICITY of Pyrazinamide in the comvination chemotherapy of pulmonary tuberculosis. La Pyrazinamide 25 ans après pp. 119. Alger.

4. PILHEU, J. A. DE SALVO, M. C. ET AL. (1981). Estudio del hígado con microscopía de luz y electrónica en pacientes tuberculosos que reciben rifampicina e isoniacida. Medicina 41: 439, Buenos Aires.

5. PILHEU, J. A. DE SALVO, M. C., KOCH, O. R, ARIAS, R. F. (1984). Acción de la pirazinamida sobre el hígado de los enfermos tuberculosos. Estudio ultraestructural. Boletín de la Unión Internacional contra la Tuberculosis. 59: 145-147.

6. RISKA, N. (1976) Hepatitis cases in Isoniazid treated groups and in a control group. Bulletin of the International Union Against Tuberculosis, 51: 203-208.

Mycobacteria of Clinical Interest. M. Casal, editor

A CONTROLLED STUDY OF 637 PATIENTS WITH TUBERCULOSIS. DIAGNOSIS AND RESULTS USING 6 AND 9 MONTH THERAPEUTIC SCHEDULES

J. de GRACIA, R. VIDAL, J. RUIZ, E. FITE, T. GONZALEZ and N. MARTIN
Division of Pneumonology and Bacteriology, Ciutat Sanitaria Vall D'Hebron, Passeig Vall D'Hebron s/n, 08035 Barcelona (Spain)

INTRODUCTION

Tuberculosis remains an important health problem in Spain. The annual morbidity rate is 50 to 60 per 100,000 persons (1). Tuberculosis prevalence at 7 years of age is 2.3% (2) and 42.9% in non-vaccinated adults from 20 to 30 years of age (3). The annual risk of infection is 0.21% (4) and the annual decline in the number of cases has remained essentially unchanged at the 7% level (4) since 1952, when isoniazid was introduced as an antituberculosis drug. All these findings, together with the lack of reliable data on diagnostic efficacy, treatment results and contact control, show the ineffectiveness of current measures of eradication.

Since 1980, we have been using in our unit a protocol of diagnosis and treatment of patients with tuberculosis and their household contacts. We report here the results of two therapeutic regimens lasting 6 and 9 months in patients with pulmonary tuberculosis and a 9-month regimen in patients with extrapulmonary tuberculosis. We also present data on the diagnostic yield of adenosine deaminase (ADA) determinations in pleural fluid in tuberculous pleural effusions.

MATERIAL AND METHODS

Patients

Between January 1980 and December 1984, 637 patients with tuberculosis were diagnosed, treated and controlled at our Division of Pneumonology. There were 417 males and 220 females with a mean (±SD) age of 38.1±19.6 years (r. 7 to 92 years).

Diagnosis

The diagnosis of tuberculosis was established by any of the following criteria: positive sputum culture for *Mycobacterium tuberculosis;* biopsy material showing granulomas with necrosis and/or caseum; and clinical and radiologic features consistent with tuberculosis.

ADA

ADA determinations in pleural fluid were performed according to the colorimetric method of Giusti (5). The samples were obtained from 34 patients with tuberculous pleural effusions diagnosed by bacteriologic or histologic methods. Test results greater than 45 U/L were considered positive.

Drug regimens

A 9-month isoniazid (H)-rifampin (R) regimen plus ethambutol (E) for the first two months of treatment (2RHE/7RH), was used in 463 patients with pulmonary tuberculosis and 58 patients with extrathoracic tuberculosis (including 43 peripheral lymphadenitis, 5 gastrointestinal, 4 cutaneous, 3 laryngeal, 2 osseous and 1 ocular). A 6-month R-H regimen plus E and pyrazinamide (Z) for the first two months of treatment (2RHEZ/4RH), was used in 106 patients with pulmonary tuberculosis.

The dosages used were as follows: H, 300 mg daily; R 600 mg daily or 450 mg if the patient weighed less than 40 kg; E, 25 mg/kg daily and Z, 30-35 mg/kg daily.

Follow-up

Each patient was evaluated after the first 3 weeks of treatment, every two months until the end of the treatment and 6, 12, and 24 months after the completion of treatment. Two sputum specimens were requested on the 6 month visit, at the end treatment and on each of the following clinical evaluations; these were examined by smear and culture.

Statistical analysis

Statistical comparisons between the two drug regimens were made by chi-square analysis with continuity correction. A p value of 0.05 or less was considered positive.

RESULTS

Site of infection

Of the 637 patients with tuberculosis, 440 (69%) had pulmonary involvement, 149 (23.3%) pleural effusions, 39 (6.1%) mediastinal lymphadenopathy, 9 (1.4%) miliary disease, 4 pericardial involvement and 4 endobronchial disease. Extrapulmonary infection occurred in 86 patients (13.5%), including 51 (8%) with peripheral lymphadenitis, 10 (1.5%) osseous involvement, 8 tuberculosis of the larynx, 8 gastrointestinal, 6 cutaneous, 2 ocular and 1 gynecologic disease. In 94 patients (14.7%) there was involvement of more than one site.

Diagnosis

The diagnosis was established bacteriologically in 453 cases (71.1%), histopathologically in 107 (16.8%) and by clinical and radiologic features in 77 (12%). In this last group, there were 35 pulmonary localizations, 20 pleural effusions, 18 mediastinal lymphadenopathies and one miliary infection.

Of the 440 patients with pulmonary tuberculosis, 404 (41.8%) were diagnosed bacteriologically (261 by sputum smear, 35 with negative cultures). Three of nine patients with miliary tuberculosis (33.3%) had tubercle bacilli in the sputum. Pulmonary biopsy was compatible with tuberculosis in all of 5 patients (100%).

Tuberculosis was demonstrated histopathologically and/or bacteriologically in 103 of 123 patients (83.7%) with tuberculous effusions but no radiologic evidence of lung involvement. Tubercle bacilli were present in 33 pleural effusions (26.8%)

including 6 cases where mycobacteriae were found in stained smears of the fluid. Histopathologic and microbiologic studies of the pleural biopsies gave a diagnosis of tuberculosis in 94 cases (75%). Bacteriologic testing of the sputum was positive in 24 cases (19.4%) and in 6 cases (5%) this was the only evidence.

ADA

ADA levels in pleural fluid were positive in 32 of 33 cases (96.9%), with microbiologic and/or histopathologic evidence of tuberculous pleural effusion.

Therapeutic results

The results of therapy are compared in Table I.

TABLE I

	2RHE/7RH	2RHEZ/4RH	Significance
No. of patients	463	106	
Lost to follow-up	21(4.5%)	2(1.8%)	
Change of treatment	16(3.4%)	13(12.2%)	0.001
Noncompliance	8(1.7%)	6(5.6%)	0.025
Adverse drug reaction	8(1.7%)*	7(6.6%)**	0.01
Treatment failures	1(0.2%)	2(1.8%)	
Controls after treatment	313(at 1 year)	66(at 6 months)	
Relapses	4(1.2%)	-	

*hepatotoxicity **2 hepatotoxicity, 2 gout, 3 uncontrolled diabetes mellitus

No treatment failures occurred among the 58 patients with extrapulmonary tuberculosis in whom treatment was administered during 9 months, and all 41 patients controlled after 12 months were free of relapse.

DISCUSSION

Bacteriologic testing is very sensitive in providing a diagnosis of pulmonary tuberculosis (91.8%, in our series). The second most common site of infection in our series was the pleural space (23.3%). Sputum and pleural fluid showed tubercle bacilli in 30% of patients with pleural involvement. Pleural biopsy was especially valuable, increasing the diagnostic yield to 83.7%. Ocana et al (6) have documented an excellent correlation between pleural ADA levels and tuberculosis infection. We have observed in our patients a correlation of 96.9%. Although false-positive results may occur with this test (7), we feel it has an important role in obtaining a prompt and non-invasive diagnosis of pleural disease.

The usefulness of a 9-month regimen containing R and H in the treatment of pulmonary tuberculosis has been well documented (8). The evidence regarding short-term therapy for extrapulmonary disease seems more problematic, although some reports have stated similar good results (9). In the current study, 41 patients responded well to treatment and remained free of relapse for a year after completion of therapy.

Our results show that regimens lasting 6 and 9 months are equally effective and are similar to data from other studies in the literature (10). Nevertheless, we have observed a significantly greater number of treatment changes in the group of patients treated for 6 months ($p < 0.001$). This was due to an increase in adverse drug reactions ($p < 0.01$) and noncompliance with the treatment schedules ($p < 0.025$). Treatment had to be modified in three patients with uncontrolled diabetes mellitus. We do not know the mechanism of this pirazinamide interference.

We feel that 6-month regimens with R, H, E and Z are adequate for most cases of pulmonary tuberculosis. The use of short-course chemotherapy is unwarranted in patients with gout; patients with diabetes mellitus should be closely monitored for complications; and treatment compliance has to be ensured if these regimens are to succeed.

REFERENCES

1. Anonymous (1985) Butlleti Epidemiologic de Catalunya 6:9-12
2. Taberner JL, Garcia A (1984) Gas San (Barc) III:50-52
3. Passarell A, Campis M, Vidal R, Gracia J de, Fite E, Vaque J (1985) III Congreso Nacional de Higiene y Medicina Preventiva Hospitalaria Salamanca (Spain)
4. March P de (1985) Rev Clin Esp 176:482-483
5. Giusti G (1974) In: Bergmeyer HU (ed) Methods of enzymatic analysis. Academic Press Inc, New York, pp 1092-1099
6. Ocaña I, Martinez JM, Segura R, Fernandez T, Capdevila JA (1983) Chest 84:51-51
7. Niwa Y, Kishimoto H, Shimokata K (1985) Chest 87:351-355
8. Briths Thoracic Association (1980) Lancet 1:1182-1183
9. Stead´s W, Dut A (1982) Am Rev Respir Dis 125:94-96
10. Briths Thoracic Association (1982) Am Rev Respir Dis 126:460-462

Mycobacteria of Clinical Interest. M. Casal, editor

CHEMOTHERAPY OF LUNG TUBERCULOSIS BY A NEW ANTIBACTERIAL SUBSTANCE OFLOXACIN (DL 8280)

MICHIO TSUKAMURA
National Chubu Hospital, Obu, Aichi 474 (Japan)

INTRODUCTION

In vitro antimycobacterial activity of a new antibacterial substance ofloxacin (DL 8280) was investigated by Tsukamura (1), and it was shown that the agent inhibits the growth of *Mycobacterium tuberculosis, M. bovis, M. kansasii, M. xenopi,* and *M. fortuitum* at concentrations of 1.25 or 2.5 μg/ml in Ogawa egg medium. *In vitro* activity against *M. tuberculosis* strain H37Rv and cross resistance between other antituberculosis agents were studied by Tsukamura (2). There was no cross resistance between this agent and other antituberculosis agents. The therapeutic effect of the agent on lung tuberculosis was studied by Tsukamura et al (3) in 'treatment failure' patients. The results of six months' administration of ofloxacin showed that the treatment with this agent is effective. The present study deals with the results of one year of administration for lung tuberculosis in treatment-failure patients and those of six months' administration for lung tuberculosis in patients not previously treated.

MATERIAL AND METHODS

Chemotherapy of treatment-failure patients

The study started with 19 patients who had chronic cavitary, treatment-failure pulmonary tuberculosis (3). However, the patients consisted of very far advanced patients and 3 patients had died due to cor pulmonare by the 8th month of observation. Accordingly, these 3 patients were omitted from the present observation. The remaining 16 patients comprised 9 males and 7 females. The average age of the patients was 57.4, and the time after initiation of tuberculous disease was 3 to 25 years. The majority of patients, except those having hypersensitivity to rifampicin, had tubercle bacilli resistant to rifampicin, isoniazid, streptomycin and ethambutol. The patients received ofloxacin, 0.3 g daily per os, together with antituberculosis agents, to which tubercle bacilli of patients had resistance. Therefore, the chemotherapy was, in practice, a single administration of ofloxacin. The administration continued for one year (5 patients received chemotherapy for 2 years).

Chemotherapy of patients not previously treated by any antituberculosis agents

All patients had the following background: (1) smear- and culture-positive in the sputum examination on admission; (2) tubercle bacilli isolated were susceptible to all antituberculosis agents; (3) the disease contains cavitary lesions and does not involve more than two lobes. Screening for non-tuberculous mycobacteria was made by a test for susceptibility to p-nitrobenzoic acid (0.5 mg/ml) in Ogawa egg medium (4). The patients who excreted non-tuberculous mycobacteria were not included in the study. Newly admitted patients were allocated to three groups: (1) the first group received a regimen of streptomycin + rifampicin + isoniazid; (2) the second group received a regimen of ethambutol + rifampicin + isoniazid; (3) the third group received a regimen of ofloxacin + rifampicin + isoniazid. The ratio of the first group to the other two groups was 3 : 2 : 1. The dosages were as follows: ofloxacin, 0.3 g daily per os (in one patient, 0.4 g); streptomycin sulfate, 0.75 g daily intramuscularly for the first 3 months and then 1 g per day, twice weekly, for the latter 3 months; rifampicin, 0.45 g daily per os; ethambutol, 0.75 g or 1.0 g daily per os; isoniazid, 0.4 g daily per os. The duration of the above chemotherapy was 6 months.

RESULTS

One year administration in treatment - failure patients

A comparison of the percentage of positive cultures before and after administration of ofloxacin is shown in Figure 1 and Table I. The positive cultures were reduced markedly by the administration of ofloxacin. There was a statistically significant difference in the percentage of positive cultures between the 12 months before administration and those after administration (Table I). Negative conversion defined as continuous negative cultures for more than 5 months occurred in 3 patients. Two other patients showed a negative culture in the last 3 months. These two had negative conversion immediately after administration, but a few positive cultures appeared in the latter 6 months. On the other hand, 11 patients did not show negative conversion. In these 11 patients, tubercle bacilli resistant to ofloxacin appeared in the 3rd or 4th month. Before administration, the tubercle bacilli of all patients were susceptible to ofloxacin, 1 μg/ml, in Ogawa egg medium, whereas tubercle bacilli persisting even after administration were resistant to 2.5 μg/ml or more.

Reduction of cavity size was observed in 5 patients, and a gain in body weight of more than 4 kg in 4 patients. No side effects were observed in any of the patients during the administration period of one year or in the following year. The following examinations were performed: blood cell counts; amounts of hemoglobin; amounts of inorganic ions in the serum; serum transaminase activities; serum alkaline phosphatase activity; amounts of bilirubin, urea nitrogen and creatinine in the serum; albumin and globulin fractions in the serum; amounts of protein, sugar and sediments in the urine; visual and hearing acuities. Ten patients received photographic examination of the retina and the cornea and were shown to have no pathologic changes.

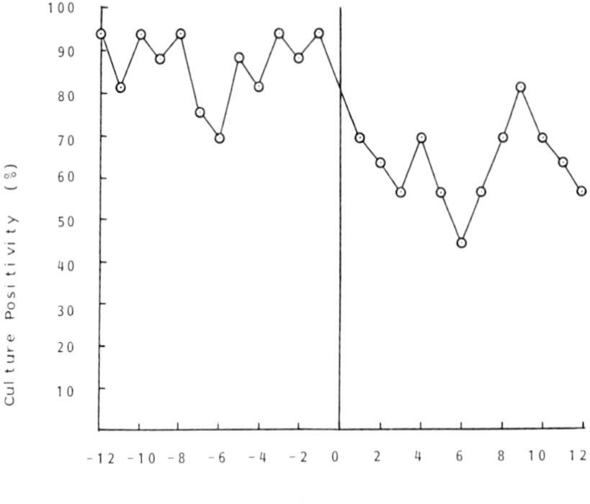

Fig. 1. Comparison of the culture positivity before and after administration of ofloxacin. Each circle indicates the culture positivity of 16 patients in each month.

Five patients received ofloxacin for 2 years, without any side effects. In the remaining 11 patients, the administration was stopped after one year of administration.

Of the 4 patients with negative conversion of sputum culture, 2 patients did not receive the drug after one year. One of these 2 patients showed a relapse in the second year. Another two patients received ofloxacin for 2 years without relapse.

Six months administration in patients not previously treated

The results are shown in Figure 2 and Table II. Negative conversion of sputum culture occurred in all patients. The time required to produce negative conversion was almost the same in the three regimens. No side effects were observed in the regimen including ofloxacin, whereas in the first regimen including streptomycin, the administration of streptomycin had to be stopped in 3 patients due to tinnitus.

DISCUSSION

In the chemotherapy of the treatment-failure patients, the administration of ofloxacin caused a significant reduction in culture positivity. The new agent ofloxacin is believed to be effective in the treatment of lung tuberculosis. Side effects of this agent were almost nil. The agent seems to be very safe. We used this drug for one year in 16 patients without marked side effects. The dosage used was 0.3 g daily, and this dose was given as a single dose in the morning

TABLE 1. COMPARISON OF CULTURE POSITIVITY BEFORE AND AFTER ADMINISTRATION OF OFLOXACIN IN SIXTEEN PATIENTS

Case No.	Culture positivity by monthly sputum examinations			
	Before administration (12 months)		After administration (12 months)	
1	7/12	58 %	2/12	17 %
2	11/12	92	0/12	0
3	7/12	58	3/12	25
4	10/12	83	9/12	75
5	10/12	83	3/12	25
6	12/12	100	12/12	100
7	12/12	100	10/12	83
8	9/12	75	6/12	50
9	11/12	92	10/12	83
10	12/12	100	9/12	75
11	12/12	100	11/12	92
12	12/12	100	11/12	92
13	12/12	100	10/12	83
14	12/12	100	12/12	100
15	12/12	100	12/12	100
16	5/12	42	0/12	0
Average		86.4 ± 18.8 (n= 16)		62.5 ± 36.9 (n= 16)

There is a statistically significant difference between the above two percentages by the t-test (P less than 0.05).

after breakfast. The above dosage was decided after taking into account the body weight of our patients (30 to 50 kg). Five patients received this agent for 2 years, and, in the second year, the dose was increased to 0.4 g daily. No side effects were observed in the patients who received ofloxacin for 2 years.

The therapeutic effect of ofloxacin seems to be inferior to that of rifampicin. The rate of negative conversion by single use of ofloxacin after one year of administration in treatment-failure patients with far-advanced disease was 19% (3/16) in the present study, while that of rifampicin in similar patients was 59% (19/35; 0.45 g daily administration) or 26% (6/23, 0.45 g per day twice weekly) (5).

TABLE 2. COMPARISON OF THE TIME REQUIRED FOR OBTAINING NEGATIVE CONVERSION OF SPUTUM CULTURES IN PATIENTS WHO RECEIVED THREE DIFFERENT REGIMENS OF ANTITUBERCULOSIS CHEMOTHERAPY

Regimen[a]	No. of patients[b]	Age in years[c]	Time in months required for causing negative conversion of sputum culture[c]	Ratio of patients in whom the cavity remained after 6 months of observation
OX+RFP+INH	17 (11M: 6F)	49.8 ± 19.6	1.76 ± 0.75	1/17 (6 %)
SM+RFP+INH	74 (55M:19F)	45.4 ± 15.9	1.75 ± 0.79	15/74 (20 %)
EB+RFP+INH	32 (26M: 6F)	51.0 ± 17.8	1.75 ± 0.79	3/32 (9 %)

a OX, Ofloxacin; RFP, Rifampicin; INH, Isoniazid; SM, Streptomycin; EB, Ethambutol.
b M, Male; F, Female.
c (Mean) $\pm$ (Standard deviation).

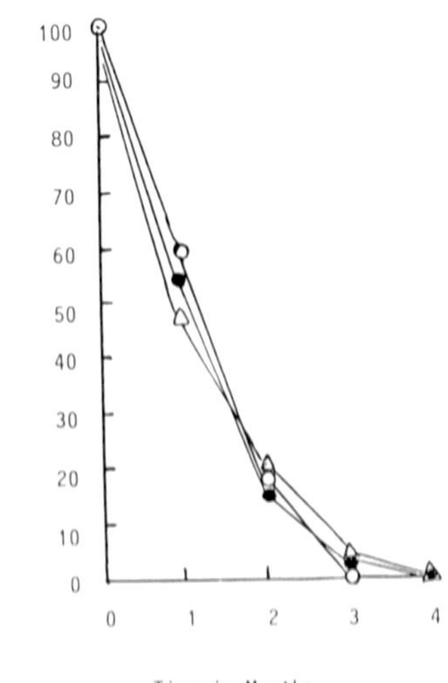

Fig. 2. Comparison of the rate of negative conversion of sputum cultures in three different chemotherapy regimens.
Open circles : ofloxacin + rifampicin + isoniazid
Closed circles : streptomycin + rifampicin + isoniazid
Open triangles : ethambutol + rifampicin + isoniazid

The chemotherapy of patients previously not treated by any antituberculosis agent with the regimen ofloxacin + rifampicin + isoniazid appeared as effective as two other triple regimens including streptomycin or ethambutol. All three regimens contained the two most effective agents, rifampicin and isoniazid, because at present it is not permitted to undertake a chemotherapy trial without these two effective agents. The inclusion of these two agents made it impossible to evaluate the effect of an additional agent (ofloxacin, streptomycin or ethambutol). However, the results of the present study which show that the three regimens are equally effective support that the regimen including ofloxacin may be used as the first choice in the treatment of lung tuberculosis, since streptomycin and ethambutol may affect hearing or visual acuity.

CONCLUSION

Ofloxacin is a moderately effective antituberculosis agent and may be used in the chemotherapy of tuberculosis. We have observed no marked side effects.

REFERENCES

1) Tsukamura, M. (1983). In vitro antimycobacterial activity of a new antibacterial substance DL8280. Differentiation between some species of mycobacteria and related organisms by the DL8280 susceptibility test. Microbiology and Immunology 27: 1129-1132.

2) Tsukamura, M. (1985). In vitro antituberculosis activity of a new antibacterial substance ofloxacin (DL 8280). American Review of Respiratory Disease 131: 348-351.

3) Tsukamura, M., Nakamura, E., Yoshii, S., and Amano, H. (1985). Therapeutic effect of a new antibacterial substance ofloxacin (DL 8280) on pulmonary tuberculosis. American Review of Respiratory Disease 131: 352-356.

4) Tsukamura, M., and Tsukamura, S. (1964). Differentiation of Mycobacterium tuberculosis and Mycobacterium bovis by p-nitrobenzoic acid susceptibility. Tubercle 45: 64-65.

5) Yokouchi, J., Tsukamura, M., Koike, K., Miwa, T. et al. (1973). Daily and intermittent chemotherapy with rifampicin in treatment-failure, far-advanced pulmonary tuberculosis. Japanese Journal of Chest Diseases 32: 117-122 (in Japanese).

DRUG DOSING IN ANTI-TUBERCULOUS CHEMOTHERAPY - LIMITS OF PRESENT PRACTICE AND POSSIBLE IMPROVEMENTS THROUGH THE USE OF A FIXED COMBINATION OF ORAL DRUGS

GIANNI ACOCELLA

Center of Reference for the Chemotherapy of Mycobacterial Infections. Institute of Phthysiology and Chest Diseases, University of Pavia, Pavia (Italy)

In modern 6-month treatment regimens for human tuberculosis four drugs are administered in the initial, intensive, 2-month phase of chemotherapy.

Of the four drugs, streptomycin (S) is usually administered at a dose of 0.75 or 1 g daily i.m.; isoniazid (H) at a daily dose of 300 mg or, in some cases, 400 mg to all patients; rifampicin (R) is administered at a daily dose of 450 mg in patients weighing <50 kg and 600 mg in those weighing >50 kg; pyrazinamide (Z) is administered at a daily dose of 1.5 g in patients weighing <50 kg, 2.0 g in patients of body weight ranging from >50 to 74 kg, and 2.5 g in patients weighing >75 kg.

The consequences in terms of complications in the administration procedure are evident, as different quantities of drugs are administered to the same patient while the amounts have to be changed as patients of varying body weight undergo treatment. The chances of errors in drug administration are therefore particularly high in countries, like the developing ones, where a very high number of cases are treated simultaneously in the same center. Errors in terms of adequacy of dosage, in the number and type of drugs to be administered and in the duration of drug administration are of particular relevance in this phase of antituberculous chemotherapy. As has been shown, this is the crucial part of the treatment because it is in this phase that the majority of sputum conversions occur, with all the obvious consequences for breaking the infectious cycle of the disease. On this basis, it is evident that any attempt at making the process of drug delivery easier and free from errors is fully justified, particularly if one takes into account the very high therapeutic potential of these regimens if correctly applied.

In this line of thinking, a fixed triple combination of H, R, and Z has been developed which presents a whole series of advantages with respect to the conventional, individual drug administration approach.

Firstly, the combination prevents monotherapy which is recognized to be probably the greatest risk in tuberculosis as it almost inevitably leads to selection of resistant mutants to the drug used and, in turn, the loss of value of the drug itself.

Secondly, the combination ensures that the drugs whose role and activity has been clearly defined in several controlled clinical trials are administered and not substituted for others of less known or unknown activity in the combined chemotherapy.

Thirdly, it prevents shortage of one or more drugs during antituberculous programs (a fact which can occur quite frequently in developing countries) as well as facilitating stockage of the drugs.

Fourthly, it prevents possible diversion of one or more drugs to indications different from tuberculosis.

In addition, as will be shown in the present paper, the combination has been developed in such a way as to ensure an almost perfect coincidence between the appropriate dosage in mg/kg body weight of each and all drugs present in the combination.

The rationale of the combination derives from the application of the concept that the dosage of a drug should be calculated on an mg/kg b.w. basis and not through the approximations made inevitable by the strength of the preparation available on the market. As far as the three oral antituberculous drugs are concerned, a large series of controlled clinical studies have indicated that the appropriate dosage of H is 5 mg/kg, that of R is 12 mg/kg and that of Z is 30 mg/kg b.w. with minor variations in excess or defect.

It was thought that a rule simple to remember by any health operator could have been that of administering one tablet of the fixed combination for every 10 kg of body weight of the patient.

In order to achieve a coincidence between the appropriate and the administered daily dosage it was enough to put into the tablet 10 times the required mg/kg dosage of each drug. The tablet contains, therefore, 50 mg of H, 120 mg of R, and 300 mg of Z. One can easily calculate that giving 3 such tablets to a patient weighing 30 kg will result in administering 150 mg of H, 360 mg of R, and 900 mg of Z which correspond exactly to the required amount of each and of the three drugs together. It is evident that such dosages cannot be made up utilizing the available preparations of H, R and Z on the market. Administering 5 such tablets to a patient weighing 50 kg will result in administering 250 mg of H, 600 mg of R and 1500 mg of Z which, again, corresponds exactly to the appropriate daily dosage of each and of the three drugs.

The exercise can obviously be repeated for any body weight class and the coincidence will always be perfect, since the whole concept of the triple fixed combination has been developed starting from the mg/kg requirements and not from a pre-established total daily dose.

An analysis of the relevant literature has in fact indicated that the application of regimens as stated at the beginning of this paper to patients of varying body weight inevitably leads to marked deviations of the administered versus the appropriate dosage.

Taking again the example of a patient weighing 30 kg, it can be easily calculated that giving 300 mg of H (and even more so if 400 mg are given) corresponds to administering 100% excess of H with respect to appropriate daily dose (300 mg versus 150 mg). Administering 450 mg of R corresponds to a 25% excess of R (450 versus 360). Administering 1500 mg of Z corresponds to a 67% excess of Z with respect to the appropriate dose based on 30 mg/kg (1500 mg versus 900 mg). Repeating this analysis (utilizing the dosage reported at the beginning of this paper) on patients of a whole range of body weights (from 30 to 80 kg) has indicated that the deviations are quite marked and include a not irrelevant underdosing in patients of body weight of 65 kg or more.

It is evident that the discrepancy between the dosage applied utilizing the individual drugs and the dosage appropriate in terms of mg/kg is now an unacceptable procedure in generic medical terms, can result in an increased number of side effects particularly when overdosing occurs and increases the cost of drug treatment by an amount corresponding, percentwise, to the excess of drug unnecessarily administered.

It must be observed that the administration of the triple combination formulation results in a reduction of the order of 50% in the amount of medication to be taken daily by the patients, a fact of some relevance in terms of compliance improvement.

Finally, the process of checking whether the patients have taken their medication can be made easier through the use of the triple combination since it contains rifampicin which makes the urine turn red. As rifampicin cannot be ingested alone in the combination, its presence in the urine is an indirect, although fairly reliable, way of making sure that all necessary drugs have been ingested by the patients.

Mycobacteria of Clinical Interest. M. Casal, editor

SHORT TERM TREATMENT (6 MONTHS) OF INITIAL PULMONARY TUBERCULOSIS. RESULTS AFTER FIVE YEARS

R. REY and F.J. GUERRA
Victoria Eugenia's Chest Diseases Hospital, Madrid (Spain)

This clinical study represents the last phase of progress in short-term therapy (1), and is based on the following premises:

- The effectiveness of this type of therapy even if the best form of chemotherapy and its ideal duration are not known exactly.
- The benefit derived by patients in being healed before they stop treatment, while making a cautious attempt to achieve the shortest possible treatment.
- The role of pyrazinamide.

MATERIAL AND METHODS

Patient selection

Admission to this trial was opened in October 1975 and was closed in September 1978. All patients can from the 'Victoria Eugenia' Chest Diseases Hospital in Madrid, and their ages ranged between 15 and 65 years.

The following conditions for inclusion in the study were: not having received previous treatment for over two weeks duration, lack of serious suspected or demonstrated hepatic disease, lack of any important ophthalmologic disease, renal failure, active gastric ulcer, pregnancy or gout. Also, patients who lacked a permanent address, were psychologically unstable or where thought unlikely to comply with the clinical visits, were not included in the study.

Therapeutic regimens

I. 6ERH. Rifampin 600 mg/day. Isoniazid 300 mg/day. Both taken in a single dose one hour before breakfast. Dosage was kept the same throughout treatment.
II.6ZRH. Rifampin and isoniazid at the same dosage as before. Pyrazinamide 35 mg/kg b.w./day in two takes. Treatment lasted six months in both series.

Examinations during treatment and follow-up

Bacteriological diagnosis was established through bacilloscopy and sample culture. Culture at this phase turned out negative in 9% of the patients in regimen I and in 7% of the patients in regimen II (Table II). During the first year, laboratory studies (microscopy and culture) of sputum were done every month; during the second and third years, these studies were done every three months, and once every six months during the last two years. During the treatment phase, hepatic function and uric acid levels and ophthalmologic evaluation were done monthly.

Bacteriological studies were systematically performed in our hospital's mycobacteria laboratory; cultures from a set number of our patients were sent to the Pasteur Institute in Paris, which kindly offered to serve us as control laboratory.

Of the 189 patients (Table I) originally selected for the study, 9 were withdrawn on account of defective selection. Another 11 patients were removed from the study at their own request, or for incorrectly following therapy or because they continued therapy longer than indicated.

Treatment was completed by 78 patients from the ERH group and 86 patients from the ZRH group.

Of all the patients who completed treatment, five (3%) suffered associated silicosis; 56% showed advanced extension of their disease in radiologic studies. The largest number of patients was found in the age group between 15 and 29 years: 72 patients (44%).

RESULTS

Culture evolution during the therapeutic phase (Table II) proved to be similar in both groups, there being no favorable difference toward pyrazinamide, as was expected.

Drug susceptibility studies performed on 105 patients showed initial resistance to isoniazid in five cases (5%) (Table III).

Mild intolerance to the drugs such as gastrointestinal problems, was observed in both groups but it disappeared with symptomatic treatment. Transaminase levels were temporarily increased in 16 patients (3ERH group and 16 ZRH group). Two patients treated with pyrazinamide had joint pain and another developed a skin rash that disappeared without having to stop medication.

Analysis of uric acid profiles showed a progressive increase in the levels during the therapeutic phase; these levels returned to normal once treatment ended.

Of the four patients withdrawn from regimen I, in three of them this was due to development of toxic hepatitis, without being able to determine the drug that caused it. A fourth patient presented retrobulbar optic neuritis during the fourth month of treatment. There was a patient in regimen II who developed a skin rash and had to be removed from the study. His condition improved when treatment was stopped.

Of the 164 patients who finished treatment, only 139 (85%) completed the five-year check-ups (Table IV). Five patients died in the follow-up phase, due to causes other than tuberculosis. Of the 20 patients who could not be followed after treatment, over half of them were lost during the first year of follow-up.

Five patients showed late relapses (Table V), four (5.8%) were on the ethambutol regimen and 1 (1.4%) on the pyrazinamide regimen. Relapses appeared

TABLE I

PATIENTS STUDIED

	6ERH	6ZRH	TOTAL Nº	TOTAL (%)
Nº patients included	94	95	189	(100)
Patients excluded due to defficient selection	5	4	9	(4.8)
Patients excluded during treatment			11	(5.8)
. Incorrect or prolonged treatment	5	1		
. Voluntary withdrawal	2	3		
. Deaths	0	0		
. Severe intolerance			5	(2.6)
Hepatitis	3	0		
Optic nerve neuritis	1	0		
Skin rash	0	1		
Nº of patients that finished treatment	78	86	164	(86,8)

TABLE II

MONTHLY NEGATIVATION OF CULTURE

Month	Regimen	Nº cases	NEGATIVE CULTURES Nº	NEGATIVE CULTURES (%)
0	6ERH	88	8	(9)
	6ZRH	86	6	(7)
1	6ERH	73	28	(38)
	6ZRH	77	36	(47)
2	6ERH	72	56	(78)
	6ZRH	71	54	(76)
3	6ERH	67	61	(91)
	6ZRH	74	71	(96)

TABLE III

DRUG SUSCEPTIBILITY TESTS

Pre-treatment susceptibility	Regimen	Nº of cases analyzed	Relapses
INH/susceptible	6ERH	51	3
	6ZRH	49	1
INH/resistant	6ERH	3	1
	6ZRH	2	0
	Total	105	5

61 cases studied by Dr. Grosset (Pasteur Institute).
44 cases studied by Drs. Ortega-Calderón and Martínez-Larriba (Victoria Eugenia, Hospital).

TABLE IV

PATIENS FOLLOW UP

Regimen	Nº of patients finishing treatment	Deceased patients non-active TP	FOLLOW UP MONTHS 7-18	19-30	31-48	49-66	Nº of patients finishing follow up
I)6ERH	78	2	4	3	1	0	68 (87)
II)6ZRH	86	3	7	5	0	0	71 (83)
	164	5	11	8	1	0	139 (85)

TABLE V

RELAPSES AT FIVE YEARS

Regimen	Nº patients finishing follow up	MONTH OF RELAPSE APPEARANCE 7-18	19-30	31-48	49-66	T O T A L Nº	(%)
I)6ERH	68	0	0	4	0	4	(5.8)
II)6ZRH	71	0	0	1	0	1	(1.4)

3 years after treatment was stopped.

One of the patients in the ERH group who relapsed also showed initial resistance to isoniazid.

It is concluded that a regimen including pyrazinamide is significantly more efficient than one including ethambutol, although it does cause more temporary intolerance.

REFERENCE

1. Guerra FJ, Rey R (1977) A clinical approach to the short-term of pulmonary tuberculosis. Excerpta Medica International Congress Series No. 437, Jun 19.

MYCOBACTERIOSIS

Mycobacteria of Clinical Interest. M. Casal, editor

DIAGNOSIS OF NON-TUBERCULOUS MYCOBACTERIOSIS

MICHIO TSUKAMURA
Respiratory Disease Department, National Chubu Hospital, Obu, Aichi 474 (Japan)

INTRODUCTION

Diagnostic criteria for non-tuberculous lung mycobacteriosis are useful for standardization in cooperative studies or for a general consensus between clinicians world-wide. Such criteria were proposed by Yamamoto et al (17) in 1967, by Tsukamura (7) in 1978, and by Ahn et al (1) in 1982. The first and third criteria were published in English and are therefore well known to many investigators. In contrast, Tsukamura's criteria were published in Japanese and are not known in other countries, though they are used very widely in Japan as diagnostic criteria for non-tuberculous lung mycobacteriosis in the Japanese National Chest Hospitals (see Table 7). The criteria were based on observations of the mode of excretion of non-tuberculous mycobacteria in tuberculosis patients and in patients with non-tuberculous lung mycobacteriosis (13).

According to the criteria of Yamamoto et al (17), the isolation of non-tuberculous mycobacteria more than 4 times and the formation each time of more than 100 colonies on isolation medium are required. The criteria do not indicate the period in which the mycobacteria are isolated 4 or more times. At present, the criteria are considered to be too strict, and it is possible to overlook 'transient infection' (6,12). So as to correct such faults, new criteria have been proposed by Tsukamura (7). These were based on observations of patients with non-tuberculous lung mycobacteriosis and of tuberculosis patients. The present paper presents the basis of Tsukamura's criteria.

MATERIAL AND METHODS

Patients hospitalized in tuberculosis departments of the National Chubu Hospital in the period of 1972 to 1976 (5 years) were subjects of the present observations. Sputum specimens from the patients were examined monthly, but on admission were examined daily for 3 to 7 days.

A morning-sputum specimen was added to an equal amount of a 4% NaOH solution and homogenized by shaking mechanically at room temperature for 15 minutes. Two loopful-samples of the sputum specimen were inoculated by a spiral loop that can deliver a 0.02 ml sample by each inoculation (15) onto Tween egg medium (14). In this medium, a neutralization process is not required since it occurs in the medium itself. The inoculated tubes were stoppered by a gum cap with a 3 mm-cut in its base and incubated at 37°C for 8 weeks. The growth of acid-fast bacteria was observed after incubation for 4 and 8 weeks.

The Tween egg medium (14) gave a slightly higher rate of positive cultures than the Ogawa egg medium, which is used widely in Japan. The composition of the Tween egg medium is as follows: basal solution (1% KH_2PO_4, 1% sodium glutamate, and 0.1% $MgSO_4.7H_2O$), 100 ml; whole eggs, 200 ml; Tween 80, 2 ml; 2% aqueous solution of malachite green, 6 ml.

The method of inoculating a non-neutralized sputum specimen with a spiral loop gave a significantly higher rate of positive cultures than inoculation of 0.1 ml of a neutralized sputum specimen by pipette onto the Löwenstein-Jensen medium (15). The reason for the superiority of spiral loop inoculation is probably as follows: (a) streaking of samples by a spiral loop on to the medium surface is effective for obtaining good growth of mycobacteria; (b) neutralization of samples occurs soon on the medium, as the amount of inoculate is small; (c) inoculation of a sample with a pipette causes part of the sample to adhere to the tube wall.

Growing colonies on the Tween egg medium were examined microscopically by the Ziehl-Neelsen method. Strongly acid-fast bacteria without mycelium were subjected to further species identification. The mycobacteria were inoculated on to Ogawa egg medium containing no agent and containing 0.5 mg of p-nitrobenzoic acid per ml (PNB medium) (16). After incubation at 37°C for 3 weeks, the mycobacteria which grew on the PNB medium were considered as non-tuberculous mycobacteria and identified according to the schedule described previously (5).

The PNB medium was prepared as follows: p-nitrobenzoic acid was dissolved in propylene glycol at a concentration of 25 mg/ml. Two volumes of this solution were added to 100 volumes of the Ogawa egg medium before sterilization. The medium was then poured in 7 ml quantities into tubes, 165 by 16.5 mm, and made as slopes by sterilization at 90°C for 60 minutes. Mycobacteria other than *M. tuberculosis, M. bovis, M. africanum* and *M. microti* are able to grow on this medium, whereas tubercle bacilli are not.

RESULTS AND DISCUSSION

Prevalence of non-tuberculous mycobacteria in sputa of tuberculosis patients. The prevalence of non-tuberculous mycobacteria and rhodococci in sputa of all patients in the tuberculosis departments was 1.5% by monthly sputum examinations. The percentage of non-tuberculous mycobacteria in all mycobacteria (*M. tuberculosis* plus non-tuberculous mycobacteria) was 9.6%. The above non-tuberculous mycobacteria contain not only mycobacteria isolated casually from tuberculosis patients but also mycobacteria from patients with lung disease due to non-tuberculous mycobacteria. The prevalence of non-tuberculous mycobacteria isolated from tuberculosis patients was 0.8% (Table 1). Therefore, the casual isolation-rate of non-tuberculosis mycobacteria from tuberculosis patients is 0.008 = 0.01 per one sputum examination. The finding that the isolates belong to the casual ones was confirmed in the follow-up of the patients.

The above casual isolation-rate of non-tuberculous mycobacteria was that obtained by monthly sputum examination. In our hospital, daily sputum examinations were carried out, as a rule, in newly admitted patients. During the observation period of 5 years, a total of 1636 new patients were admitted. Of these, 21 patients excreted every day non-tuberculous mycobacteria according to daily sputum examinations which were carried out after admission. Excluding these 21 patients, the remaining 1615 patients had a total of 78,528 daily sputum examinations and a total of 486 strains of non-tuberculous mycobacteria and rhodococci were isolated. The casual isolation-rate is therefore 0.006 by daily sputum examination. This value is not so very different from the rate obtained by monthly sputum examination. We may consider the background isolation-rate is ca. 0.01 (1%).

TABLE 1

PREVALENCE OF NON-TUBERCULOUS MYCOBACTERIA IN PATIENTS HOSPITALIZED IN TUBERCULOSIS DEPARTMENTS OF THE NATIONAL CHUBU HOSPITAL IN THE PERIOD 1972-1976 (5 YEARS)

	All hospitalized patients in tuberculosis departments	Excluding patients with lung disease due to non-tuberculous mycobacteria[a]
No. of sputum specimens examined monthly : A	36,243	36,008
No. of mycobacterial strains isolated, including *M. tuberculosis*: B	5,600	5,365
No. of non-tuberculous mycobacterial strains: C	569	304
Ratio (C/B) x 100%	9.6	5.7
Ratio (C/A) x 100%	1.5	0.8

a - Diagnostic criteria tentatively used are as follows: (a) Isolation on 3 or more occasions of non-tuberculous mycobacteria belonging to the same species by monthly sputum examinations and at least one culture should have more than 100 colonies on the isolation medium; (b) co-appearance of clinical symptoms, including the presence of cavities.

What is the unusual prevalence of non-tuberculous mycobacteria?

The unusual prevalence of non-tuberculous mycobacteria is considered as that of non-tuberculous mycobacteria higher than the prevalence of casual isolates of non-tuberculous mycobacteria. It is known that tuberculosis patients who have a cavity (cavities) or bronchiectasis excrete non-tuberculous mycobacteria more frequently than normal persons (2). A prevalence of non-tuberculous mycobacteria

Table 2. Abnormal (unusual) prevalence rate of non-tuberculous mycobacteria by monthly or daily sputum examinations

	No. of positive cultures of non-tuberculous mycobacteria	No. of sputum specimens examined
Unusual prevalence rate	2	2
of non-tuberculous	2 or more	3
mycobacteria[a]	2 or more	4
	2 or more	5
	2 or more	6
	2 or more	7

[a]These rates are significantly higher than the casual isolation-rate of non-tuberculous mycobacteria in tuberculosis patients (0.006-0.015 per one sputum examination) by the χ^2-test ($P<0.05$).

Table 3. Mode of casual excretion of non-tuberculous mycobacteria from tuberculosis patients

No. of isolation	No. of patients	No. of isolates
Once	222	1 x 222 = 222
Twice	18	2 x 18 = 36
Three times	14	3 x 14 = 42
Four times	1	4 x 1 = 4
Total	255	304

higher than that of tuberculosis patients is, therefore, regarded as unusual. Prevalence higher than the background-isolation rate (the isolation rate of non-tuberculous mycobacteria in tuberculosis patients) has been calculated (Table 2). These are 'unusual' prevalence values for non-tuberculous mycobacteria. Unusual prevalence values indicate merely the presence of an unusual state and do not always indicate the presence of disease due to non-tuberculous mycobacteria.

Diagnostic criteria for non-tuberculous lung mycobacteriosis

From long-term follow-up of non-tuberculous mycobacteriosis, Tsukamura (2-4, 10) classified the disease into two types: primary infection type and secondary infection type. Primary infection means infection by non-tuberculous mycobacteria of 'healthy' lungs and shows fresh cavitary or caseo-infiltrative lesions without a tendency to fibrosis on roentgenography. Secondary infection means infection by non-tuberculous mycobacteria of tuberculosis cavities, from which tubercle bacilli have usually disappeared, and usually shows a cavity with a sclerotic wall or cavities in sclerotic lesions.

I. Primary infection type

In this case a fresh cavitary lesion is observed in the roentgenographic picture. Evidence of unusual excretion of non-tuberculous mycobacteria is sufficient to diagnose the disease. Simultaneous occurrence of unusual excretion of non-tuberculous mycobacteria and a fresh cavitary lesion in the same patient indicates the existence of disease. These two events are rare. If they occur independently, it is improbable that they will occur simultaneously in the same patient.

The criteria for diagnosing primary infection type disease are as follows:

a) presence of a fresh cavitary (or caseo-infiltrative) lesion;

b) occurrence of non-tuberculous mycobacteria belonging to the same species in sputum on two or more occasions on 2 to 7 daily sputum examinations or 2 or 3 monthly sputum examinations.

II. Secondary infection type

Here non-tuberculous mycobacteria are seen in sputa of patients who have a cavity with sclerotic walls or cavities in sclerotic lesions. Cavitary lesions already exist. Therefore, the presence of a cavity does not serve for diagnosis. The diagnosis must be made only from the bacteriological findings.

On the basis of such criteria, the mode of excretion of non-tuberculous mycobacteria was observed in tuberculosis patients for 5 years. During the observation period, non-tuberculous mycobacteria were isolated from a total of 255 tuberculosis patients. These isolations were not accompanied by any change in the clinical symptoms. They were therefore considered as casual. The mode of isolation is shown in Table 3. Eighteen patients excreted non-tuberculous mycobacteria on two occasions. Of these 18, 12 excreted the same organism twice (Table 4). As shown in Table 4, several patients excreted the same organism twice within 3 months.

In 15 patients, non-tuberculous mycobacteria were seen on 3 or more occasions. The mode of excretion is shown in Table 5. One patient excreted *M. avium*-*M. intracellulare* 3 times in 16 months. In all 3 cultures of this case, the number of colonies on isolation medium remained less than 22. In another two cases, three cultures of *M. gordonae* and of *Rhodococcus bronchialis* were isolated in periods of 5 and 3 months, respectively. In these cases, the number of colonies did not exceed 10 (Table 5). In a further 28 patients, *M. avium*-*M. intracellulare* were excreted 3 to 60 times according to monthly sputum examination (Table 6). In these patients, the appearance of non-tuberculous mycobacteria accompanied some change in the clinical symptoms (increase in cough and sputum, fever, deterioration of roentgenographic features, etc.). Mycobacteria belonging to the same

TABLE 4. PATIENTS FROM WHOM NON-TUBERCULOUS MYCOBACTERIA BELONGING TO THE SAME SPECIES WERE ISOLATED TWICE BUT WHO WERE NOT CONSIDERED AS HAVING LUNG DISEASE DUE TO THESE ORGANISMS AND WHO DID NOT HAVE ANY SIGNIFICANT CHANGE IN CLINICAL SYMPTOMS

Case No.[a]	Species	Number of colonies[b]		Interval between two isolations
1	*Rhodococcus bronchialis*	25	15	2 months
2	*M.avium-M.intracellulare*	+	+	2 months
3	*R.bronchialis*	+	15	1 month
4	*Rhodococcus sp.*	1	1	2 months
5	*M.fortuitum*	10	6	2 months
6	*R.bronchialis*	2	2	1 month
7	*M.avium-M.intracellulare*	25	10	2 months
8	*M.avium-M.intracellulare*	+	5	1 month
9	*M.fortuitum*	25	5	1 month
10	*M.gordonae*	1	6	28 months
11	*R.rubropertinctus*	1	1	15 months
12	*M.avium-M.intracellulare*	2	1	4 months

a - Another 6 patients excreted two isolates belonging to different species

b - The symbol + indicates more than 100 colonies, and the arabic numbers show the actual number of colonies on isolation medium

Remark. All patients in whom non-tuberculous mycobacteria were isolated twice had a cavity (cavities) in a sclerotic lesion

species were isolated at least 3 times within 6 months, according to monthly sputum examinations. The number of colonies on isolation medium was more than 100 at least in one or more cultures. Only two patients did not show excretion of more than 100 colonies, but did excrete the organism every month for 8 months. The 28 patients who had a change in their clinical symptoms were considered as having lung disease due to non-tuberculous mycobacteria, i.e. *M. avium-M. intracellulare*. In a search for positive conditions in patients who have disease due to non-tuberculous mycobacteria and for negative conditions in tuberculosis patients, we obtained the following criteria for the diagnosis of secondary infection type disease:

1) excretion of 3 or more isolates belonging to the same species (*M. avium-M. intracellulare*) found at monthly sputum examination during an observation period of 3 to 6 months, and evidence of at least one culture showing more than 100 colonies on isolation medium;

2) deterioration of clinical symptoms which appear in parallel with the isolation of mycobacteria;

TABLE 5. PATIENTS IN WHOM THREE OR FOUR CULTURES OF NON-TUBERCULOUS MYCOBACTERIA WERE FOUND ON MONTHLY SPUTUM EXAMINATIONS MADE OVER 5 YEARS AND WHO WERE NOT CONSIDERED AS HAVING LUNG DISEASE DUE TO THESE ORGANISMS, AND WHO DID NOT SHOW ANY SIGNIFICANT CHANGES IN CLINICAL SYMPTOMS

Case	Species and number of colonies that grew on isolation medium[a]				Period in which bacteria were isolated 3 or 4 times
1	Rb (1)		G (1)	G (1)	19 months
2	R (1)		R (1)	R (3)	22 months
3	I (+)		I (+)	F (1)	14 months
4	I (2)		I (4)	Rt (1)	35 months
5	I (+)		I (2)	I (1)	41 months
6	I (22)		I (20)	I (5)	16 months
7	R (1)		R (10)	I (4)	35 months
8	R (3)		G (2)	Rr (2)	12 months
9	G (1)		G (10)	G (1)	5 months
10	Rb (2)		F (10)	I (12)	8 months
11	I (3)		R (6)	R (10)	11 months
12	G (2)	R (8)	G (5)	R (10)	10 months
13	F (1)		Rb (1)	Rb (1)	5 months
14	Rb (2)		Rb (1)	Rb (1)	3 months
15	R (2)		Rr (2)	R (4)	10 months

a - Rb, *Rhodococcus bronchialis*; G, *M. gordonae*; R, *Rhodococcus sp.*; I, *M. avium-M. intracellulare* complex; F, *M. fortuitum*; Rt, *Rhodococcus terrae*; Rr, *Rhodococcus rubropertinctus*. The number in parentheses indicates the number of colonies. Symbol + indicates more than 100 colonies.

Remark. All patients had cavities in a sclerotic lesion

3) presence of cavitary lesions.

Secondary infection is usually observed in disease due to *M. avium-M. intracellulare* (2-4,8,9-11).

In the presence of a fresh cavitary lesion without a sclerotic process, only the evidence of the positive culture of non-tuberculous mycobacteria belonging to the same species is important; the number of colonies on isolation medium is not important for diagnosing the disease. Previously, we reported that a number of patients showed positive cultures of non-tuberculous mycobacteria and soon showed negative conversion of sputum cultures and disappearance of cavities within 3 to 6 months. These cases we called 'transient infection' (6,12). To find these, 3 to 7 daily sputum examinations at admission of patients are more important than monthly sputum examinations. However, evidence of two or more positive cultures on three monthly sputum examinations in early stages of admission certainly

TABLE 6. MODE OF EXCRETION OF *M. AVIUM-M. INTRACELLULARE* COMPLEX IN PATIENTS WHO SHOWED THREE OR MORE POSITIVE CULTURES WITHIN 6 MONTHS AT MONTHLY AND/OR DAILY SPUTUM EXAMINATIONS

Case	Age and sex	Number of isolations					
		Number of colonies on isolation medium					Total number of isolations
		1-10	11-50	51-100	more than 100	partially confluent	
1	42 M	3			2		5
2	80 F	4			4		8
3	63 F	2			3		5
4	81 F	3	1		1	1	6
5	54 M	10			1		11
6	61 F	6	2				8
7	82 F	8					8
8	80 F	5	1		6		12
9	51 M	43	17		27		87
10	49 M	5	3		14	1	23
11	42 M	4	2		8		14
12	69 F	13	4		6		23
13	63 F	7	2		22		31
14	41 F	8	6		53		67
15	73 M	3	1		4		8
16	62 M	2	3		3		8
17	77 M		1		2		3
18	59 M	85	66	1	245		397
19	32 M	6	2		33		41
20	71 M				17	2	19
21	61 M	5	4		3		12
22	67 M	7	1		2		10
23	60 M	2			6		8
24	76 F	2	2		21		25
25	82 F	3	3		3		9
26	50 M		1		3		4
27	63 M	2			1		3
28	39 F	24	21		239	3	287
Total		262 22.9%	143 12.5%	1 0.1%	729 63.8%	7 0.6%	1142 100.0%

The observation period differed in each patient and ranged from 3 months to 5 years. The data in this Table include both monthly and daily sputum examinations. All patients showed cavities, and the appearance of non-tuberculous mycobacteria paralleled deterioration in clinical symptoms (enlargement of cavities, increase in cough and sputum, and fever). Patients other than case Nos. 7, 8, 10-13, 18, 21, 24 and 25 showed a new cavity (cavities) with appearance of non-tuberculous mycobacteria.

supports the presence of disease. In the presence of old cavitary lesions (cavities in a sclerotic lesion or a cavity with a sclerotic wall), the number of colonies is also important for diagnosis.

Diagnostic criteria for lung mycobacteriosis of the primary infection type may be applied in all cases of non-tuberculous lung mycobacteriosis. On the other

TABLE 7. DIAGNOSTIC CRITERIA OF NON-TUBERCULOUS MYCOBACTERIOSIS OF THE JAPANESE NATIONAL CHEST HOSPITALS (SECOND EDITION)

I. In the roentgenographic feature, a fresh cavity or caseous lesion has newly appeared.

a) Two or more positive cultures belonging to the same species of pathogenic mycobacteria should be shown by three daily sputum examinations, which are made within one month, or

b) two or more of positive cultures belonging to the same species of pathogenic mycobacteria should be shown by monthly sputum examinations within 3 months.

New appearance of the cavitary lesion and appearance of the mycobacteria should occur simultaneously. The amount of colonies on isolation medium may be not more than 100.

II. In the roentgenographic feature, a cavity (cavities) with sclerotic wall or in sclerotic lesion (or bronchiectasis which can be considered as the source of excretion of non-tuberculous mycobacteria) already exists.

Three or more positive cultures belonging to the same species of pathogenic mycobacteria should be shown by monthly sputum examinations, which are made within 6 months, and one or more of the above cultures should be 100 or more colonies on isolation medium. The above excretion of non-tuberculous mycobacteria should be accompanied by change of clinical symptoms (change in the roentgenographic feature, fever, hemoptysis, increase of cough and sputum, etc.).

The following mycobacteria are considered as pathogenic non-tuberculous mycobacteria: *M.kansasii*, *M.szulgai*, *M.scrofulaceum*, *M.xenopi*, *M.avium-M.intracellulare*, *M.simiae*, *M.shimoidei*, *M.nonchromogenicum*, *M.fortuitum* and *M.chelonae*.

The Mycobacteriosis Research Group of the Japanese National Chest Hospitals (chairman: M. Tsukamura). 1985. Diagnostic criteria of non-tuberculous lung mycobacteriosis. Kekkaku 60:51.

hand, the diagnostic criteria for non-tuberculous mycobacteriosis of the secondary infection type have been made, based on observations of the disease due to *M.avium-M.intracellulare*. However, here we have used these criteria for mycobacteriosis due to other mycobacteria (Table 7). The criteria shown in Table 7 are those which are used currently by the Mycobacteriosis Research Group of the Japanese National Chest Hospitals as well as by most investigators in Japan.

SUMMARY

Diagnostic criteria for lung disease due to non-tuberculous mycobacteria are shown based on observations of the mode of excretion of non-tuberculous mycobacteria from tuberculosis patients and from patients considered as having lung disease due to these organisms.

In the presence of a fresh cavitary (or caseo-infiltrative) lesion, the isolation of two or more positive cultures belonging to the same species at 3 to 7 daily or 3 monthly sputum examinations is sufficient to demonstrate the presence of lung disease due to these organisms. In the presence of a cavity with a sclerotic wall or cavities in a sclerotic lesion, the isolation of three or more cultures belonging to the same species at monthly sputum examinations carried out within 6 months and the demonstration of one or more cultures showing more than 100 colonies on isolation medium are required to indicate the presence of infection.

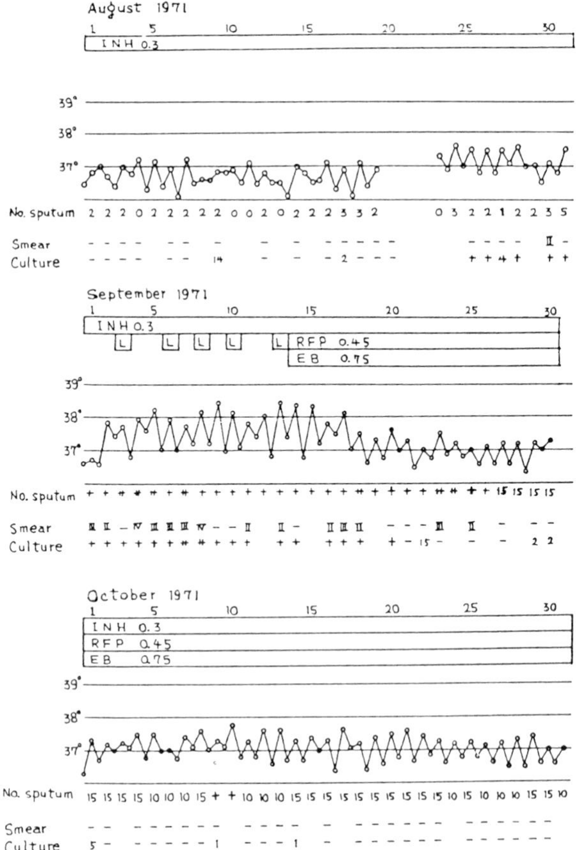

Fig. 1. The mode of excretion of *Mycobacterium intracellulare* in sputum of a female patient, 66 years old.
RFP, rifampicin; INH, isoniazid; EB, ethambutol; L, lividomycin, 1 g per day, intramuscularly.

REFERENCES

1. Ahn, C.H., McLarty, J.W., Ahn, S.S., Ahn, S.I., and Hurst, G.A. 1982. Diagnostic criteria for pulmonary disease caused by Mycobacterium kansasii and Mycobacterium intracellulare. American Review of Respiratory Disease 125: 388-392.
2. Tsukamura, M. 1973. Background factors for casual isolation of Mycobacterium intracellulare from sputa of patients with tuberculosis. American Review of Respiratory Disease 108: 679-683.
3. Tsukamura, M. 1973. A clinical study of pulmonary infection due to Mycobacterium intracellulare. Japanese Journal of Chest Diseases 32: 23-36 (in Japanese).
4. Tsukamura, M. 1975. Roentgenographic features of lung disease due to Mycobacterium intracellulare (primary and secondary infection). Kekkaku 50: 17-30 (in Japanese with English summary and tables and figures).
5. Tsukamura, M. 1975. Identification of mycobacteria. p. 1-75, The National Chubu Hospital, Obu, Aichi, Japan 474.
6. Tsukamura, M. 1976. Infection of the lungs due to Mycobacterium intracellulare improved in a short period. Kekkaku 51: 35-39 (in Japanese with English summary, tables and figues).
7. Tsukamura, M. 1978. A trial of standarization of diagnosing lung disease due to mycobacteria other than tubercle bacilli. Kekkaku 53: 367-376 (in Japanese with English summary, tables and figures).
8. Tsukamura, M. 1979. Secondary infection of Mycobacterium avium-intracellulare complex to open-negative cavities of lung tuberculosis. Kekkaku 54: 71-74 (in Japanese with English summary, tables and figures).
9. Tsukamura, M. 1980. Mode of excretion of Mycobacterium intracellulare at the onset of lung infection. Diagnosis of secondary infection of tuberculosis cavity. Kekkaku 55: 215-218 (in Japanese with English summary and figures).
10. Tsukamura, M. 1981. Characteristics of the x-ray feature of lung disease due to Mycobacterium avium -Mycobacterium intracellulare complex. Kekkaku 56: 22-33 (In Japanese with English summary, tables and figures).
11. Tsukamura, M. 1981. A comparative study of the fate of cavities between lung tuberculosis and lung disease due to Mycobacterium avium-Mycobacterium intracellulare complex. Kekkaku 56: 361-368 (in Japanese with English summary, tables and figures).
12. Tsukamura, M., Kita, N., Shimoide, H., and Kawakami, K. 1981. "Transient infection" of the lungs due to Mycobacterium avium- Mycobacterium intracellulare complex. Kekkaku 56: 309-317 (in Japanese with English summary and tables).

13. Tsukamura, M., Mizuno, S., and MUrata, H. 1978. Frequency of isolation and kind of species of acid-fast organisms isolated from patients hospitalized in the National Chubu Hospital in 1972 to 1976 (five years). Kekkaku 53: 307-312 (in Japanese with English summary and tables).

14. Tsukamura, M., Toyama, H., and Fukaya, Y. 1979. "Tween egg medium" for isolating mycobacteria from sputum specimens. Microbiology and Immunology 23: 833-838.

15. Tsukamura, M., Toyama, H., and Fukaya, Y. 1979. Use of spiral loop-inoculation for isolating mycobacteria from sputum specimens. Iryo 33: 509-513 (in Japanese with English summary and tables).

16. Tsukamura, M., and Tsukamura, S. 1964. Differentiation of _Mycobacterium tuberculosis_ and _Mycobacterium bovis_ by p-nitrobenzoic acid susceptibility. Tubercle 45: 64-65.

17. Yamamoto, M., Ogura, Y., Sudo, K., and Hibino, S. 1967. Diagnostic criteria for disease caused by atypical mycobacteria. American Review of Respiratory Disease 96: 773-778.

APPENDIX

A CASE REPORT OF SECONDARY INFECTION DUE TO MYCOBACTERIUM INTRACELLULARE

A 63-year-old female suffered from lung tuberculosis since January1968 and received antituberculosis agents (the regimen used is unkown). The treatment was stopped in November 1969. Since January 1970she complained increase of cough and sputum, and, on the 31th March, she was admitted to this hospital. On admission, she had a large cavity in sclerotic lesion in the left upper lobe. By daily sputum examination in April 1970, three cultures of acid-fast organism were isolated and all three were identified as M.tuberculosis. Tubercle bacilli disappeared by administration of isoniazid plus ethambutol. In August 1970, however, an acid-fast organism was isolated and this was identified as M.intracellulare. At this time, the cavity still remained almost unchanged, and she was included to a subject of the daily sputum examinations for 6 months. The study was begun from the 1st June 1971. During the first two months, no acid-fast organism was isolated from this patient. In August, she went to home and stayed there for four days. After returning to hospital, she had slight fever. Before this onset of clinical manifestation, two cultures of M.intracellulare were isolated on the 9th and 17th August. With appearance of fever, M.intracellulare (serotype Boone) began to appear in sputum every day. After administration of rifampicin plus ethambutol plus isoniazid, the amount of sputum was reduced and the fever decreased. The mode of excretion of M.intracellulare is shown in Fig.1. In this case, isolation of M.intracellulare was in accordance with appearance of clinical symptoms, fever and cough and sputum. Most cultures showed 100 or more colonies on isolation medium.The case is considered as secondary infection of M.intracellulare to tuberculosis cavity, from which tubercle bacilli disappeared. The observation of the case shows that the causative organism has appeared in sputum two weeks earlier than the appearance of clinical symptoms, and the amount of the organism increases with the manifestation of clinical symptoms.

Mycobacteria of Clinical Interest. M. Casal, editor

CHROMATOGRAPHIC ANALYSIS IN MYCOBACTERIAL TAXONOMY AND IDENTIFICATION

FRANCISCO MARTIN-LUENGO and PEDRO L. VALERO-GUILLEN
Department of Microbiology, Faculty of Medicine, University of Murcia, Murcia (Spain)

INTRODUCTION

Mycobacteria are rich in lipids. These compounds, which are found in the cell wall and plasmatic membrane and which have very diverse structure, play a part in the biological and biophysical properties of the cell envelope.

Lipids of mycobacteria were initially examined from a chemical viewpoint by Anderson (14) and later studied in further detail by several authors (1,14). Present information on lipids of these bacteria shows a complex composition, most compounds being specific (1). The differential lipid pattern between mycobacteria and related organisms make lipid analysis a powerful criterion in the taxonomy and diagnosis of the genus *Mycobacterium*; some important constituents can, moreover, be easily analysed by chromatographic procedures.

Thin-layer chromatography (TLC) was first applied by Marks and Szulga (13) in a study on glycolipids as key to the diagnosis of mycobacteria. The possibilities of this technique have increased in several laboratories for a variety of compounds (6) and in conjunction with gas-liquid chromatography (GLC) (4) and high performance liquid chromatography (HPLC) (3,9) enable an isolate of *Mycobacterium* to be characterized by chemical criteria alone. Some of these analyses are, however, difficult to carry out in a normal mycobacterial laboratory, being both time consuming and highly sophisticated. This review will deal only with studies of mycolic acids, fatty acids and alcohols that can be easily examined by TLC and GLC.

MYCOLIC ACIDS

Mycolic acids are defined as alpha-branched beta-hydroxylated long-chain fatty acids. Their general structure is

$$R_1\text{-CHOH-}\underset{\displaystyle R_2}{\underset{|}{\text{CH}}}\text{-COOH}$$

where R_1 and R_2 are aliphatic chains. The length of the R_1 chain is ca 40-70 carbon atoms and that of R_2 20-24 C. Different mycolic acids related to the presence of several structural variations in R_1 have been described in *Mycobacterium* (4,6), and are summarized in Table 1. Mycolic acids play a structural role in the cell wall, being linked to arabino-galactan and trehalose, and probably contribute to maintaining the fluidity of the cell envelope (14).

Mycolic acid structure has been established by means of spectroscopic, mass-spectrometry and chromatographic techniques, and detailed procedures can be found in a variety of references (1,14); a method based on GLC-mass spectrometry has, moreover, been reported as successful (22), but these techniques are sophisticated. The whole mycolic acid pattern of mycobacteria can, nevertheless, be elucidated by TLC (4,6,17), the results being, in general, consistent and reproducible.

TABLE I

STRUCTURAL CHARACTERISTICS OF MYCOLIC ACIDS OF MYCOBACTERIA.

Mycolic acid type	Cyclopropane rings	Double bonds	Oxygen functions	Number of carbon atoms
alfa*,I**	0,1,2	0,1,2***		65-90
alfa',II		1		58-64
methoxy,III	0,1	1	$-OCH_3$	65-90
keto,IV	0,1	1	-C=O	65-90
epoxy,V	0,1	1	-C–O–C- (epoxide ring)	65-80
carboxy,VI	0,1	1	-COOH	65-80

*According to Minnikin 1982. **According to Daffé et al. 1983.

***In *M. fallax* tri- and tetra-unsaturated have been described (19).

Two basic procedures employing TLC are at present being applied in analysis of the mycolic acid pattern of mycobacteria (4,6). Minnikin and associates (15) used mono- and bidimensional TLC methods after acid hydrolysis of bacteria; alkaline hydrolysis followed by esterification of liberated acids is, however, a more convenient procedure (6), as previously discussed by Daffé et al (4) who follow a combination of single dimensional TLCs. Solvents employed by Minnikin et al (15) are basically petroleum ether-acetone (95:5 v/v)x3 in the first direction and toluene-acetone (97:3 v/v)x1 in the second direction, while Daffé et al (4) use dichlomethane (x1) and petroleum ether-ether (90:10 v/v)x4. Some examples of separation of mycolic acids by some of these procedures are shown in Figure 1.

When TLC is applied in the elucidation of the mycolic acid pattern of mycobacteria some problems arise in the interpretation of the results, the inclusion of some well-established patterns in all analyses thereby being necessary. At the same time, alpha'-mycolates and methoxy-mycolates, as also epoxy-mycolates and keto-mycolates (Fig. 1A) show a similar chromatographic behaviour in the systems of Minnikin et al (6,15); this problem can be resolved by chromatographing the extracts in dichloromethane, but in this solvent methoxy and keto-mycolates co-chromatograph together (Fig. 1B), being, however, separated in petroleum ether-acetone (Fig. 1A) or in petroleum ether-ether (4). In the system of Daffé et al (4) epoxy-mycolates migrate closely with alcohols, but this problem can be resolved by acid methanolysis, this compound giving four characteristic polar components (15). Acid methanolysis also attacks other mycolic acids, mainly at their double bonds and cyclopropane rings, artefacts appearing in the chromatograms R_f of which are related to several mycolic acids (6,15,21). A procedure that implies a transformation of the -OH group to tert-butyldimethylsilyl ether with ulterior TLC has recently been developed by Dobson et al (6), thus enabling the above problems to be overcome.

The mycolic acid composition of the *Mycobacterium* species has been reviewed in detail by Daffé et al (4) and Dobson et al (6). Table 2 summarizes this composition. Note that some species contain a specific combination (i.e. *M. chelonae* and *M. simiae*) and some closely related species can be differentiated by this criterion (i.e. *M. triviale* and *M. nonchromogenicum*).

Another possibility for analysis of mycolic acids consists of study of their pyrolysis products. Mycolic acids pyrolyse to meroaldehyde and a fatty acid according to the reaction:

$$R_1\text{-CHOH-}\underset{\displaystyle R_2}{\underset{|}{\text{CH}}}\text{-COOCH}_3 \longrightarrow R_1\text{-CHO} \quad + \quad R_2\text{-CH}_2\text{-COOCH}_3$$

This pyrolysis can be carried out in the injector of the gas-chromatograph (8) at a temperature of 300-350°C, then analysing the methyl ester produced. Mycolic acids of mycobacteria pyrolyse saturated fatty acids of 22, 24 and 26 C; this characteristic can also be used as a diagnostic key in these microorganisms (12). Table 2 also shows the major pyrolysis products of the different species of *Mycobacterium*.

FATTY ACIDS AND ALCOHOLS

Mycobacteria contain simple fatty acids from ca 12 to 26 C, saturated and mono-unsaturated together with some 10-methyl branched fatty acids, such as 10-methyl octadecanoic acid (tuberculostearic acid). Moreover, a group of 2-methyl branched fatty acids has been detected in some species (14).

Tuberculostearic acid is present in most of the species of *Mycobacterium* and only in *M. gordonae* is its amount, in general, below the limits of detection of gas-liquid chromatography. 2-methyl branched fatty acids can be considered as complex; their number of carbon atoms ranges from ca 15 to 40, in some cases being multi-methyl branched and sometimes also hydroxylated. These acids have been given several names according to some of their structural features (see for revision reference 14). A special fatty acid that can be included in this group

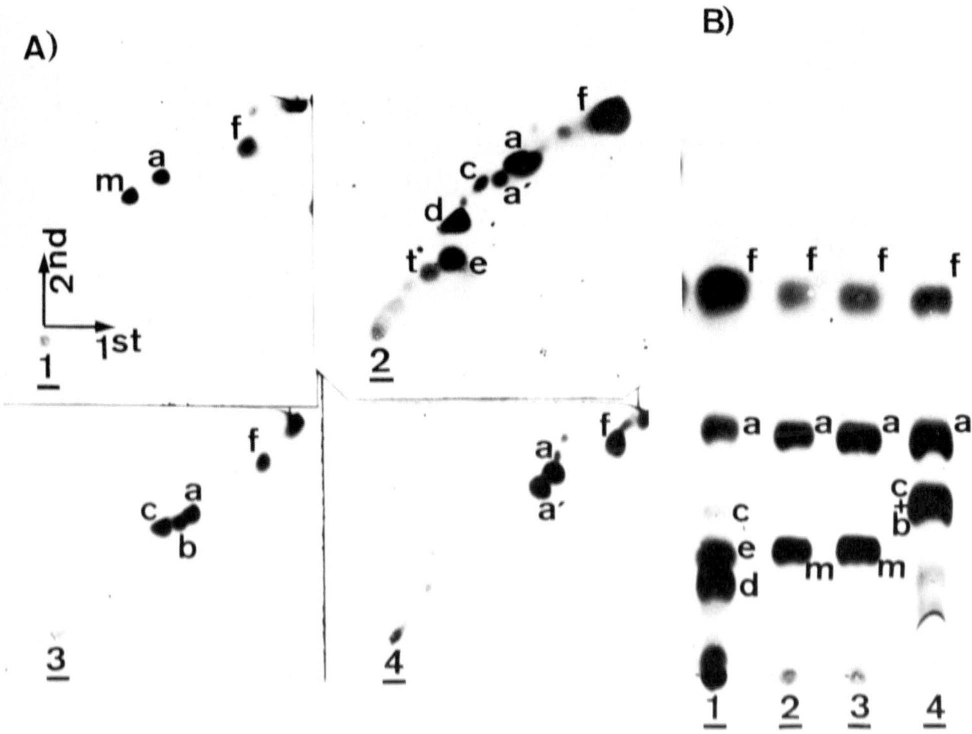

Fig. 1. TLC of mycolic acids of mycobacteria. A) Two-dimensional TLC of 1, *M. fortuitum* ATCC 6841; 2, *M. vaccae* CCM 206; 3, *M. gordonae* CHC 21808, and 4, *M. chelonae* TMC 1542. Solvents: first direction (1st), petroleum-ether-acetone 95:5x3; second direction (2nd), toluene-acetone 97:3x1. B) Single dimensional TLC of 1, *M. nonchromogenicum* CCM 56; 2, *M. senegalense* CCM 226; 3, *M. fortuitum* ATCC 6841, and 4, *M. szulgai* ATCC 20886. Solvent: dichloromethane.
a = alpha-mycolates, a' = alpha'-mycolates, b = methoxy-mycolates, c = keto-mycolates, m = epoxy-mycolates, d = carboxy-mycolates, e = 2-eicosanol+2-octadecanol, t = n-tetradecanol, f = fatty acids.
Plates revealed with molybdophosphoric acid (5% in ethanol).

is that found in *M. gordonae* and described as 2-methyl-3-hydroxy-eicosanoic acid (4).

Secondary alcohols have long been described in some mycobacteria belonging to the structural types 2-eicosanol and 2-octadecanol; moreover, preliminary analyses carried out in our laboratory have shown that some new species can also contain primary alcohols such as n-tetradecanol. Table 2 also shows the composition of some characteristic fatty acids and alcohols of *Mycobacterium* species.

Fatty acids (as methyl esters) and alcohols can be examined by GLC preferably using apolar columns, either packed or capillary. Capillary columns enable a better resolution to be obtained and, in some cases, show a certain grade of complexity in the unsaturated fatty acids of 16 and 18 C. With both types of columns, however, care should be taken to eliminate non-esterified fatty acids after saponification (11) because of possible irreversible absorption of the alcohols to the stationary phase. A little tailing of these compounds should, in any case, be expected (and also of hydroxylated fatty acids), although this problem is overcome with trimethylsilyl or trifluoroacetyl derivatives (4,11).

TABLE 2

MYCOLIC ACIDS, FATTY ACIDS AND ALCOHOLS OF MYCOBACTERIUM SPECIES.

Species	Mycolic acids	Pyrolysis	TBS*	2-methyl branched**	Alcohols#
M.tuberculosis	a,b,c##	26	+	24-32	-
M.bovis	a,b,c	26	+	24-32	-
M.africanum	a,b,c	26	+	24-32	-
M. microti	a,b,c		+	27-32	-
M. leprae	a,c	22	+/-	30-34	-
M.lepraemurium	a,c,d	24	+		s
M.ulcerans	a,b,c	24	+	27-29	-
M.paratuberculosis	a,c,d				s
M.gordonae	a,b,c	24	-	15,22	-
M.szulgai	a,b,c	24	+	15,23,25	-
M.marinum	a,b,c	24	+	25-30	-
M.kansasii	a,b,c	24	+	16,27-32	-
M.xenopi	a,c,d	24	+	25	s
M.scrofulaceum	a,c,d	24	+		s
M.simiae	a,a',c	26	+	29-32	-
M.asiaticum	a,b,c		+		-
M.avium	a,c,d	24	+	25	s
M.intracellulare	a,c,d	24	+	25	s
M.nonchromogenicum	a,c,d		+		s
M.novum	a,c,d	24	+		s
M.terrae	a,c,d	24	+	25	s
M.triviale	a	24	+	25	-
M.fallax	a	24	+		-
M.gastri	a,b,c	24	+	29-32	-

TABLE 2 (Continued)

Species	Mycolic acids	Pyrolysis	TBS	2-methyl branched	Alcohols
M.shimoidei	a,a',c,d	24	+		+
M.fortuitum	a,a',m	24	+	19	-
M.farcinogenes	a,a',m	24	+		-
M.senegalense	a,a',m	24	+		-
M.smegmatis	a,a',m	24	+	15	-
M.chitae	a,a',m	24	+		-
M.chelonae	a,a'	24	+		-
M.agri	a,a',b,c		+		-
M.diernhoferi	a,c,d	24	+		s
M.thermoresistibile	a,a',b,c	24	+		s
M.flavescens	a,c,d	24	+		s
M.phlei	a,c,d	24	+		s
M.aurum	a,c,d	24	+	18-22	s
M.neoaurum	a,c,d	22	+		s
M.gadium	a,c,d	22	+		s
M.aichiense	a,c,d	24	+		s
M.rhodesiae	a,c,d	24	+		s
M.sphagni	a,c,d	24	+		s
M.tokaiense	a,c,d	24	+	22	s
M.duvalii	a,a',c,d	24	+		s
M.chubuense	a,a',c,d	24	+		p,s
M.gilvum	a,a',c,d	22	+		s
M.parafortuitum	a,a',c,d	22	+	19?	s
M.obuense	a,a',c,d	24	+		p,s
M.vaccae	a,a',c,d	22	+		p,s
M.komossense	a,b,c,d		+		s

This Table is based on references 2,4,5,6,7,10,12,16,18,19,20 and also on data from the laboratory of the authors.

*TBS = Tuberculostearic acid

**Range of total carbon atoms is given. s= secondary; p = primary

See Figure 1 for assignment of mycolic acids.

REFERENCES

1. ASSELINEAU C, ASSELINEAU J (1978) Ann Microbiol Inst Pasteur 129A:49
2. ASSELINEAU C, CLAVEL S, CLEMENT F, DAFFE M, DAVID H, LANEELLE MA, PROME JC (1981) Ann Microbiol Inst Pasteur 132A:19
3. COLLINS MD (1982) J Appl Bacteriol 52:457.
4. DAFFE M, LANEELLE MA, ASSELINEAU C, LEVY-FREBAULT V, DAVID H (1983) Ann Microbiol Inst Pasteur 134B:241
5. DAFFE M, LANEELLE MA, ROUSSEL J, ASSELINEAU C (1984) Ann Microbiol Inst Pasteur 135A:191
6. DOBSON G, MINNIKIN DE, MINNIKIN SM, PARLETT JH, GOODFELLOW M, RIDELL M, MAGNUSSON M (1985) In: Goodfellow M, Minnikin DE (eds) Chemical Methods in Bacterial Systematic. Academic Press, London, pp 237-265
7. DRAPER P, DOBSON G, MINNIKIN DE, MINNIKIN SM (1982) Ann Microbiol Inst Pasteur 133B:39
8. GERRANT GO, LAMBERT MA, MOSS CW (1981) J Clin Microbiol 13:899
9. HALL RM, RATLEDGE C (1984) J Gen Microbiol 130:1883
10. KUSAKA T, KOSHAKA M, FUKUNISHI Y, AKIMORI H (1981) Int J Lepr 49:406
11. LARSSON L (1983) Acta Path Microbiol Immunol Scand Sect B 91:235
12. LECHEVALIER MP, HORAN C, LECHEVALIER H (1971) J Bacteriol 105:313
13. MARKS J, SZULGA T (1965) Tubercle 46:400
14. MINNIKIN DE (1982) In: Ratledge C, Standford JL (eds) The Biology of the Mycobacteria, Vol 1. Academic Press, London, pp 95-184
15. MINNIKIN DE, HUTCHINSON IG, CALDICOTT AB, GOODFELLOW M (1980) J Chromatogr 188:221
16. MINNIKIN DE, PARLETT JH, MAGNUSSON M, RIDELL M; LIND A (1984) J Gen Microbiol 130:2733
17. MINNIKIN DE, MINNIKIN SM, PARLETT JH, GOODFELLOW M (1985) Zbl Bakt Hyg A 259:446
18. RAFIDINAVIRO E, SAVAGNAC A, LACAVE C, PROME JC (1982) Biochim Biophys Acta 711:266
19. RAFIDINAVIRO E, PROME JC, LEVY-FREBAULT V (1985) Chem Phys Lipids 36:215
20. TISDALL PA, ROBERTS GD, ANHALT JP (1979) J Clin Microbiol 10:506
21. VALERO-GUILLEN PL, MARTIN-LUENGO F (1985) J Appl Bacteriol 59:113
22. YANO I, TORIYAMA S, MASUI M, KUSUNOSE E, KUSUNOSE M (1978) Jap J Bacteriol 33:173

MYCOBACTERIUM AVIUM COMPLEX INFECTIONS IN IMMUNOCOMPROMISED PATIENTS

TIMOTHY KIEHN, DONALD ARMSTRONG, EDWARD BERNARD, PATRICIA BRANNON, ARTHUR BROWN, ROBERT CAMMARATA, FITZROY EDWARDS, JONATHAN GOLD, CATHERINE HAWKINS, ESTELLA WHIMBEY, STEPHEN WINTER AND BRIAN WONG

Memorial Sloan-Kettering Cancer Center, New York, New York

Before 1980 there were few reports of disseminated infection in humans caused by members of the *Mycobacterium avium* complex (MAC). These organisms have been isolated from several environmental sources including soil, water, and house dust and they produce diseases in birds and animals. In humans, MAC causes infections of the lungs, lymph nodes, skin, bones, soft tissues, and the genitourinary tract. These organisms are slow-growing, usually non-pigmented, and have been placed in the Runyon Group III. Niacin and nitrate tests are negative, there is low catalase production, Tween 80 is not hydrolyzed, urease is not produced, tellurite is usually reduced and by in vitro testing methods these mycobacteria are resistant to many antimicrobials.

Since 1981 we have followed, at Memorial Sloan-Kettering Cancer Center, approximately 350 patients with the acquired immune deficiency syndrome (AIDS) and 67 of these patients have had disseminated MAC infection. During this time we have also seen MAC infection in 4 patients with hairy cell leukemia and 2 patients with severe combined immunodeficiency disease. MAC infection has been a common finding in the AIDS patients who died. Of the 79 AIDS patients who died and had autopsies, 41 had evidence of disseminated MAC infection.

Isolation of MAC from an antemortem specimen did not always indicate MAC infection. Of the total AIDS patients, 93 have had positive cultures for MAC. These patients were divided into those with MAC infection, colonization only with the mycobacterium, or of indeterminate results where infection or colonization could not be determined. Infection was documented by a positive culture from blood or tissue obtained by biopsy or autopsy. Colonization was documented by a positive culture from respiratory secretions, urine or stool and negative cultures and histology from autopsy. The indeterminate group included patients who had positive cultures from respiratory secretions, urine or stool, who were living, and in whom cultures from sterile sites had not been positive; or who died and did not have an autopsy. Also included in the indeterminate group were patients who had only one or two colonies of MAC isolated in culture from one biopsy or autopsy site. Of the 93 AIDS patients who had positive cultures; 67 had infection, 7 were colonized with MAC and 19 patients were in the indeterminate group. Of the 67 with infection; 60 were male homosexuals, 5 were intravenous drug users and 2 were children of intravenous drug users.

Body sites from which MAC were frequently isolated by antemortem culture were: blood (54 of 56 attempts), bone marrow (14/16), bronchial or tracheal secretions (16/23), sputum (14/22), and urine (13/31). Acid-fast stains of stool and subsequent stool cultures were positive in 13 of 31 attempts and cultures of cerebrospinal fluid were positive only twice in 19 attempts.

Examination of fecal material for acid-fast organisms was not a routine procedure in our mycobacteriology laboratory before the onset of the AIDS epidemic. During the period of this study, a slide preparation of fecal material was stained with auramine O and when positive, fecal material was processed through a routine sputum decontamination procedure using N-acetyl-L-cysteine and sodium hydroxide. The sediment was inoculated onto conventional mycobacterial agar media which included Lowenstein-Jensen agar slants and Middlebrook 7H11 and Mitchison selective 7H11 agar plates. We recently expanded the stool culture procedure and it now includes inoculation of BACTEC 12A broth media used in the radiometric detection system from Johnston Laboratories. Preliminary results indicate that this new 12A media may be the most sensitive method of detecting low-levels of MAC in stool specimens. This new culture system detected MAC from stool specimens when the direct acid-fast stain of the feces was negative.

The new Isolator lysis blood culture system was used to detect MAC in blood cultures and to determine the number of mycobacteria per milliliter of blood. Blood was collected in the tube that contains an anticoagulant and the lytic agent, saponin. After one hour, thus allowing for lysis of blood cells, the tube was centrifuged and the sediment was inoculated directly onto 7H11 agar plates. Colonies of MAC were usually visible by 6 to 14 days of incubation and the number of colony forming units were determined. When colony counts were expected to be high, >300 CFU/ml, blood was collected in a 1.5 ml Isolator tube and processed in the laboratory. This procedure did not require a centrifugation step and the entire 1.5 ml of blood-lysate was inoculated directly onto Middlebrook 7H11 agar plates after 10 fold dilutions of the blood-lysate were made in 7H9 broth.

Colony counts of mycobacteria in the blood ranged from as low as <1 CFU/ml of MAC to as high as 28,000 CFU/ml. Early in the AIDS epidemic we typically saw patients with advanced MAC infection and colony counts >1000 were seen to decrease into the 100s and less as the patients received treatment. Recently we have diagnosed the disease earlier, with colony counts <100 CFU/ml, and these levels of mycobacteremia have often persisted for several months despite treatment. Blood from one patient with high-grade mycobacteremia was fractionated, and most of the circulating mycobacteria were found within the leukocytes. Lytic agents, such as saponin or desoxycholate, increased the colony counts 2 to 5 fold when compared to unlysed blood. The patient at autopsy had more than 10 CFU per gram of mycobacteria in the spleen, liver, and an abdominal lymph node. These findings are similar to those previously observed in patients with lepromatous leprosy with continuous mycobacteremia and organisms found predominantly in leukocytes.

Organisms from AIDS patients with disseminated MAC infection often had two common characteristics; many of them were the same serotype and they often were pigmented. Fifty-eight MAC isolates, from patients with disseminated disease, were serotyped at National Jewish Hospital and Research Center, Denver, Colorado, and at the Centers for Disease Control, Atlanta, Georgia. Forty-three (74%) were serovar 4, six were serovar 8, five were serovar 1, and one each were serovars 9, 10, 17, and 19. Forty-four of the 58 strains (76%) produced a deep yellow pigment early in culture. The remaining strains exhibited the more traditional buff color usually associated with MAC isolates.

All isolates were susceptible, in vitro, to standard concentrations of clofazimine, cycloserine, and ansamycin. Most strains were resistant to isoniazid, streptomycin, and lower concentrations of ethambutol, ethionamide, and rifampin. Because of the in vitro susceptibility results, many of the patients were treated with clofazimine, ansamycin, and ethionamide or ethambutol. When colony forming units in sequential blood cultures were used to measure response to therapy, we found that in about an equal number of cases, colony counts decreased, increased, or remained unchanged. Of 31 patients who were treated for their MAC infection and not lost to follow-up, all are either alive with persistent bacteremia or have died, and MAC was found in all 17 of those with autopsies. None of our patients with disseminated MAC infection have, to our knowledge, been cured of their infection. At autopsy the organs which were most often positive for MAC in 41 autopsies were the spleen, liver, lymph node, lung, and kidney.

In an effort to diagnose MAC infection earlier, an ELISA test was developed to measure antibodies to MAC. Patients with AIDS and disseminated infection had antimycobacterial antibody levels that did not differ significantly from those in uninfected patients with AIDS or controls. Infected patients with hairy cell leukemia had significantly higher levels of antimycobacterial antibody than did uninfected patients with hairy cell leukemia or controls. These results further demonstrate that AIDS involves a functional defect in humoral immunity, in addition to impairment of cellular immune function.

In summary, *Mycobacterium avium* complex has been seen as a common cause of widely disseminated infection among AIDS patients. The origin of the infection appears to be the gastrointestinal tract and stool smears and cultures for acid-fast bacilli and blood cultures have been sensitive diagnostic tests. Quantitative blood culture methods have been useful in enumerating organisms in the peripheral blood and organisms have been shown to be present in leukocytes. Patients have been treated with a variety of antimicrobials but without a sustained remission of MAC disease. These patients' inability to produce high levels of MAC antibody is further evidence of the effect that AIDS has on the immune system.

REFERENCES

Wong B, Edwards F, Kiehn T, Whimbey E, Donnelly H, Bernard E, Gold J, Armstrong D (1985) Amer J Med 78:35-40

Kiehn T, Edwards F, Brannon P, Tsang A, Maio M, Gold J, Whimbey E, Wong B, McClatchy J, Armstrong D (1985) J Clin Microbiol 21:168-173

Winter S, Bernard E, Gold J, Armstrong D (1985) J Inf Dis 151:523-527

Hawkins C, Kiehn T, Whimbey E, Gold J, Brown A, Armstrong D (1985) Int Conf on AIDS, Atlanta, Georgia, USA

CUTANEOUS INFECTIONS CAUSED BY *MYCOBACTERIUM CHELONEI* SUBSP. *CHELONEI* AND SUBSP. *ABSCESSUS*

A. REZUSTA, J. GIL, M.A. VITORIA, G. MARTINEZ* and C. RUBIO-CALVO

Department of Microbiology, *Department of Dermatology, School of Medicine, University of Zaragoza, 50009 Zaragoza (Spain)

INTRODUCTION

The 'rapidly-growing' mycobacteria of Runyon's group IV are free-living organisms found in soil and water throughout the world (1). *Mycobacterium chelonei* is a member of this group and has been increasingly recognized as a pathogen in a variety of human diseases (2,3,4).

We report two cases of posttraumatic subcutaneous abscesses caused by *Mycobacterium chelonei*. The first, a posttraumatic colonization produced by subsp. *abscessus* was found in a 60-year-old man who had previously been diagnosed as having chronic bronchitis and diabetes. At the time of study he presented with multiple painful erythematous subcutaneous nodular lesions with progression to abscess formation on his left fingers and arm. Several days previously he had suffered numerous cutaneous cuts and scratches. The second case was a 40-year-old man who presented an abscess with fistulation from his amputated leg. *Mycobacterium chelonei* subsp. *chelonei* was isolated in pure culture.

MATERIAL AND METHODS

The smears on microscopic examination showed many polymorphonuclear leucocytes and acid-fast bacilli by Ziehl-Neelsen staining.

The exudates were inoculated on: Löwenstein-Jensen (LJ), LJ with pyruvate (LJP), Sabouraud Agar (SA), Blood Agar (BA) and Schaelder Agar (ScA) were incubated at 32° and 37°C, 27°C, 37°C aerobically and 37°C in an anaerobic atmosphere respectively, to rule out any other etiological agents (fungi, aerobic and anaerobic bacteria). The acid-fast bacilli grew on LJ, LJP, SA and BA. Growth of *Mycobacterium chelonei* subsp. *abscessus* occurred at 32° and 37°C, *Mycobacterium chelonei* subsp. *chelonei* only at 32°C. They were identified according to the method of Runyon et al (5).

Susceptibility was tested to Penicillin (P), Cloxacillin (Cx), Ampicillin (Am), Carbenicillin (Cb), Ticarcillin (Tic), Piperacillin (Pip), Azlocillin (Az), Mezlocillin (Mz), Cephalothin (Cr), Cefamandole (Ma), Cefoxitin (Fox), Cefotaxime (Ctx), Moxalactam (Mox), Cefoperazone (Cfp), Gentamicin (Gm), Tobramycin (Tm), Dibekacin (DKB), Amikacin (An), Netilmicin (Net), Erythromycin (E), Lincomicin (L), Clindamycin (cc), Rifampicin (Ra), Tetracycline (Te), Minocycline (Min), Colistin (Cl), Chloramphenicol (Cm), Fosfomycin (Fo), Trimethoprim-sulfamethoxazole (Sxt), Pipemidic acid (Pipe), and Vancomycin (Va) was performed by the disk-plate method (Wallace) (6) on Mueller-Hinton Agar (MHA) and MHA supplemented with 10% OADC.

RESULTS AND DISCUSSION

Mycobacterium chelonei subsp. *abscessus* and *M. chelonei* subsp. *chelonei* show similar sensitivity patterns to antibiotics. Both are resistant to P, Cx, Am, Cb, Tic, Pip, Az, Mz, Cr, Ma, Fox, Mox, Cfp, Gm, Tm, L, cc, Ra, Te, Min, Cl, Cm, Fo, Sxt, Pipe and Va, and sensitive to An, Net and E. *Mycobacterium chelonei* subsp. *abscessus* was sensitive to Ctx and resistant to DKB; subsp. *chelonei* was sensitive to DKB and resistant to Ctx.

In the first case, Amikacin therapy was started (1 g/day given intramuscularly) and interrupted 15 days later due to increased BUN. Cefotaxime treatment (2 g/day for 10 days) was initiated and his cutaneous lesion responded with resolution. One month later subcutaneous nodules appeared at the same zone. No acid-fast material has been observed, and all cultures have been sterile. Lesions resolved without therapy. Four months later he relapsed with the reappearance of similar cutaneous nodules on the same zone. Cefotaxime therapy was instituted (2 g/day for 15 days given intramuscularly) in addition to procodazol (300 mg/day orally for 4 months). This time a cure was obtained.

In the second case, Erythromycin was administered orally (2 g/day for 30 days) in conjunction with surgical removal of the lesion.

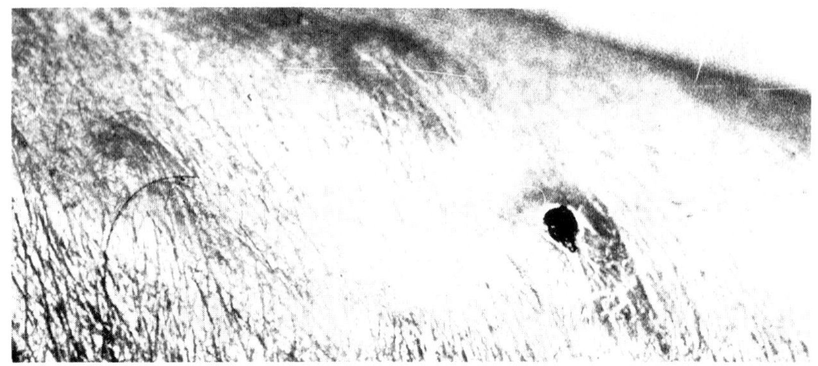

Fig. 1. Cutaneous nodular lesions with ulcerations on left arm caused by *Mycobacterium chelonei* subsp. *abscessus*.

REFERENCES

1. Hand W.L., Sanford J.P. Mycobacterium fortuitum - a human pathogen. Ann. Inter. Med. 1970; 73: 971-7

2. Jauregui L., Arbulu A., Wilson F. Osteomyelitis, pericarditis, mediastinitis and vaculitis due to Mycobacterium chelonei. Am. Rev. Respir. Dis. 1977; 115: 699-703

3. Graybill J.R., Silva J.Jr., Fraser D.W., London R., Rogers E. Disseminated mycobacteriosis due to Mycobacterium abscessus in two recipients of renal homografts. Am. Rev. Respir. Dis 1974; 109: 4-10

4. Center for Disease Control. Follow-up on mycobacterial contamination of porcine heart valve prosthesis. Morbid Mortal Weekly Rep. 1978; 27: 96-8

5. Runyon E.H., Karlson A.G., Kubica G.P., Wayne L.G. In Lennette E.H., Balows A., Hausler W.J., Truant J.P. (ed), Manual of Clinical Microbiology 3rd. American Society for Microbiology. Washington D.C.

6. Wallace R.J., Jr., Dalovisio J.R., Pankey G.A. Disk diffusion testing of susceptibility of Mycobacterium fortuitum and Mycobacterium chelonei to antibacterial agents. Antimicrob. Agents Chemother., 1979, pp. 611-614.

Mycobacteria of Clinical Interest. M. Casal, editor

SUBCUTANEOUS ABSCESS DUE TO A RAPIDLY GROWING, SCOTOCHROMOGENIC MYCOBACTERIUM

HAJIME SAITO[1), HARUAKI TOMIOKA[1), KENJI ASANO[1), AND SHOSO YAMAMOTO[2)
Department of Microbiology and Immunology, Shimane Medical University, Izumo, Shimane 693[1)] and Department of Dermatology, Hiroshima University School of Medicine, Hiroshima, Hiroshima 730[2)] (Japan)

INTRODUCTION

Among the rapidly growing mycobacteria, two species have been recognized as pathogenic for humans: *Mycobacterium fortuitum* and *M. chelonae* (1). A recent study has reported on pulmonary infections attributable to *M. thermoresistibile* (2). The purpose of the present paper is to present a new acid-fast organism of rapidly growing, scotochromogenic mycobacteria that has been isolated from a subcutaneous abscess of the left chest in a patient of a 37-year-old housewife.

MATERIALS AND METHODS

Detection of acid-fast bacilli: The pus discharged from the fistula and biopsied specimens of the lesions were stained by the Ziehl-Neelsen method and examined under a microscope. Gauze smeared with the pus was immersed and shaken in 1 ml of saline to free the attached organisms. The sample was treated with an equal amount of 4% H_2SO_4 at room temperature for 30 min and centrifuged at 3,000 rpm for 30 min. The sediment obtained was washed with saline, inoculated onto 1% Ogawa egg medium (Ogawa medium), and incubated at 37°C and 25°C.

Bacterial strains: The new isolate (strain Okumoto) and strains of 18 species of rapidly growing, scotochromogenic mycobacteria were used. Strain Okumoto, *M. komossense*, and *M. spahgni* were cultured at 33°C, while the other strains were cultured at 37°C.

Colony characteristics: Tests were performed for the following properties after incubation using Ogawa medium at 22°C and at either 37°C or 33°C: the photoactivity of the colony after incubation for 3 days (3), the colony morphology after incubation for 7 days in the dark, and the photoactivity of the colony after incubation for 7 days under exposure to light from a 15-watt fluorescent lamp at a distance of 30 cm.

Growth at different temperatures: One loopful (3 mm in diameter) of bacteria grown in Dubos Tween albumin liquid medium for 3-5 days was inoculated onto Ogawa medium. The ability of the bacteria to grow was determined after incubation for 7 days at 28°C, 33°C, 37°C, 42°C, 45°C, and 52°C.

Resistance tests: The growth of bacteria on Ogawa medium containing 3% sodium chloride; 125, 250, or 500 μg of $NH_2OH{\cdot}HCl$ per ml; or 20 μg of sodium azide per ml was determined by the same method as above.

Biochemical characteristics: The following tests were performed by the procedure outlined in the references listed: iron uptake (4); degradation of salicylates (5), *p*-aminobenzoate (PABA)(6), and *p*-aminosalicylate (PAS)(7); acid phosphatase (8); pH 5.0/70°C acid phosphatase (9); arylsulfatase (3); nitrate reduction (3); Tween 80 hydrolysis (3); semiquantitative catalase (3); pH 7.0/68°C catalase (3); and Bönicke's amidase series (10). The production of acid from 16 carbohydrates and utilization of 11 organic acids were determined after 2 weeks'incubation at 33°C (11).

Numerical analysis: Based on the comparison of the 69 properties listed above, the similarity (S) values between the strains were calculated by dividing the number of similar positive matches by the number of similar positive matches plus the number of dissimilar matches (12). The S values are given as percentages in the present study.

Pathogenicity test: Strain Okumoto (2.2×10^6 organisms/mouse) was inoculated into the tail vein of 5-week-old female ddY mice. After infection, the superficial lesions were observed daily and 20 animals each were sacrificed on days 7, 14, 21, and 28. Any gross lesions in the lungs, liver, spleen, or kidneys were observed, and the viable organisms in the respective organs were cultured as described previously (13).

RESULTS

Detection of acid-fast bacteria: The results of detection of acid-fast bacilli from specimens taken from the lesion are as follows: No acid-fast rods were observed microscopically from the tissue sample obtained on September 26, 1983 and March 3, 1984. Of the 14 pus samples examined between November 17, 1983 and July 29, 1985, acid-fast bacilli were demonstrated in 9 samples by smear and in 6 samples by culture. Evaluation of results was impossible in 3 samples that showed positive smears because of contamination of the culture medium. The isolates were all shown to be rapidly growing, scotochromogenic mycobacteria with same colony morphology.

Description of strain Okumoto: Microscopic observation of cells grown on Ogawa medium revealed acid-alcohol-fast, short rods. However, the cells were much longer when grown in Dubos Tween-albumin medium. At 28°C and 33°C, growth was observed after 3 days. At 37°C, 42°C, 45°C, and 52°C, however, the organism failed to grow. There was some growth at 22°C as well, but it was less eugonic than at 33°C. At 22°C, the colonies were non-pigmented. At 33°C, they were pale yellowish, smooth and glistening. The colonies produced at either temperature showed no photoactivity, but the colonies cultured under exposure to light for 7 days showed positive photoactivity and a dark yellowish color. The organism failed to grow on Ogawa medium containing 125, 250, or 500 μg of $NH_2OH{\cdot}HCl$ per ml, 20 μg of NaN_3

per ml, or 3% NaCl. Degradation of salicylate, PABA, and PAS was negative. Iron uptake was positive. Heat-stable acid phosphatase, nitrate reductase, arylsulfatase (3-day), and Tween 80 hydrolysis (7-day) were negative, while acid phosphatase, catalase, heat-stable catalase, arylsulfatase (2-week), and Tween 80 hydrolysis (2-week) were positive. The organism showed a positive urease, nicotinamidase, and pyrazinamidase, and a negative acetamidase, benzamidase, isonicotinamidase, salicylamidase, allantoinase, succinamidase, and malonamidase. Acid was produced from arabinose, xylose, glucose, laevulose, mannose, trehalose, mannitol, and inositol, but not from rhamnose, galactose, sucrose, lactose, maltose, raffinose, dulcitol, or sorbitol. Utilization of organic acids was positive for acetate, citrate, pyruvate, fumarate, propionate, and lactate, but negative for succinate, benzoate, tartrate, malonate, and oxalate.

In the pathogenicity test conducted on mice, none of the infected animals showed any superficial or organic lesions. No viable organnisms were recovered from the lungs or kidneys 7 days after infection nor from the liver or spleen 14 days after infection.

Similarity values: The similarity values between strain Okumoto and strains of 18 species of rapidly growing scotochromogenic mycobacteria were calculated. Strain Okumoto exhibited low inter-taxon S values for all species, the highest being *M. aurum* (74%), followed by *M. austroafricanum* (68%), *M. parafortuitum* (68%), *M. obuense* (64%), *M. tokaiense* (64%), *M. rhodesiae* (62%), *M. chubuense* (62%), *M. gadium* (62%), *M. vaccae* (61%), *M. neoaurum* (61%), *M. spahgni* (59%), *M.aichiense* (58%), *M. phlei* (57%), *M. komossense* (57%), *M. flavescens* (54%), *M. duvalii* (50%), *M. gilvum* (43%), and *M. thermoresistibile* (41%).

Comparison of strain Okumoto and *M. parafortuitum*: The similarity between strain Okumoto and strains of *M. parafortuitum* was pointed out by M. Tsukamura, National Chubu Chest Hospital, Aichi, Japan. Therefore, a comparative study was made on 69 properties (see Materials and Methods) in strain Okumoto and 4 strains of *M. parafortuitum* (ATCC 19686, ATCC 19687, ATCC 19688, and ATCC 25807). Positive results were obtained for all the strains of *M. parafortuitum* and negative results for strain Okumoto in regard to the following properties: pigmantation of colonies cultured at 22°C in the dark; photoactivity of colonies cultured at 22°C and 37°C; growth at 37°C and 42°C; acetmaidase; acid production from sucrose; utilization of succinate, benzoate, and malonate; and resistance to 3% NaCl. The internal S value of *M. parafortuitum* was 97%, while the inter-taxon S value with strain Okumoto was 68%.

DISCUSSION

The acid-fast bacilli that are pathogenic for humans, have a low optimal temperature for growth (30°C and 33°C), and do not develop at 37°C in primary culture are *M. marinum*, *M. ulcerans*, and *M. haemophilum* (slow growers) and *M. chelonae* subsp. *chelonae* (rapid growers). The former 3 species characteristically cause superficial lesions in humans and do not invade the viscera. The organism described in this paper is a rapidly growing, scotochromogenic mycobacterium with temperature limits for growth and was repeatedly isolated from the pus specimens taken from the patient's skin lesions. The isolate was different from the type strain (ATCC 23366) of *M. aurum*, which is the most closely related species among the rapidly growing, scotochromogenic mycobacteria, in the following points: pigmentation in the dark at 22°C; photoactivity; growth at 37°C; acid phosphatase; acid production from galactose and sucrose; utilization of succinate; and resistance to 3% NaCl. The organism is also easily differentiated from *M. parafortuitum* (see Results). The organism produced neither organic nor superficial lesions in mice. Furthermore, according to recent studies on lipid analysis performed by I. Yano, Niigata University School of Medicine, Niigata, Japan, the isolate showed a unique lipid pattern.

In view of the above results, the isolate is considered to belong to a new species of rapidly growing mycobacteria. However, this organism has so far been obtained from only one patient. Therefore, we hope in the future to isolate the same mycobacterium from clinical or environmental samples and to propose its existence as a new species.

REFERENCES

1. Wolinsky E (1979) Am Rev Respir Dis 119:107-159
2. Weitzman I, Osadczyi D, Corrado ML, Karp D (1981) J Clin Microbiol 14:593-595
3. Vestal AL (1977) Procedures for the isolation and identification of mycobacteria U.S. Public Health Service Publication 75-8230 Center for Disease Cntrol, Atlanta
4. Saito H, Yamaoka K, Kiyotani K (1976) Int J Syst Bacteriol 26:111-115
5. Tsukamura M (1965) J Gen Microbiol 41:309-315
6. Tsukamura M, Mizuno S, Tsukamura S (1966) Med Biol 72:266-269
7. Tsukamura M (1961) Jap J Tuberc 9:70-79
8. Urabe K, Saito H, Tasaka H, Matsubayashi H (1966) Med Biol 72:127-130
9. Saito H, Hosokawa H, Tasaka H (1968) Am Rev Respir Dis 97:474-476
10. Bönicke R (1962) Bull Int Union Tuberc 27:13-68
11. Gordon RE, Smith MM (1953) J Bacteriol 66:41-48
12. Sneath PHA (1957) J Gen Microbiol 17:201-226
13. Saito H, Sato K, Jin BW (1984) Antimicrob Agents Chemother 26:270-271

Mycobacteria of Clinical Interest. M. Casal, editor

CLINICAL SIGNIFICANCE OF ENVIRONMENTAL MYCOBACTERIA (1975-1984)

N. MARTIN CASABONA, T. GONZALEZ FUENTE, R. VIDAL PLA, J. REINA PRIETO and J. DE GRACIA ROLDAN

Service of Microbiology, Service of Pneumology, Ciudad Sanitaria 'Vall d'Hebron', Barcelona, Generalitat de Catalunya, Departament de Sanitat i Seguretat Social (Spain)

INTRODUCTION

From 1975-1984 inclusive, the percentage of environmental *Mycobacteria* (EM) isolated each year in our laboratory was between 10.1 and 29.7, with a mean of 15.

Epidemiologically, *Mycobacteria* detection is directly related to the decline in tuberculosis. However, the percentage of pathogenic EM strains is low in our environment in comparison to all *Mycobacteria* isolated.

This article is an up-date of an article published previously on *Mycobacteria* in our environment, on the number of cases and their development (1).

MATERIAL AND METHODS

During the period mentioned above, 6700 mycobacterial strains were isolated; of these 5664 were identified as *M. tuberculosis*, 30 as *M. bovis*-BCG, 1 as *M. africanum* and 1005 as EM.

Two hundred and eighty-five case histories of patients with EM isolates were studied according to principal and secondary criteria. Principal criteria were: (A) repeated and abundant isolates; (B) tuberculosis suspected clinically in absence of *M. tuberculosis*; (C) exeresis isolates with compatible histology; (D) clinical response to antibiotic therapy, as indicated by sensitivity in vitro of the isolated strain. Secondary criteria were: (a) isolation of a species which is probably a human pathogen (2) or a saprophyte in an aseptic product; (b) one isolate with more than 100 colonies or various isolates with less than 100; (c) histological lesions compatible with a chronic granulomatous process in samples different from that of the isolate; (d) response to antibiotic treatment without previous antibiotic testing; (e) a PPD-5UT reaction of more than 10 mm, without previous contact with *M. tuberculosis*. Minimum criteria for mycobacteriosis are 2 principal and 3 secondary criteria.

RESULTS

Forty-four cases fulfilled the minimal criteria (Table 1). Specimens were as follows: For pulmonary diseases, sputum, gastric aspirates and transtracheal puncture. For skin samples, back of hand biopsy, exudates from abscessed nodules on legs and arms. For lymphadenitis, biopsy or exeresis. For pleural infections, biopsy or pleural cavity liquid. For disseminated infections, blood, urine, nephrectomy sections, renal cavity liquid, exudates from inguinal fistula. For subarachnoid abscesses, CSF, skin exudates from points along decompression catheter. For sternal osteitis, pus from cutaneous abscesses in the thoracotomy zone, and drainage contents of the same zone. For urinary infections, urine specimens.

All patients with a pulmonary process had a history of pneumopathy. The patient with a disseminated infection had renal insufficiency treated by hemodialysis. In the case with CSF infection, the patient had a cerebral hemangioma for which a peritoneal drainage had been set in position; the patient with sternal osteitis had post-operative valve replacement with a Björk-Shiley prosthesis.

In 14 of the 21 patients with pleural or pulmonary infection, the clinician, suspecting a tuberculous process, started triple tuberculostatic treatment. In the other cases, tuberculostatics or antibiotics were given as recommended in the literature, or found to be effective *in vitro*.

In patients in whom the evolution of the disease was followed, positive cultures were found for a long time in pulmonary cases despite improvement in clinical and radiological symptoms, apart from patients with *M. kansasii* and one patient with *M. avium*, in whom bacilloscopy and culture became negative after 2 months (case no. 43).

DISCUSSION

Pulmonary processes caused by *M. kansasii* responded well under triple tuberculostatic treatment which was mostly given prior to identification and despite the fact that of the drugs used only EB was found to be active *in vitro* in 7 of the cases. Pulmonary infection due to *M. kansasii* has been described as being easy to treat and it is now recommended that RP, INH and EB be used for 2 years (3).

Pulmonary infections have often been described for *M. avium* or *M. fortuitum*, but *M. gordonae* is rarely found, and is generally recognised as a saprophyte although it has occasionally been found to be the pathologic agent in pulmonary pathology (4,5). These 3 species do not react in a uniform way to treatment; of the 9 patients with pulmonary infection due to these species, 6 had cultures which were repeatedly positive despite clinical improvement.

The clinical but not the bacteriological resolution has been described previously (6,7), especially in cases with *M. avium*. It did not prove possible, since the number of patients involved was small, to establish any relationship between therapy employed and sputum conversion or clinical development.

We found no criteria given in the literature as to how to proceed in cases

TABLE I

Species Sites	Case No.	Criteria		Treatment	Development Clinically	Bacteriologically
M.kansasii						
pulmonary	1, 2, 3, 4	AB	abde	RP, INH, EB	?	?
	7	AB	abde	RP, INH, EB	good	negative
	6	ABD	abde	RP, INH, EB	good	negative
	8	ABCD	abde	RP, INH, EB	good	negative
	5	ABD	abdec	RP, EB, Er	good	negative
M.marinum						
cutaneous	33	BCD	abc	RP, Er, Tr	good	N.R.
M.gordonae						
thyroid abscess	36	BC	bcde	Q+RP,INH,EB	good	N.R.
maxilar osteitis	37	ABCD	bcde	Q+RP,EB,SM,Er	good	N.R.
pulmonary	42	AB	bde	RP,INH,EB,SM	good	on-going (1 year) persistent + cult.
M.scrofulaceum						
adenopathy	19, 23	BC	ace	Q	good	N.R.
	22	AB	abe	Q	good	N.R.
	44	ABC	abcde	Q	good	N.R.
	20, 21	BC	acde	Q+RP,INH,EB	good	N.R.
pleural	18	BC	acde	RP, INH, EB	good	N.R.
M.avium						
pulmonary	9	AB	abde	RP, INH, EB	exitus	1 year + culture
	10	AB	abde	RP, INH, EB	good	on-going (6 m.) + culture
	41	AB	abd	RP,INH,Dx,Cl	good	on going (16 m.) + culture
	43	AB	abcd	RP,INH,EB,Pz	good	positive (2 m)

where treatment should be stopped, although bacteriologically still positive.

In the four cases with pleural infection, three samples were biopsy specimens and it is not known if EM isolates have been found before in this type of sample. We only know of four EM isolates in CSF (8,9,10,11); in our case, the patient died of a subarachnoid abscess which was found only on autopsy.

In the patient with sternal osteitis (publication in press) the infection was not controlled by any therapeutic measures recommended and could be eliminated only by radical surgery. This has been described in epidemics in cardiac surgery (12).

We have now established a protocol for this type of infection, in terms of guidelines for treatment and development.

TABLE I (continued)

Species Sites	Case No.	Criteria		Treatment	Development Clinically	Bacteriologically
M.avium						
adenopathy	24,25,26	BC	abce	Q	good	N.R.
	28	AB	abe	Q	good	N.R.
	29,31	BC	ace	Q	good	N.R.
	27	BC	acde	Q+RP,INH,EB	good	N.R.
submandibular abscess	30	BC	abcde	Q+RP,INH,EB	good	N.R.
M.fortuitum						
pulmonary	11	ABD	abcde	RP,INH,EB,Cf,Ak	good	?
	12	AB	abde	RP, INH, EB	?	?
	13	AB	abde	RP,INH,EB,Tr,Er	good	on-going (2 years) + culture
	14	AB	abde	EB, Er, Tr	good	4 months control + culture
pleural	15,17	BC	abcde	RP, INH, EB	good	N.R.
	16	BC	acde	RP, INH, EB	good	N.R.
adenopathy	32	BC	abcde	Q+RP,INH,EB	good	N.R.
disseminated	38	ABCD	abcde	Q+RP,INH,EB,Cf Ak	good	negative
subarachnoid abscess	39	ABC	abc	RP,INH,EB,Ak	exitus	always + culture
sternal osteitis	40	ABCD	abcde	Q+RP,INH,Ak,Cf, Dx	good	negative after sternectomy
M.chelonei sub.**abscessus**						
urinary	35	AB	abde	RP, INH, EB	good	?
cutaneous	34	ABC	ac	Tr, Ak, Er	good	?

Case No. corresponds to case Nr. given in the previous publication (1) (No. 1-38) to which 6 cases have been added over the last year (No. 39-44)
Q = surgical ; RP = Rifampicin ; INH = Isoniazide ; EB = Ethambutol ; SM = Streptomycin Er = Erythromycin ; Ak = Amikacin ; Tr = Tetracycline ; Dx = Doxycycline ; Cl = cefaclor ; Pz = Pirazinamide ; Cf = Cefoxitin ; N.R. = not realized ; ? = unkown ;

REFERENCES

1. Martin Casabona N, González Fuente T, Fernández Pérez F (1985) Med Clin (Barc) 84 : 651-654
2. Wolinsky E (1979) Am Rev Respir Dis 119 : 107-159
3. Bailey WC (1983) Chest 84 : 5 625-628
4. Clague H, Hopkins CA (1985) Tubercle 66:61-63
5. Gernez-Rieux C, Tacquet A (1959) Bull Int Union Tuberc 29:330-342
6. Yeager H, Raleigh JW (1973) Am Rev Respir Dis 108:547-552
7. Dutt AK, Stead WW (1979) Am J Med 67:449-453
8. Hand WL, Stanförd YP (1970) Ann Int Med 73 : 971-977
9. Dalovisio JR, Pankey GA (1978) Am Rev Respir Dis 117:625-630
10. Santamaria Jauregui J, Sanz Hospital J, Berenguer J, Muñoz D, Gómez Mampaso E, Bouza E (1984) Am Rev Respir Dis 130:136-137
11. Dalavisio JR, Pankey GA, Wallace RJ, Jones DB (1981) Rev Infect Dis 3:1068-1074
12. Robicsek F, Daugherty HK and al (1978) J Thorac Cardiovasc Surg 75:91-96

Mycobacteria of Clinical Interest. M. Casal, editor

DISEASES DUE TO MYCOBACTERIA OTHER THAN *M. tuberculosis* IN HUMAN POPULATIONS. ARGENTINA, 1982-1984.

LUCIA BARRERA[1], ISABEL N. de KANTOR[2], ANDREA SALINAS[1]

[1]Instituto Nacional de Microbiología "Dr. Carlos G. Malbrán" and

[2]Centro Panamericano de Zoonosis (CEPANZO, OPS/OMS), C.C. 3092,Correo Central, 1000 Buenos Aires, Argentina

INTRODUCTION

The relative importance of nontuberculous mycobacteria as a cause of human disease has been analyzed in studies performed in the city of Buenos Aires, covering the periods 1963-67 (1), and 1977-80 (2), and in Santa Fe (3), for the years 1977-80. Percentages of cases due to atypical mycobacteria as compared with total mycobacterial cases were 0.20, 0.37, and 1.6%,respectively.

It has also been established that 4.5% of the country's cattle is infected with *Mycobacterium bovis* (4). These infected animals are undoubtedly a source of disease for man.

Of the total pulmonary cases, those caused by *M. bovis* were 5.7% in 1946 (5), and 1.5% in 1971 (6) in Buenos Aires. To date, no national survey on the frequency and distribution of diseases due to mycobacteria other than *M. tuberculosis* has been carried out in Argentina.

MATERIAL AND METHODS

Mycobacterial strains were isolated by fifteen public health laboratories,from five major regions of the country (Provinces of Buenos Aires and Santa Fe, Central, Northwestern and Southern regions).

The study was initiated on January 1, 1982 and concluded on December 31, 1984. During this period the participant laboratories submitted their cultures for identification when at least one of the following conditions was observed: a) growth on Stonebrink medium and scant of no growth on Lowenstein-Jensen medium; b) growth in less than ten days; c) colonial morphology or pigmentation different from those of *M. tuberculosis*; d) resistance to several antituberculous drugs in strains isolated from patients who had not been previously treated; e) negative niacin test.

Laboratories were requested to report annually: a) number of positive cultures; b) number of patients with positive cultures; c) number of cultures other than *M. tuberculosis*; d) number of patients with cultures different from *M.tuberculosis*; e) number with mycobacteriosis or bovine tuberculosis.

Diseases due to nontuberculous mycobacteria were diagnosed on the basis of the following criteria: a) clinical evidence of disease consistent with mycobacteriosis; b) repeated isolation of the same strain from naturally contaminated specimens or isolation of the mycobacteria at least once from closed lesions under sterile conditions; c) no isolation of *M. tuberculosis*; d) growth of more than 5 colonies in each isolate; e) low response to standard antituberculous chemotherapy regimens.

Mycobacterial identification was performed by the Instituto Nacional de Microbiología (INM) in collaboration with CEPANZO. The methods employed have been described in detail (7).

RESULTS

During the three-year period of the study, the participant laboratories reported 13 544 mycobacterial positive cultures from 7 672 patients. Of these, 437 were classified as nontuberculous mycobacteria and 49 (0.36%) as *M.bovis*(Table I).

Twenty seven patients (0.35%) were mycobacteriosis cases, 36 (0.47%) were tuberculosis cases due to *M.bovis* and the rest were *M.tuberculosis* cases(TableII).

M. avium-intracellulare-scrofulaceum complex (MAIS) was the etiologic agent in 26 of the 27 mycobacteriosis cases (25 pulmonary and 1 lymphadenitis).The remaining case was due to *M. chelonei*. The number of cultures from these patients ranged from two to seven. Nontuberculous mycobacteria were isolated once only from 214 patients and were considered of no clinical significance.

Information on cases due to *M. bovis*, was collected on 32 out of 36 cases. All of them suffered from pulmonary disease and 2 also had nodal and skin complications. At least 19 had been in contact with cattle. One case was caused by a *M. bovis* BCG strain isolated from endocranial granulomatous tissue and skin lesions biopsies from a three-year old immunocompromised boy.

DISCUSSION

The present study confirmed that diseases due to non tuberculous mycobacteria cannot be considered as a major sanitary problem in Argentina. The average national percentage of atypical mycobacteriosis obtained in the present survey (0.35%) is similar to the percentage previously reported in Buenos Aires (2), but surprisingly, *M.kansasii* , the most frequent nontuberculous pathogenic strain reported in Buenos Aires during the period 1977-1980, was not detected in this survey. This could be due to failure to recognize the cultures of this species as nontuberculous mycobacteria or to the fact that *M. kansasii* infection is circumscribed to that city.

The highest percentages of tuberculosis cases due to *M. bovis* were recorded in the provinces of Santa Fe and Buenos Aires. Cases in the remaining regions of the country were low or nil. A correlation was observed between this regional variation of human cases and the rate of cattle infection (4).

According to our results, pulmonary *M. bovis* disease in adults is not an important epidemiologic problem either. It would be restricted to high risk population such as rural or slaughterhouse workers. However, there is not enough information on *M. bovis* infection in children. Hygienic and preventive measures such as inspection of cattle at slaughter, pasteurization or boiling of milk would protect effectively the population in spite of the high rate of tuberculous infection in cattle.

TABLE I

ATYPICAL MYCOBACTERIA AND *M. bovis* STRAINS ISOLATED FROM MAN IN ARGENTINA (1982-1984) REGIONAL DISTRIBUTION

Region	Positive cultures	Atypical mycobacteria	*M. bovis*
1) Buenos Aires	5 936	58 (0.98%)	29* (0.49%)
2) N.E.	690	7 (1.01%)	0 (0.0%)
3) Santa Fe	1 953	251 (12.85%)	16 (0.82%)
4) Center	1 312	83 (6.33%)	3 (0.23%)
5) N.W.	3 476	31 (0.89%)	1 (0.03%)
6) South	177	7 (3.95%)	0 (0.0%)
Total	13 544	437 (3.22%)	49 (0.36%)

* One BCG strain.

TABLE II

DISEASE DUE TO ATYPICAL MYCOBACTERIA OR TO *M. bovis* IN MAN, ARGENTINA (1982-1984) REGIONAL DISTRIBUTION

Region	Patients with positive cultures	Mycobacteriosis cases	*M. bovis* cases
1) Buenos Aires	3 010	6 (0.20%)	15 (0.50%)[a]
2) N.E.	588	0 (0.0%)	0 (0.0%)
3) Santa Fe	863	8 (0.93%)	16 (1.85%)
4) Center	821	11 (11.34%)	3 (0.37%)
5) N.W.	2 288	2 (0.09%)	1 (0.04%)
6) South	102	0 (0.09%)	0 (0.0%)
Total	7 672	27 (0.35%)	36 (0.47%)

[a] One case due to BCG.

ACKNOWLEDGEMENTS

The study group wishes to acknowledge the following special contributions:

- Servicio Nacional de Tuberculosis y Enfermedades Respiratorias/Ministerio de Salud y Acción Social, Argentina, for its support with the participating laboratories.

- Dr. Adalbert Laszlo (National Reference Centre for Tuberculosis of Canada/ Laboratory Centre for Disease Control), who provided a collection of mycobacterial reference strains and some chemical products.

REFERENCES

1. Cetrángolo A, Kantor IN de (1969) Medicina (B. Aires)29:186-189
2. Di Lonardo M, Isola N, Ambroggi M, Fulladosa G, Kantor IN de (1983) Bol Of San Panam 95: 134-141
3. Latini O, Inst Nac Epid Santa Fe, 1977-81 (1982) Informe N° 7, Santa Fe, Argentina, pp. 2-3
4. Comisión Nacional de Zoonosis (1982) La tuberculosis bovina en la R. Argentina, B. Aires, pp. 11-15
5. Arena A, Cetrángolo A (1946) An Cát Pat Tuberc (B. Aires), 8: 36-40
6. Cetrángolo A, Marchesini L, Isola NC, Kantor IN, Di Lonardo M (1971) Actas XIII Congreso Argentino de Tisiología y Neumonología, M. del Plata, pp. 845-846
7. Kantor IN de (1979) Bacteriología de la tuberculosis humana y animal. Buenos Aires, Centro Panamericano de Zoonosis, OPS/OMS (Ser Monog Cient Técn 11)

© 1986 Elsevier Science Publishers B.V. (Biomedical Division)
Mycobacteria of Clinical Interest. M. Casal, editor

MYCOBACTERIA IDENTIFIED IN THE PASTEUR INSTITUTE (PARIS) DURING 1978-1984

HUGO L. DAVID, VERONIQUE LEVY-FREBAULT, ANTOINETTE FEUILLET and JEANNINE GRANDRY
Unité de la Tuberculose et des Mycobacteries, Institut Pasteur, Paris (France)

INTRODUCTION

The occurrence and increase in the number of diagnosed cases of nontuberculous mycobacterial disease in man is well documented, and the subject has been reviewed by Wolinsky (12). The most common clinically significant mycobacteria have been thoroughly described (6), and most are environmental bacteria (7,10). It has recently been shown that the isolation of clinically significant mycobacteria from clinical specimens correlates reasonably well with their geographic distribution (2,3) and may contribute to a better understanding of their occasional role as pathogens (4,5). Also, it has been suggested that subclinical infection by these bacteria may interfere with antituberculous vaccination (11). From the above considerations we decided to review the data on mycobacteria identified in the Pasteur Institute. A previous report in France was published by Boisvert (1), and similar surveys have been published in the USA (4,5).

MATERIAL AND METHODS

During 1978-1984, 7715 isolates from 7671 patients were identified. The strains were submitted by public and private laboratories located in each of the 13 health regions into which France is divided. The strains were isolated from the pathological secretions of patients suspected of having tuberculosis, or a search for tubercle bacilli was considered necessary in the differential diagnosis of other pathological conditions. For the purpose of this report, when more than one strain was received for one patient, only the first isolate was entered in the statistical analysis, while mixed cultures were entered as independent isolates.

Because the distribution of mycobacteria referred to a reference laboratory may not reflect the findings in clinical laboratories, a questionnaire was sent to corresponding laboratories to ascertain the distribution of isolates and their frequency in respect to the total number of specimens examined. Seventy-one of the laboratory directors responded to the questionnaire, and their data are presented in this report.

For identification, the strains were inoculated onto 7H10 agar plates for screening mixed cultures; pure cultures were identified using methods recommended by the Pasteur Institute (8,9).

RESULTS AND DISCUSSION

During the survey period, 44 mixed cultures were identified. As shown in

Table 1, 47.73% of these cultures were of tubercle bacilli mixed with nontuberculous mycobacteria; and one culture contained *M. tuberculosis* and *M. bovis*. The clinically significant mycobacteria more often mixed with tubercle bacilli were the *M. fortuitum*-complex (33.33%) and *M. kansasii* (14.29%).

The distribution of 7715 isolates identified and recorded by the geographic location of the submitting laboratory is shown in Table 2. The more frequent clinically significant nontuberculous mycobacteria were *M. fortuitum*-complex (7.66%), and *M. kansasii* (6.53%). Among the tubercle bacilli (5171 isolates) there were 158 *M. bovis* (3.06%).

A comparison of the frequency distribution of clinically significant mycobacteria identified in the Pasteur Institute and clinical laboratories is shown in Table 3. According to the information received, the clinical laboratories performed 221,259 examinations; 7609 (3.44%) were positive by microscopy alone or by microscopy and culture; and 5964 cultures (2.70%) yielded mycobacterial isolates. *M. tuberculosis* was more often represented in the clinical laboratories than in the Pasteur Institute (95.03% and 73.97%, respectively), while *M. bovis* was more often identified in the Pasteur Institute (2.36% and 0.61%); other isolates more frequently represented in the Pasteur Institute data were BCG, *M. kansasii*, *M. xenopi*, *M. avium* complex, and *M. fortuitum*-complex.

A definitive diagnosis of nontuberculous mycobacterial disease was obtained in 71 patients (Table 4). Because of a lack of patient follow-up, the relative frequency of disease caused by nontuberculous mycobacteria could not be established.

Table 1. Mixed mycobacterial cultures identified.

Mycobacterium	tuberc	bovis	avi	xeno	kans	fort	gast	ter	gord	flav
tuberculosis	-									
bovis	2	-								
avium-complex	2	0	-							
xenopi	1	1	0	-						
kansasii	3	0	0	0	-					
fort.complex	6	1	1	0	0	-				
gastri	0	0	0	0	1	0	-			
terrae	2	0	0	0	1	1	0	-		
gordonae	3	0	4	0	1	2	2	1	-	
flavescens	3	0	0	0	0	4	0	1	1	-
Total	22	2	5	0	3	7	2	2	1	

Table 2. Isolates of mycobacteria referred to the Pasteur Institute, and recorded by the geographic location of the submitting laboratory.

Sanitary region	Number of clinically significant isolates tuberc	bovis	afric	BCG	mari	kans	sim	scrof	szul	xeno	avi	ulce	fort	Total	Non-signif.[1]	Total isolates
1	2631	30	44	76	11	275	2	10	0	143	55	1	277	3555	407	3962
2	171	3	1	0	0	2	0	0	0	7	4	0	13	211	30	231
3	88	5	2	8	0	0	0	0	0	5	2	0	13	123	36	159
4	103	5	1	6	0	18	0	2	0	5	11	0	21	172	44	216
5	235	15	2	5	0	115	1	1	2	25	12	0	22	435	180	615
6	315	19	2	5	1	8	1	3	0	10	24	0	53	441	124	565
7	30	3	1	4	0	4	0	0	0	3	1	0	4	50	12	62
8	118	1	1	6	1	0	0	0	0	2	3	0	27	159	18	177
9	68	29	1	3	0	0	0	2	1	1	6	0	14	125	8	133
10	194	5	0	0	0	0	0	0	0	3	1	0	26	229	43	272
11	502	19	2	11	0	5	1	0	1	5	8	0	34	588	59	647
12	348	6	1	6	0	4	0	2	2	8	12	0	5	394	44	438
13	151	18	1	4	0	6	0	0	0	20	21	0	4	225	13	238
TOTAL	4954	158	59	134	13	437	5	20	6	237	160	1	513	6697	1018	7715
%[2]	73.97	2.4	0.9	2.0	0.2	6.5	0.1	0.3	0.1	3.5	2.4	0.01	7.7	100	-	-
%[3]	64.21	2.0	0.8	1.7	0.2	5.6	0.1	0.3	0.1	3.1	2.1	0.01	6.6	86.8	13.2	100.0

[1] M. gordonae(441);M. flavescens(102);M. terrae(220);M. gastri(136);M.thermoresistibile(1);various rapidly growing(118).

[2] Frequency of clinically significant isolates; [3] Frequency of total isolates.

NB:"Clinically significant nontuberculous mycobacteria"="opportunists"="potentially pathogenic".

Table 3. Clinically significant mycobacteria identified in the Pasteur Institute and 71 clinical laboratories.

Mycobacterium	Pasteur Institute (1978-1984)		Clinical laboratories (1980)	
	Number	%	Number	%
tuberculosis	4954	73.97	5450	95.03
bovis	158	2.36	35	0.61
africanum	59	0.88	11	0.19
BCG	134	2.00	24	0.42
marinum	13	0.19	0	-
kansasii	437	6.53	52	0.91
simiae	5	0.07	0	-
scrofulaceum	20	0.30	6	0.10
szulgai	6	0.09	0	-
xenopi	237	3.54	70	1.22
avium-complex	160	2.39	22	0.38
ulcerans	1	0.01	0	-
fortuitum-complex	513	7.66	65	1.13

Table 4. Nontuberculous mycobacterial disease reported to the Pasteur Institute.

Location of infection	Mycobacterium										
	mari	kans	sim	scro	szul	xeno	avi	ulc	for	other	No.
Lungs		1			6	2					9
Nodes[1]		4		1		2	22		1	1[2]	31
Bone& joints						1	1		4	1[3]	7
Skin	13							1	2		16
Dissemi-nated		2	1								3
Peritoni-tis							2		2		4
Number	13	7	1	1	6	6	25	1	9	2	71

[1] For a total of 209 infections of lymph nodes, 129 were caused by M. tuberculosis(61.72%); 8 by M. bovis(3.83%); 38 by BCG(18.8%); 3 by M. africanum(1.44%); and the remaining 31 (14.83%) by the nontuberculous mycobacteria indicated in the table.

[2] M. thermoresistibile; [3] M. terrae.

The data presented in this report give some useful information about the geographic distribution of nontuberculous mycobacteria in France, and indicate that a controlled investigation should be considered in view of their uneven distribution. In clinical laboratories, the frequency of nontuberculous clinically significant mycobacteria was 3.74%, which was significantly less than in the USA (31.84%, ref. 3).

Judging from the data shown in this report, the frequency of nontuberculous mycobacteria disease in France was conservatively established at 1.35% of all cases of mycobacterial disease. Even if the referral of cultures and the lack of patient follow-up did not follow a standardized reporting system, the data do give sufficient information on which to base a controlled national survey.

ACKNOWLEDGEMENTS

We wish to thank our colleagues who referred their isolates, and those who contributed clinical information, too many to be listed herein.

REFERENCES

1. Boisvert,H.(1973)Ann. Soc. Belge Med. Trop. 53,233-245.
2. FalkinhamIII,J.O.,Parker,B.C.,Gruft,H. (1980) Am. Rev. Resp. Dis.121,931-937.
3. George,K.L.,Parker,B.G.,Gruft,H.C.,FalkinhamIII,J.O. (1980)Am.Rev. Resp. Dis. 122,89-94.
4. Good,R.C.(1980) J. Infect. Dis. 142,779-783.
5. Good,R.C.,Snider Jr.,D.E.(1982)J. Infect. Dis.146,829-833.
6. Kubica,G.P., David,H.L.(1980), in"Gradwohl's Clinical Laboratory Methods and Diagnosis",C.V.Mosby Co.,USA.
7. Kubin,M.(1984) in"The Mycobacteria:a source book", Marcel Dekker Inc., USA.
8. Levy-Frébault,V.,Grandry,J.,David,H.L.(1982)J. Med. Micro. 15,575-577.
9. Meyer,L.,David,H.L. (1979) Ann. Microbiol.(I. Pasteur) 130B,323-332.
10. Tsukamura,M.(1984) in "The Mycobacteria: a source book", Marcel Dekker Inc., USA.
11. Vaccination against Tuberculosis, WHO Technical Report series No.651, WHO-Geneva, 1980.

LEPROSY

Mycobacteria of Clinical Interest. M. Casal, editor

RECENT ADVANCES IN THE BACTERIOLOGY OF *MYCOBACTERIUM LEPRAE*

HUGO L. DAVID and NALIN RASTOGI

Service de la Tuberculose et des Mycobactéries, Institut Pasteur, 75015 Paris (France)

INTRODUCTION

Since 1971, there has been a significant increase in reports describing the properties of *Mycobacterium leprae*. This was a consequence of the discovery that the nine-banded armadillo was highly susceptible to experimental leprosy (1), which was followed by the decision of the World Health Organisation to establish a *M.leprae*-bank, making infected tissues and bacteria available for investigation (2). The present review is concerned with the knowledge gained on *M. leprae* recovered from armadillos since the Paris meeting in June 1984. For further information, the reader is referred to 8 review articles (and the references included) that appeared between 1971 and 1984 (3,4,5,6,7,8,9,10).

RECOVERY AND PURIFICATION OF *M. LEPRAE* SUSPENSIONS

M. leprae is recovered and purified from the tissues of infected animals since it could not be cultivated in artificial media. Recovery and purification are two distinct procedures in the preparation of bacterial cells for examination. The addition of chemicals (acids, alkalis, detergents, etc.) and enzymes (hyaluronidase, collagenase, proteases, nucleases) to the ground host tissues may help to liberate the bacteria but are not sufficient to eliminate the host components attached to their surface. Thus, in general, it may be stated that the final purification step needs to be applied to bacterial suspensions, already largely liberated from host material. The final purification step depends on the aim of the investigator who, nonetheless, is constrained to select appropriate controls because no method of purification can ensure the removal of all unwarranted host components and/or may affect the bacterial components in which the investigator is mostly interested.

Various procedures for recovering and purifying *M. leprae* are summarized in Table 1. The tissues used may have been non-sterilized or may have been sterilized using γ-radiations. The tissues, usually frozen, are chopped into small fragments and are ground into a paste either manually in the presence of an abrasive (alumina, sterilized sand, etc.), or with the help of a tissue grinding device. Usually, a buffer is added to facilitate grinding and the release of the bacteria.

Regarding evaluation of the quality of the final product, the methods used may be classed as general and specific. The general methods include culturing for contamination by mycobacteria other than *M. leprae* and other micro-organisms,

Table 1 : Methods used for purification of *M. leprae* from infected tissues.

GRINDING	RECOVERY	PURIFICATION	REFERENCES
1. Homogenization in Tris-HCl	Differential centrifugation and washing of 10,000 x g pellet with HEPES-NaCl (+0.1% Tween)	DNAase treatment (pH 7.2). Percoll density gradient. 2 phase separation on dextran-PEG (pH 6.9)	P.DRAPER, Protocol 1/79, IMMLEP, WHO.
2a. Grinding with sterilized alumina in Spizizen salt solution.	Differential centrifugation repeated 2-3 times.	High speed pellet. 4% H_2SO_4 and/or 4% NaOH (10 min at 20°) and repeated washings with salt solution.	H.L.DAVID *et al.* Ann. Virol. (Inst. Pasteur) (1980). 131E : 167-184.
2b. Idem	Idem	Idem, followed by DNAase treatment & washed with Tris-EDTA-NaCl buffer	H.L.DAVID *et al.* Acta Leprologica (Nouvelle Série) (1984). 2:129-136.
3a. Homogenization in 0.25 mM EDTA + 0.3M sucrose (pH 7.2)	Differential centrifugation.	Pellet washed in buffred sucrose → 0.1M dipotassium EDTA → 0.1% Tween → separated by dextran-PEG partition	C.C.SHEPARD *et al.* Brit. J. Exp. Pathol. (1980). 61 : 376-379.
3b. Homogenization in 0.01M dipotassium EDTA.	Differential centrifugation.	Pellet → 0.1M NaOH (2H 20°) → Tris-HCl washed → Collagenase (24H) → Trypsin + Chemotrypsin (24H) → Dextran-PEG 2 Phase partition.	
4 Homogenization in Sorvall Omnimixer → 1H (37°) treatment with Pronase + SDS in Tris-HCl	Differential centrifugation.	High speed pellet → DNAase (1H, 37°) → 1H (37°) Pronase treatment → Bacilli recovered after washing.	T.IMAEDA *et al.* J. Bacteriol. (1982) 150 : 414-417.

Table 1 Continued on next page →

Table 1 continued.

	GRINDING	RECOVERY	PURIFICATION	REFERENCES
5a.	Homogenization in Polytron homogenizer in 0.3M sucrose + 0.1% Tween & 10 mM Tris base (pH 9.5)	7,000 x g cenfugation → pellet washed in Tris-NaCl or PBS.	DNAase treatment → dextran-PEG 2 phase partition → sedimentation in sucrose gradient using a unit gravity chamber.	P.K.DAS & A.TULP Ann. Microbiol. (Inst. Pasteur). (1982) 133B : 389-400.
5b.	Homogenization in Tris-HCl → Tween 80 → Collagenase + Elastase treatment	Idem	Idem	
6.	Homogenization in Sorvall Omnimixer in HEPES-NaCl with $MgSO_4$ → 12,000 x g centrifugation → high speed pellet further homogenized*	12,000 x g centrifugation and pellet washed.*	Pellet → Collagenase → DNAase → Percoll density gradient centrifugation → 2 phase dextran-PEG partition.	J.CLARK-CURTISS et al. J. Bacteriol. (1985).61 : 1093-1102.

*1 mM benzamidine always present as inhibitor of armadillo cellular proteases.

optical microscopy using soluble blue as counterstain, immunodiffusion using antibodies raised against armadillo tissues, search for nonmycobacterial lipids such as cholesterol and phosphatidylcholine, and electron microscopy. The specific methods relate to the purposes of the individual investigator and may require fairly large amounts of the product. Examples of specific methods (enzyme activities, chemical composition, etc.) are given throughout this review.

MORPHOLOGY AND ULTRASTRUCTURE OF *M. leprae*

M. leprae are acid-fast, rod-shaped bacteria. Their morphology and staining properties as observed using the optical microscope have often been reviewed and need not be recalled here. Instead, we shall review their structure as observed in the electron microscope. It should however be indicated that most of the ultrastructural data described before 1971 were from sources other than armadillos.

Observed *in situ* in sections of infected tissues, most bacilli appear located inside phagocytic cells (Fig. 1A). In the host cells, the bacilli are often in groups, but isolated bacilli are also seen. The bacteria are located in phagosomes, and are surrounded by a substance that separates the bacterial wall from the phagosomal membrane as if the bacteria were surrounded by a 'capsule'. A similar observation in *M. lepraemurium* was made by Chapman et al (11) who designated the substance as 'capsule substance'; however the designation 'electron-transparent zone' (ETZ) used by Draper & Rees (12) has gained general acceptance. In the tissues, most of the bacteria seen are often extensively damaged (Fig. 1).

M. leprae separated from the tissues are not surrounded by ETZ. However, when the bacteria are phagocytized *in vitro*, they are rapidly surrounded by a ETZ, whether they are living or dead (13,14), as if the ETZ formation was induced by some compound located on the bacterial cell wall.

As in other mycobacteria, the cell wall of *M. leprae* has a trilaminar structure. This statement must be qualified because often the mycobacterial cell walls are described as being composed of a basal peptidoglycan layer and a slightly electron-dense outer layer. The reason for the apparent contradiction is that the outer layer does not stain using the standard lead citrate coloration, and special methods are required to demonstrate a mycobacterial outer layer (15,16). Using the special cytochemical stainings, it can be concluded that the mycobacterial cell wall is made up of a basal peptidoglycan layer, a middle electron-transparent layer, and a polysaccharide-rich outer layer.

Underneath the cell wall is located the cytoplasmic membrane, which has a symmetric geometry. When first recognized, this geometry was interpreted as an unusual feature among mycobacteria, but the typical asymmetric geometry has now been reported (17). As in all other mycobacteria, the cytoplasmic membrane of the leprosy bacilli, but not the cell wall, stains using the cytochemical method of Thiéry (6,18).

Fig. 1. Ultrastructure of *M. leprae*
(A) Bacteria in armadillo liver are often seen in clumps inside phagocytic cells. Engulfed by the phagosomes, an electron-transparent zone (ETZ) separates the bacterial surface from the phagosomal membrane. (B and C) Intact bacteria with a central fibrillar, nucleic acid area. The cytoplasmic membranes are distinctly symmetric, and are often highly developed as in C. (D) In an intact bacterium from nude mice footpad, a central nucleic acid core and uniformly distributed ribosomes are apparent. A uniform, continuous cytoplasmic membrane is distinctly symmetric. (E) A degraded bacterium from nodes of an experimentally infected nude mouse. (F) Longitudinal section of an intact bacterium with highly developed symmetrically stained, cytoplasmic membrane structure and a central nucleic acid area. The cytoplasm however is devoid of ribosomes and is homogenous (Type II cell according to a previous morphological classification of *M. leprae* bacteria, for further details, see reference 18).

Bar marker represents 100 nm

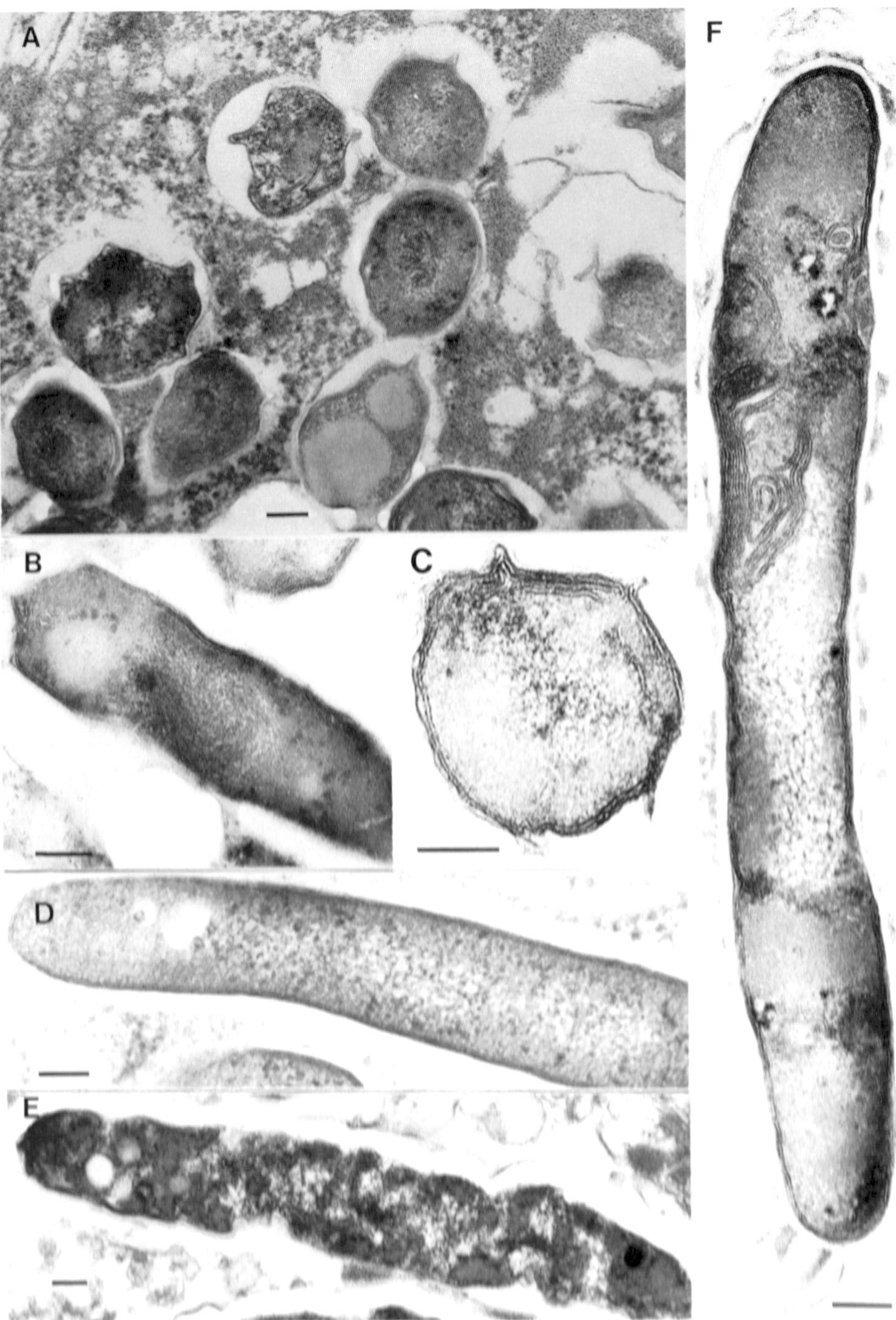

Fig. 1. Ultrastructure of *M. leprae*.

Electron microscopic studies in bacteria *in situ* or of bacterial suspensions show large numbers in various stages of autolysis. In general, the data published in recent years confirm early studies (19) correlating ultrastructure and optical microscopy, and ultrastructure and viability. The occurrence of large numbers of damaged bacteria in purified suspensions must be taken into account when evaluating chemical, biochemical, genetic, and immunological data (6).

The outer surfaces of *M. leprae* were examined in the electron microscope using negative staining and freeze-fracture methods (for a review, see 4,5, and included references); and recently using cytochemical markers: ruthenium red (which reacts with acidic polysaccharide residues); concanavalin A and wheat germ agglutinin (which bind to α-D-glucose, α-D-mannose, or N-acetyl glucosamine residues); and colloidal iron and cationized ferritin (revealing negatively charged groups). While wheat germ agglutinin did not react with the bacterial surfaces, all the remaining compounds more or less reacted with the bacterial surface (20).

CHEMICAL COMPOSITION

Very little work has been done to establish the chemical composition of *M. leprae*, mainly because of the difficulty in liberating the bacilli from contaminating host components. Most of the published work concerned the occurrence of complex lipids already known to be characteristic in the genus *Mycobacterium* as listed in Table 2. Most of these lipids cumulate in the tissues of diseased animals (21,22), and their isolation does not require highly purified bacterial suspensions because they can be separated from contaminating host lipids. As indicated in Table 2, some of the complex lipids were isolated and purified, and their molecular structure was extensively examined; other lipids were tentatively identified using thin layer chromatography only.

The presence of diaminopimelic acid (DAP) in the cell walls of *M. leprae* appears to be firmly established (23,24), but whether it is an LL- or DL-enantiomorph remains to be demonstrated (DL-DAP is characteristic in all mycobacterial species examined so far). The walls contain muramic acid and glucosamine, and the presence of an arabinogalactan has been suggested (23).

The structure of DNA from *M. leprae* was reported by Imaeda et al (25). These investigators indicated that the DNA had a G + C content of about 55-56%, and a molecular weight of about 1.3×10^9. Although these data were questioned by David et al (26), they were confirmed by Curtiss et al (27), who however found a genome size of 2.2×10^9 dalton. The G + C content of armadillo DNA was determined to be about 36-38% (26,27).

Homologies between the DNA from *M. leprae* and *M. tuberculosis* and *Corynebacterium tuberculostearicum* were, respectively, 21% and 20% (28). Curtiss et al (27) indicate some homology, and thus conserved sequences, between *M. leprae*

Table 2 : Chemistry of M. leprae

COMPOUNDS IDENTIFIED	REFERENCES
1. Mycolic acids : α-mycolate keto-mycolate methoxy-mycolate ?	Etemadi & Convit (1974). Infect. Immun. 10 :236-239 Asselineau et al.(1981).Ann.Microbiol. (Inst. Pasteur) 132A : 19-30 ; Draper et al. (1982). Ann. Microbiol.(Inst.Pasteur).133B : 39-47 ; Kusaka et al.(1982).Int.J.Lepr. 49 : 406-419 ; Young (1980). J.Gen.Microbiol.121 :249-253 ; Minnikin et al. (1985).J.Gen.Microbiol.131 : 2013-2021.
2. Tuberculostearic acid	Anderson et al.(1981).Ann.Microbiol.(Inst.Pasteur) 132B : 29-37 ; Asselineau et al.(1984).Acta Leprologica (nouvelle série) 2 : 121-127 ; Kusaka & Izumi (1983). Microbiol. Immunol. 27 : 409-414.
3. Phthiocerol dimycocerosate and phthiodiolone.	Draper et al.(1983).J.Gen.Microbiol.129 :859-863. Minnikin et al.(1984).Acta Leprologica (Nouvelle série) 2 : 113-120 ; Asselineau et al.(1984).Acta Leprologica.(Nouvelle série) 2 : 121-127 ; Minnikin et al.(1985).J.Gen.Microbiol. 131 : 2007-2011.
4. Mycoside type A (phenol glycolipids)	Brennan & Barrow (1981).Int.J.Lepr. 48 :382-387. Brennan (1983).Int.J.Lepr. 51 : 387-396.
5. Phospholipids* : phosphatidylethanolamine phosphoionositides	Minnikin et al.(1985).J.Gen.Microbiol. 131:2007-2011 ; David et al.(1985).Ann.Microbiol. (Inst. Pasteur).136A : 303-310 ; Young (1982).Ann.Microbiol.(Inst.Pasteur).133B : 53-58 ; Minnikin et al. (1984).Acta Leprologica (Nouvelle série) 2 :113-120 ; Young (1981).Int.J.Lepr. 49 : 198-204.
6. DAP in cell wall* : DD-DAP (mixture of LL and DL ?)	Draper (1976).Int. J. Lepr. 44 : 95-98.
meso-DAP	David & Rastogi. (1983). Curr.Microbiol. 9:269-274
7. Muramic acid in cell wall*	Draper (1976). Int. J. Lepr. 44 : 95-98.
8. Arabinogalactan* ?	Draper (1976). Int. J. Lepr. 44 : 95-98.

* Tentative identification based on TLC patterns only.

and *M. vaccae* and a strain designated 'lufu', but unfortunately they did not provide any data. Because significant levels of DNA-DNA homologies may be surprising between DNA's largely distinct in their G + C content, further studies on these lines are necessary.

PHYSIOLOGICAL AND BIOCHEMICAL ACTIVITIES

The overall physiological activity of *M. leprae* has been examined using various radiolabeled substrates, measurement of ATP content, leakage of potassium ions, and reduction of potassium tellurite as a measure of respiratory activity (5). In general, the data reported indicated that the bacteria were poorly efficient in respect to each of the parameters measured and in comparison with the activities reported for other laboratory-grown mycobacteria. This situation is thought to be related to the large numbers of highly damaged and dead bacteria present in the tissues from which they were recovered (6).

Various enzymes from *M. leprae* have now been identified from which certain metabolic pathways were inferred, as shown in Table 3. Because many of the reported enzymes are widespread among living cells, in most of the work reported, specific methods for control were described. It should be recognised that these methods may be applied to evaluate the quality of products used in other areas of endeavour such as immunological and genetic investigations.

Although the specific activities of *M. leprae* enzymes were usually lower than corresponding enzyme activities in cultivable species *M. phlei*, this was not always the case (for example succinyl-CoA-synthetase, succinate dehydrogenase, and fumarase; reference 7). Considering that a large proportion of *M. leprae* bacteria are in a state of autolysis, these data seem interesting and merit further examination.

GENETIC STUDIES

Genetic studies in *M. leprae* were approached in two ways: one consists of using *M. leprae* as a recipient for mycobacterial genes from other mycobacteria, and the other of using other bacteria as recipients of *M. leprae* genes.

Regarding the first of these approaches, the adsorption of mycobacteriophages onto *M. leprae*, and the injection of their DNA into the bacteria were demonstrated; however, the incoming DNA was rapidly degraded (29,30). Although the mechanism of DNA degradation has not been elucidated, it was attributed to autolysis of *M. leprae* when suspended in artificial media. Nonetheless, the preparation of appropriate mutants of these phages may eventually permit the packaging of mycobacterial genes (cloned into mycobacterial plasmids).

The second approach has been more successful since Curtis et al (27) reported the construction of genomic libraries of *M. leprae*. These investigations cloned

DNA fragments obtained by endonuclease digestion into plasmids, and used these hybrid plasmids to transform or transduce *E. coli* minicells. Using SDS-PAGE electrophoresis, these authors demonstrated the synthesis of mycobacterial polypeptides in these minicells. Whether the *E. coli* protein synthetic machinery recognized the mycobacterial promotors was not established.

DISCUSSION AND CONCLUSIONS

Studies of *M. leprae* have been plagued by the heterogeneity of the purified bacterial suspensions containing high proportions of bacterial ghosts and because of contamination of these preparations by host components. The discovery of a method to sort out living bacteria from these suspensions should resolve many of the discrepant data in the literature, and efforts to accomplish this object are necessary. As far as the purification methods are concerned, it seems clear that there is no best method to recommend, and consequently investigators will use the method they consider most appropriate for their purposes.

Since the Paris meeting in June 1984, the construction of a genomic library of *M. leprae*, and the cloning of its genes, appears as the most significant advance in the field. In the meantime, two other reports related to the subject appeared in the literature. In one of these reports, it was shown that promotors in mycobacterial plasmid DNA were recognized by the *E. coli* protein synthesis machinery (31). In the other paper concerning the construction of a genomic library of *M. tuberculosis*, it was reported that polypeptides formed in the *E. coli* recipients were recognized by monoclonal antibodies (32). These initial reports indicate the great possibilities ahead, particularly the development of diagnostic reagents for leprosy and other mycobacterial diseases.

Regarding the evaluation of potential armadillo-derived vaccines, investigations of the outer surface of bacteria using chemical, cytochemical (20) and ultrastructural methods such as freeze-fracture and -etching (33) should be pursued.

Table 3 : Biochemical activities reported for *M. leprae*.

Activity	Reference
Hydrolytic enzymes :	
N-acetyl- -D-glucosaminidase[ab]	[a]E.Matsuo & O.K.SKINSNES. Int.J.Lepr. (1974). 42 : 399-411.
β-glucuronidase[ab]	[b]P.R.WHEELER *et al*.J.Gen.Microbiol.(1982). 128 : 1063-1071.
Acid phosphatase[b]	
Oxygen free radicals :	
Catalase[abd]	[a]K.PRABHAKARAN. Int.J.Lepr.(1967).35:34-41
Peroxidase[ab]	[b]P.R.WHEELER & D.GREGORY.J.Gen.Microbiol. (1980).121 : 457-464.
Superoxide dismutase[bc]	[c]E.KUSUNOSE *et al*. FEMS Microbiol. Lett. (1981).10 : 49-52.
	[d]V.M.KATOCH *et al*.Ann.Microbiol.(Inst. Pasteur).133B : 407-414.
Respiratory activities :	
NADH oxidase and presence of cytochromes a + a_3, b, c, & o.	M.ISHAQUE *et al*.Int.J.Lepr.(1977). 45 : 114-119.
Reduction of potassium tellurite (by electron microscopy).	H.L.DAVID *et al*. Ann.Microbiol.(Inst. Pasteur).133B : 129-139.
Carbohydrate metabolism :	
- Key enzymes of EMP pathway & HMP shunt. (Hexokinase, Phoshexokinase, Fructose-1,6-diphosphate aldolase (class II), Pyruvate kinase, Glucose-6-PO_4-dehydrogenase, 6-phosphogluconate dehydrogenase. - Glycerol-3-PO_4 dehydrogenase Glycerol dehydrogenase	P.R.WHEELER. J.Gen.Microbiol. (1983). 129 : 1481-1495.
- Lactate dehydrogenase	A.M. SAOJI *et al*. Int.J.Lepr.(1980).48 : 425-430 ; P.R.WHEELER *et al*. Int.J.Lepr. (1984). 52 : 371-376.
- Lactate oxygenase	K.PRABHAKARAN.Int.J.Lepr.(1967).35:34-41.

Table 3 continued.

<u>Carbohydrate metabolism (continued)</u> :	
- NAD-dependent malate dehydrogenase and FAD-dependent malate-vitamin K-reductase.	P.R.WHEELER & V.P.BHARADWAJ. J.Gen. Microbiol. (1983). <u>129</u> : 1321-2325.
Key enzymes of Kreb's TCA cycle :	
- Pyruvate dehydrogenase, Citrate synthase, Aconitase, Isocitrate dehydrogenase, Succinyl-CoA-synthetase, Succinate dehydrogenase, Fumarase.	P.R.WHEELER. J.Gen.Microbiol. (1984). <u>130</u> : 381-389.
<u>ATP content</u> : -12.4 pg/mg (dry weight)	A.M.DHOPLE & J.H.HANKS. Int.J.Lepr. (1981). <u>49</u> : 57-59.
-0.4 fg/bacterium	H.L.DAVID <u>et al</u>. Ann.Microbiol.(Inst. Pasteur). <u>133B</u> : 129-139.
<u>Amino acid metabolism</u> :	
-3,4-dihydroxyphenylalanine (DOPA) oxidase.	K.PRABHAKARAN & W.F.KIRCHEIMER. J.Bacteriol.(1966).<u>92</u> : 1267-1268.
-glutamic acid decarboxylase	K.PRABHAKARAN & B.M.BRAGANCA. Nature (1962).<u>196</u> : 589-590.
- γ-glutamyl transpeptidase	K.T.SHETTY <u>et al</u>. Int.J.Lepr.(1981). <u>49</u> : 49-56.
-Incorporation of ^{14}C-amino acids	H.K.PRASAD & R.C.HASTINGS. J.Clin. Microbiol.(1985).<u>21</u> : 861-864.
<u>Nucleic acid metabolism</u> : -Thymidine incorporation	M.SATISH & I.NATH.Int.J.Lepr.(1981). <u>49</u> : 187-193.
-Uracil incorporation & Incorporation of various purines	S.R.KHANOLKAR & P.R.WHEELER. FEMS Microbiol.Lett.(1983).<u>20</u> : 273-278.
-^{32}P-orthophosphate incorporation in DNA.	H.L.DAVID <u>et al</u>. Acta Leprologica. (1984).<u>2</u> : 359-367.
<u>Phospholipid metabolism</u> : -^{32}P-orthophosphate incorporation	H.L.DAVID <u>et al</u>. Ann.Microbiol.(Inst. Pasteur) 1985, 136A : 303-310.
Iron uptake (exochelin mediated).	R.M.Hall et al.Int.J.Lepr.(1983).51. 490-494.

REFERENCES

1. Kirchheimer WF, Storrs EE (1971). Attempts to establish the armadillo (Dasypus novemcinctus Linn.) as a model for the study of leprosy. 1. Report of lepromatoid leprosy in experimentally infected armadillo. Int.J.Lepr. 39 : 693-702.
2. Sansarricq H (1982) The WHO leprosy programme. Ann.Microbiol.(Inst.Pasteur). 133B : 5-12.
3. Draper P (1982). Bacteriology of Mycobacterium leprae : state of the art paper. Ann.Microbiol.(Inst.Pasteur) 133B : 13-14.
4. Draper P (1983) The bacteriology of Mycobacterium leprae. Tubercle 64 : 43-56
5. Draper P (1984) Host-grown Mycobacterium leprae : a credible microorganisms. Acta Leprologica (Nouvelle série) 2 : 99-112.
6. David H.L. (1984) Classification and identification of Mycobacterium leprae. Acta Leprologica (Nouvelle série) 2 : 137-151.
7. Wheeler PR (1983) Oxidation of carbon sources through the tricarboxylic acid cycle in Mycobacterium leprae grown in armadillo liver. J.Gen.Microbiol. 130 : 381-389.
8. Wheeler PR (1984) Metabolism in Mycobacterium leprae : its relation to other research on M. leprae and to aspects of metabolism in other mycobacteria and intracellular parasites. Int.J.Lepr. 52 : 208-230.
9. Brennan PJ (1983) The phthiocerol-containing surface lipids of Mycobacterium leprae : a perspective of past and present work. Int.J.Lepr. 51 : 387-396.
10. Kato L (1977) The jaunus-face of Mycobacterium leprae : characteristics of in vitro grown M. leprae are not predictable. Int.J.Lepr. 45 : 175-182.
11. Chapman GB, Hanks JH, Wallace JH (1958) An electron microscope study of the disposition and fine structure of Mycobacterium lepraemurium in mouse spleen. J.Bacteriol. 77 : 205-211.
12. Draper P, Rees RJW (1970) Electron transparent zone of mycobacteria may be a defence mechanism. Nature (London) 228 : 860-861.
13. Rastogi N, David HL (1984) Phagocytosis of Mycobacterium leprae and M. avium by armadillo lung fibroblasts and kidney epithelial cells. Acta Leprologica (Nouvelle série) 2 : 267-276.
14. Ryter A, Frehel C, Rastogi N, David HL (1984) Macrophage interaction with mycobacteria including M. leprae. Acta Leprologica (Nouvelle série) 2 :211-226
15. Rastogi N, Frehel C, Ryter A, Ohayon H, Lesourd M, David HL (1981) Multiple drug resistance in Mycobacterium avium : is the wall architecture responsible for the exclusion of antimicrobial agents ? Antimicrob Agents Chemother. 20 : 666-677.
16. Takade A, Takeya K, Taniguchi H, Mizuguchi Y (1983) Electron microscopic observations of cell division in Mycobacterium vaccae V_1. J.Gen.Microbiol. 129 : 2315-2320.
17. Silva MT, Macedo PM, Portaels, Pattyn SR (1984) Correlation viability - morphology in Mycobacterium leprae. Acta Leprologica (Nouvelle série) 2 : 281-291
18. Rastogi N, Frehel C, Ryter A, David HL (1982) Comparative ultrastructure of Mycobacterium leprae and M. avium grown in experimental hosts. Ann.Microbiol. (Inst.Pasteur) 133B : 109-128.
19. Rees RJW, Valentine RC, Wong PC . (1960) Application of quantitative electron microscopy to the study of Mycobacterium lepraemurium and M. leprae. J.Gen. Microbiol. 22 : 443-457.

20. Picard B, Frehel C, Rastogi N (1984). Cytochemical characterization of mycobacterial outer surfaces. Acta Leprologica (Nouvelle série) 2 : 227-235.

21. Young DB (1982) Mycobacterial lipids in infected tissue samples. Ann.Microbiol. (Inst.Pasteur) 133B : 53-58.

22. Asselineau C, Daffé M, David HL, Lanéelle MA, Rastogi N (1984) Lipids as taxonomic markers for bacteria derived from leprosy infections. Acta Leprologica (Nouvelle série) 2 : 121-127.

23. Draper P (1976) Cell walls of Mycobacterium leprae. Int.J.Lepr. 44 : 95-98.

24. David HL, Rastogi N (1983) Partial characterization of the cell walls of Mycobacterium leprae. Curr.Microbiol. 9 : 269-274.

25. Imaeda T, Kirchheimer WF, Barksdale L (1982) DNA isolated from Mycobacterium leprae : genome size, base ratio, and homology with other related bacteria as determined by optical DNA-DNA reassociation. J.Bacteriol. 150 : 414-417.

26. David HL, Lévy-Frébault V, Dauguet C, Grimont F (1984) DNA from Mycobacterium leprae. Acta Leprologica (Nouvelle série) 2 : 129-136.

27. Clark-Curtiss J, Jacobs WR, Docherty MA, Ritchie LR, Curtiss R (1985) Molecular analysis of DNA and construction of genomic libraries of Mycobacterium leprae. J.Bacteriol. 161 : 1093-1102.

28. Barksdale L, Kim KS (1984) Propionibacterium, Corynebacterium, Mycobacterium and lepra bacilli. Acta Leprologica (Nouvelle série) 2 : 153-174.

29. David HL, Clément F, Clavel-Sérès S, Rastogi N (1984) Abortive infection of Mycobacterium leprae by the mycobacteriophage D_{29}. Int.J.Lepr. 52 : 515-523.

30. David HL, Sérès-Clavel S, Clément F, Rastogi N (1984) Further observations on the mycobacteriophage D29-mycobacterial interactions. Acta Leprologica (Nouvelle série) 2 : 359-367.

31. Labidi A, David HL, Roulland-Dussoix D (1985) Cloning and expression of mycobacterial plasmid DNA in Escherichia coli. FEMS Microbiol.Lett. In Press.

32 Young RA, Bloom BR, Grosskinsky CM, Ivanyi J, Thomas D, Davis RW (1985). Dissection of Mycobacterium tuberculosis antigens using recombinant DNA. Proc.Natl. Acad.Sci. USA 82 : 2583-2587.

33. Benedetti EL, Dunia I, Ludosky MA, Van Man N, Trach DD, Rastogi N, David HL (1984) Freeze-etching and freeze-fracture structural features of cell envelopes in mycobacteria and leprosy derived corynebacteria. Acta Leprologica (Nouvelle série) 2 : 237-248.

Mycobacteria of Clinical Interest. M. Casal, editor

SOME IMMUNOLOGICAL ASPECTS OF LEPROSY, AND YET MORE HYPOTHESES!

J.L. STANFORD

School of Pathology, Middlesex Hospital Medical School, Riding House Street, London, W1P 7LD.

INTRODUCTION

Since the recognition of the immunopathological spectrum of leprosy there have been a number of hypotheses put forward to explain this spectrum. These theories have been based on supposed defects in either macrophages or T lymphocytes, congenital or acquired. Congenital defects imply genetic differences and a number of studies have sought to detect genotypes of humans particularly susceptible to leprosy in its various forms. It has also been postulated that the bacillus itself is different in the two polar extremes of the spectrum.

The immunology of leprosy is not just a fascinating subject in its own right, but the key to tomorrows therapy and eventual control of the desease worldwide. Immunology provides the basis for both prophylaxis, and therapy. The anti leprosy drugs, multiple or otherwise, suffer from the same deficiencies as do the antituberculosis drugs. First, they do not correct the state of immunity of the patient so that the abnormal immune mechanisms initiated by the process of disease continue after the infection has been overcome. Second, they require administration for a very long time leading to problems of non compliance, in which the patient stops taking the drugs when he feels better and his disease subsequently relapses. The failure of a return to a state of protective immunity when most of the invading bacilli have been killed leads to the phenomenon of bacterial persistance. Organisms metabolising at a low rate, either because they are short of oxygen, as in dead tissue, or because they are in a state of stasis induced by the cell in which they are residing, are not susceptible to antibacterial agents. It is this long term susvival of viable organisms that necessitates the long period of chemotherapy required in multibacillary disease, and the problems of noncompliance.

The purpose of immunoprophylaxis is to induce, or consolidate and make long-lasting the state of immune protection (fig. 3). We know that BCG can do this in some places at least, and that it can do it for both tuberculosis and leprosy, making it highly likely that protective immunity is the same for both diseases. For treatment, immunotherapy offers great hope for the future. By reintroduction of protective immune mechanisms in place of those of disease, persisting live bacilli can be identified and removed by cellular immune mechanisms. Chemotherapy can then be stopped, perhaps only after a few weeks, dramatically reducing the cost of treatment and the problems of non compliance at the same time.

Despite all the work on the immunology of mycobacterial infections, we still do not clearly know the mechanism of immune protection or the mechanisms of the abnormal immune states of disease. There appear to be at least three of the latter, that of tuberculosis, that of lepromatous leprosy and that of tuberculoid leprosy. Although it is possible that the same immunotherapeutic measure could correct all three anomalies it is too soon to know if this will be so.

T cells, B cells, macrophages, and perhaps other cell types are the participants of cell mediated immunity, and their study provides the scientific basis for complete understanding of immune mechanisms. However, details of their interactions are superfluous to a clinical consideration of immunity. Similarly, detailed chemistry of bacterial products will be required before the processes of disease are fully understood, yet a clinical consideration of this does not require the chemistry. The account I shall give will dwell on clinical observations and skin tests - the only test of immunity that directly reflects how the individual responds to mycobacterial antigens.

Fig 1.

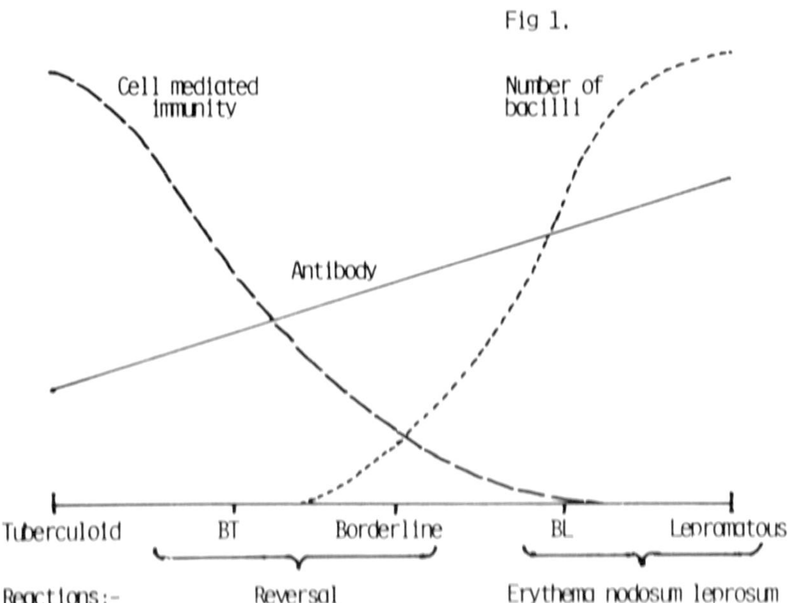

A BRIEF REVIEW OF THE DOGMA

It is generally accepted that antibodies are not important in immunity to mycobacterial disease, although the evidence upon which this is based is open to question. All the immunity is said to be of cell mediated type (c.m.i.) in which it is thought that a protected individual has macrophages capable of killing a phagocytosed bacterium under the guidance of T lymphocytes, and the diseased individual has not. Yet the failure in immunity differs with the kind of leprosy present. Whether this failure precedes infection, or is a consequence of it is unknown.

The classical immunopathological spectrum of leprosy, based on the work of Ridley and Jopling is shown in the diagram (fig. 1). At the tuberculoid end of the spectrum a few bacilli are surrounded by marked granuloma formation containing many lymphocytes and macrophages without necrosis, the Lepromin test is positive by both Ferandez' and Mitsuda's reactions, and the *in vitro* correlates of c.m.i. are positive. At the lepromatous end of the spectrum the numerous bacilli are surrounded by granulomas of the macrophage series with few lymphocytes, and without necrosis. The Lepromin test is negative, and *in vitro* tests of c.m.i. are negative to the antigens of leprosy bacilli. Borderline categories, into which most patients fall, show a mixture of the characteristics of both polar extremes. Because of this, almost all immunological research is done on the polar forms of disease. Superimposed on the spectrum are two kinds of reaction. One of these known as erythema nodosum leprosum (e.n.l.), occurs in lepromatous and borderline lepromatous disease. It is believed to be a phenomenon of formation of immune complexes of antibody and antigen in the affected parts, and is often associated with the early phase of treatment. The other type of reaction occurs in borderline and borderline tuberculoid patients and is known by a number of names, of which Reversal reaction is used in this paper. In this type of reaction immune recognition of bacterial antigen appears to increase rapidly, and this is associated with a sudden increase in infiltration of lymphocytes damaging skin, and particularly nerves. Of these two reactions e.n.l. is clinically important, but not fundamental to the immune mechanism of the disease. The reversal reaction probably represents an important change in c.m.i. to the disease, and is in urgent need of better understanding. Prior to effective chemotherapy of leprosy there was another type of devasting reaction that could occur in advanced lepromatous disease. In this reaction massive necrosis of lepromas occurred, which often led to the patients' death. Although not seen today this necrotic reaction is of immunological interest, since it suggests that the survival of the patient with lepromatous disease may depend on the maintenance of suppressor mechanisms.

SKIN TESTING IN LEPROSY

The Fernandez response to Lepromin has already been referred to as positive in tuberculoid, and negative in lepromatous leprosy. Use of soluble skin test reagents made from *M. leprae* harvested from armadillos,and from cultures of other mycobacterial species has shown several types of response in leprosy patients, in tuberculosis patients and in healthy subjects. Although there are a variety of early responses to skin tests these have not yet been fully evaluated, and most data is available for responses at 48-72 hours. Soluble antigens of *M. leprae* have been prepared by Convit and his colleagues in Venezuela, and by Rees and his colleagues in London. These reagents are prepared somewhat differently from each other. They produce the same negative results in BB, Bl and LL disease, but may differ in their reactions in BT and TT disease; Convit's antigen being the more positive. Rees' antigen, known as Leprosin A, is positive in only about one half of BT cases, but may be dividing them in a meaningful way, the cases with most neural involvement being the most positive. Amongst healthy persons both reagents show numbers of positive reactors depending to some extent on their contact with leprosy. All positive reactions to Leprosin A are rather illdefined, soft, and itching rather than painful, similar to Tuberculin test responses in the majority of BCG recipients in the United Kingdom. A necrotic element is never seen in Leprosin A responses. Tuberculin is an interesting reagent to test in leprosy patients. In tuberculoid (BT and TT) patients results are similar to those in the general population of the place where the patient lives. In lepromatous (BL and LL) patients many Tuberculin responses are larger than normal and often show signs of some central necrosis, similar to that seen in tuberculosis patients. However, in most cases simultaneous leprosy and tuberculosis can be ruled out. Thus this appearance is either due to abnormal immunoregulation, or reflects the status of individuals liable to develop lepromatous disease. Untreated lepromatous patients are often Tuberculin negative, only becoming positive after some months of chemotherapy. There is a particular type of response to Tuberculin in lepromatous leprosy known as the "giant reaction". This type of reaction may be 40-100mm in diameter, is surrounded by marked oedema and is sometimes accompanied by lymphadenopathy. Quite different from anything seen in tuberculosis, giant reactions may occur in 25% of patients soon after chemotherapy has been started and persist indefinitely in about 3% of patients. These giant reactions probably represent some instability in immune regulation.

Soluble preparations of mycobacteria contain antigens of at least three specificities. There are those shared by all mycobacteria, the common or group i antigens, those limited to individual species, the group iv antigens, and those shared amongst slowly growing species, group ii antigens, or amongst rapidly growing species, group iii antigens. Leprosin A is known to contain antigens of the first two specificities. It is also known that positive skin test responses can be to common or species specific antigens. Thus a positive response to Leprosin A in an healthy person can be directed against the species specific antigens if he has made immunologically effective contact with the leprosy bacillus, or against the common antigens if he has not. By simultaneous testing with four different reagents, one or more of them having been produced with species rare in the individual's environment, persons responding to common antigens can be identified with a fair degree of accuracy. Three categories of responders can be recognised by quadruple skin testing. Category 1 are positive to all four, recognising common antigens. Category 2 are negative to all four, combining those who have not met the four species and those who regulate out, or suppress, their responses. Category 3 persons respond to group iv antigens, being positive to some reagents and negative to others.

The key for those categories amongst healthy persons, tuberculosis patients and leprosy patients is shown in fig. 2.

RESPONDER CATEGORIES IN HEALTHY PERSONS AND THOSE WITH MYCOBACTERIAL DISEASES

Fig 2

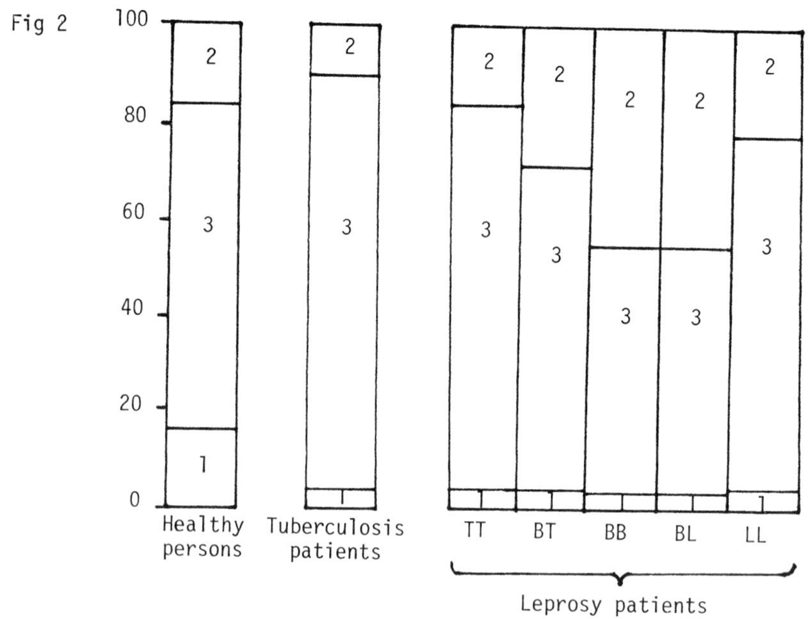

It can be seen from the figure that category 1 responders are reduced in patients and that category 2 non-responders are increased in the borderline types of leprosy. This is especially so in patients whose disease has been recently diagnosed.

THE IMMUNE STATUS PRIOR TO CLINICAL LEPROSY

Although much is now known about the immune status of leprosy patients once they have developed the disease, their state of immunity prior to this remains conjectural. Unlike most infectious diseases leprosy does not seem to be at all easily acquired and it is generally accepted that a long period, often many years, has to be spent in an endemic country before the visitor is likely to develop the disease. This period is often in excess of that required for indigenous children to become leprous, yet there does not seem to be any particular racial predisposition to the disease. Leprosy is also a disease of the poor, why? Even amongst the poor the disease is very unevenly distributed and its geographical distribution has never been explained. Who too does leprosy have a spectrum? Why is one person susceptible to one form of the disease and another to a different type? How is it that multibacillary leprosy is common in one place and paucibacillary leprosy in another?

For such a complicated set of conditions an explanation is unlikely to be dependant on any single factor. Indeed 3 factors appear to be necessary to satisfy the observations. These are, 1) there must be cases of untreated multibacillary disease to act as the source of infection; 2) there must be an environmental factor, and 3) there is likely to be a genetic factor.

Prior to effective chemotherapy, the biblical principle of separating the leprous from the healthy probably prevented many cases of the disease by reducing the chances of meeting the leprosy bacillus. Today, the same can be achieved with adequate chemotherapy. The often quoted success in eradicating the disease from Norway prior to chemotherapy was probably not only due to isolation of cases, but to a combination of this and an improvement in socioeconomic conditions.

Since it has yet to be proven that there are major environmental reservoirs of leprosy bacilli, the environmental factor is likely to be indirect. Factors that might act in this way include diet, proximity of cohabitation, reduced penetration of ultraviolet light, and the effects of environmental mycobacteria other than leprosy bacilli.

Regarding diet, claims were made at the very beginning of this century that leprosy was associated with fish eating, but since that time there have been no serious claims for its effect. Nevertheless, intriguing recent data on the effects of vitamin D (fish oil is a major source) on cell mediated immunity may make the suggestion worthy of reconsideration. It is likely that poor protein nutrition is associated with reduced cell mediated immunity. However, there is no clear association between leprosy and starvation zones.

If, as in tuberculosis, infection takes place through aerosols, living close together and poor ventilation prevent adequate dilution of droplet nuclei in the air, thus increasing the challenge dose in a confined space. The principal cause of death of mycobacteria in the air is u.v. light, and when this is cut off by shadow or high water vapour content of the air, leprosy bacilli survive longer, again increasing the chances of infection.

Contact with other mycobacterial species by inhalation, ingestion, or wound contamination also has a very significant effect on protection from, and susceptibility to, both leprosy and tuberculosis. The mechanisms by which this works are only just beginning to be elucidated. Naturally acquired protection is easily explained, but increased susceptibility is rather more difficult.

The evidence that BCG can protect man from both tuberculosis and leprosy firmly associates protection with those antigens that the vaccine shares with both pathogens. Crossprotection studies in animals showing that any mycobacterial species can be used to vaccinate from challenge with any other also show that the protective antigens are those that all mycobacteria share. These are the common mycobacterial antigens or group i antigens. That these are protective is supported further by the absence of circulating T cells responding to common antigens in patients with mycobacterial disease, and the inability of such patients to produce positive skin tests to common antigens. Since these antigens are by definition present in any mycobacterial species, sensitisation to them can arise through contact with any of the species. Such naturally acquired protection can be consolidated by BCG vaccination (fig. 3).

Although early contact with mycobacteria leads to sensitisation and to a degree of protection, excessive contact can lead to a return of susceptibility to infection. There may be several mechanisms by which this can occur, perhaps depending on the genetic composition of the individual.

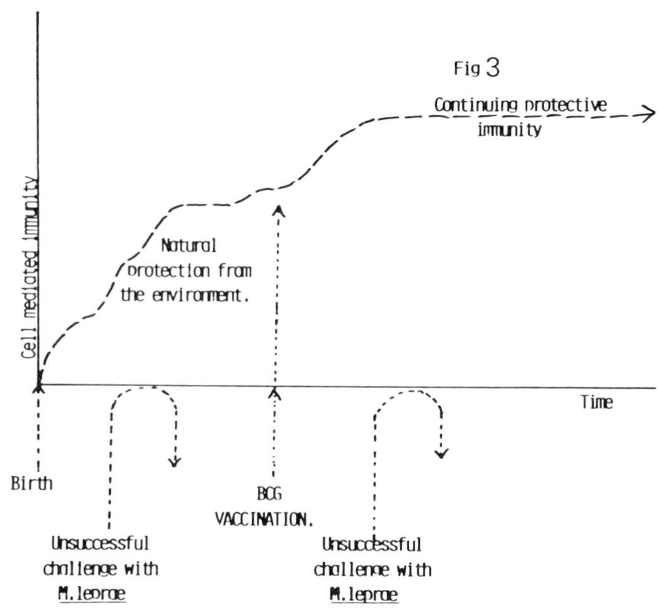

Natural protection appears to work by T lymphocyte-macrophage interaction. Macrophages, perhaps of a particular type, phagocytose invading pathogenic mycobacteria and express some of the bacterial constituents on their surface. The common antigens are recognised by T cells which release mediators instructing the macrophage to prevent the multiplication of, and perhaps kill, the mycobacterium. At the same time, species specific antigens also expressed on the macrophage surface may initiate replication of T cells that can recognise them, leading to additional sensitisation. Excessive sensitisation to some antigens, including the species specific antigens, but apparently not the common ones, leads to initiation of a second cell-mediated pathway in which the macrophage, or any other cell expressing the relevant antigens, is killed leaving the mycobacterium alive inside it. The mechanism by which it works is not understood, but it appears to lead to increased susceptibility to infection. However, it may be a valuable mechanism for the control of established disease, and is expressed as necrosis at skin test sites.

Other mechanisms leading to susceptibility may involve immune suppression. In some cases this may work by initial sensitisation occurring through the gut, leading to suppression of responses to later percutaneous challenge. Studies employing a complicated system of mixed skin test reagents have shown that mycobacteria have several determinants which can be recognised by some individuals as suppressive. At least one of these works locally and prevents the skin test response to other antigens injected with the suppressor determinant. Another works distantly so that antigens injected with the suppressor determinant at one site do not produce the expected response when injected simultaneously without the suppressor determinant at a distant site. Recognition of the determinants as suppressive is itself associated with earlier environmental sensitisation to the species possessing the determinant, and only occurs in a proportion of individuals. It is at this point that genetics may play a part. Studies on skin test responsiveness to tuberculins of healthy persons has shown a significant relationship between inability to react to routine concentrations of tuberculins (category 2 non-responders) and the lack of the HLA DR3 marker. Lack of this same marker has been shown to be associated with lepromatous leprosy.

To make an hypothesis linking recognition of certain determinants of mycobacteria as suppressive, and susceptibility to the development of lepromatous leprosy is very tempting. Such an hypothesis might propose the following :-

1) A small number of leprosy bacilli fall onto a susceptible suface, let us say the intranasal epithelium. They release a minute quantity of the suppressor determinant that acts locally, so that when they are phagocytosed there is no immune recognition of their presence.

2) As a result they begin to multiply, and with increasing numbers they produce more suppressor determinants, perhaps now a determinant that acts at a distance. Bacilli entering the blood stream can now settle anywhere without immune recognition, and in the sites most suitable to them they give rise to lepromatous disease.

3) Suppression of recognition of *M. leprae* includes recognition of common mycobacterial antigens, only leaving responses to species specific antigens of each species unaffected.

If such a hypothesis were true, then persons genetically liable to recognise certain mycobacterial determinants as suppressive become susceptible after sufficient environmental mycobacterial bombardment has induced the condition. Although suppressor determinants may explain the early phase of multibacillary infection, direct suppressor activity associated with phenolic glycolipid and perhaps other substances must play a part as infection becomes established. The phenolic glycolipids may be responsible for the nonspecific tuberculin unresponsiveness seen in newly diagnosed multibacillary patients. However, it is difficult to see how the glycolipid can explain the specific unresponsiveness to leprosy bacilli that is uncovered during the course of treatment, and which persists for the rest of the patients life. Recognition of *M. leprae* specific suppressor determinants could explain persistance of the specific defect, but experimental proof is lacking.

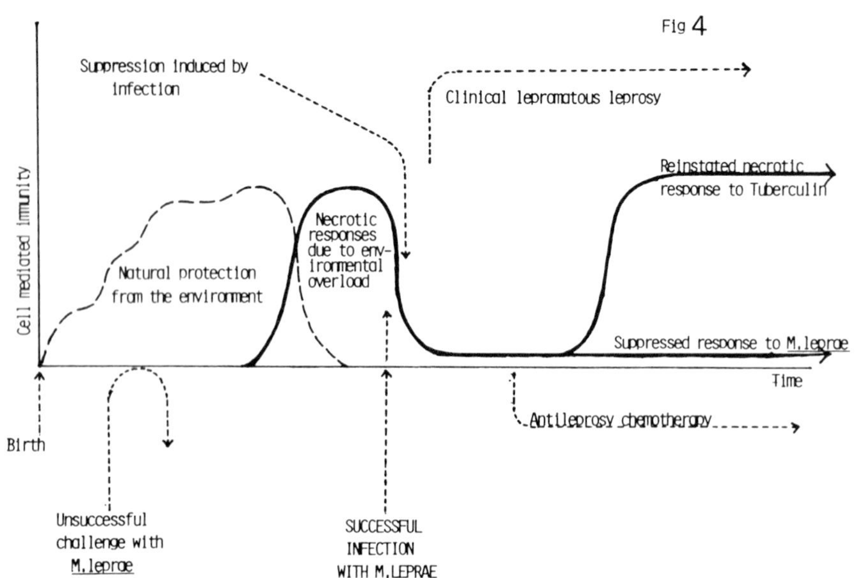

Such a hypothesis based on suppression by *M. leprae* of necrotic responses induced by over exposure to environmental mycobacteria in some people, or by further suppression of category 2 non-responder individuals, may offer an explanation for the inception of lepromatous disease. However, an alternative explanation is required for tuberculoid forms of leprosy.

Skin test data on persons with TT or BT disease show that they produce positive reactions of non-necrotic type to soluble species specific antigens of various mycobacterial species, frequently including those of *M. leprae*. But, they do not produce responses to the common mycobacterial (? protective) antigens, possibly due to continuous oral bombardment with these antigens, which in some individuals may induce suppression of skin responsiveness. Such persons might be susceptible to develop tuberculoid disease when challenged with leprosy bacilli, as shown in fig. 5.

Fig 5.

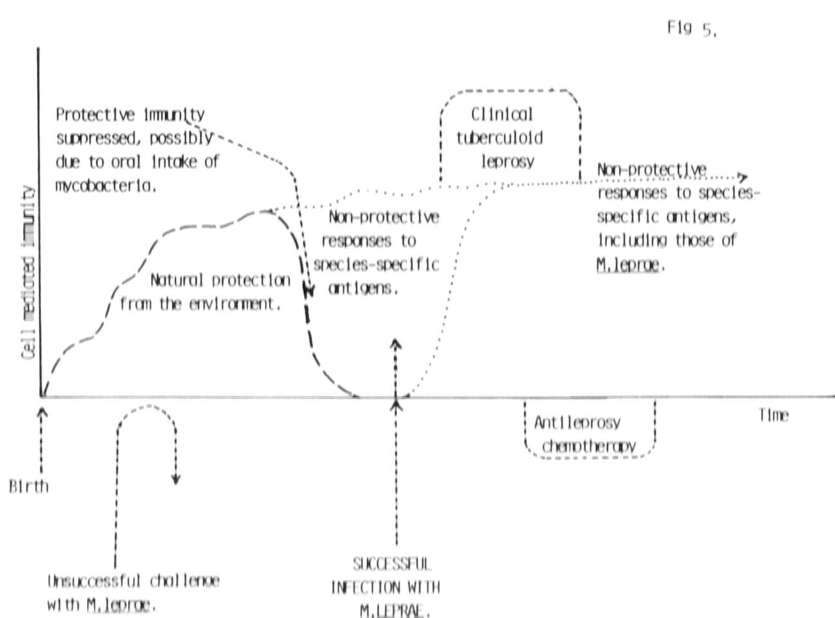

Spontaneous cure without treatment occurring in polar TT leprosy may perhaps be due to desuppression and a return to protective immunity, but much more data is needed to clarify this obscure area.

POSSIBILITIES FOR IMMUNOTHERAPY

From the above it can be seen that leprosy is a disease in which successful infection is probably associated with pre-existing environmentally induced immune deficiencies in the genetically susceptible. Once established, the infection causes a further perturbation of immunity which does not resolve, at least in the case of multibacillary disease, when the great majority of organisms are killed by multiple drug therapy. This leads to the problem of bacterial persistance with its requirement for long term chemotherapy, the concommitant likelihood of non-compliance with treatment, and subsequent relapse of disease. Even if a new drug was developed effective in killing persisters in a short time, the immune defect would lead to the patients susceptibility to reinfection.

Thus what is needed is a form of immunotherapy designed to return defective immune mechanisms to a state of protective immunity in which the persisters are rapidly killed and the patient left resistant to both reactivation and reinfection, as shown in fig. 6.

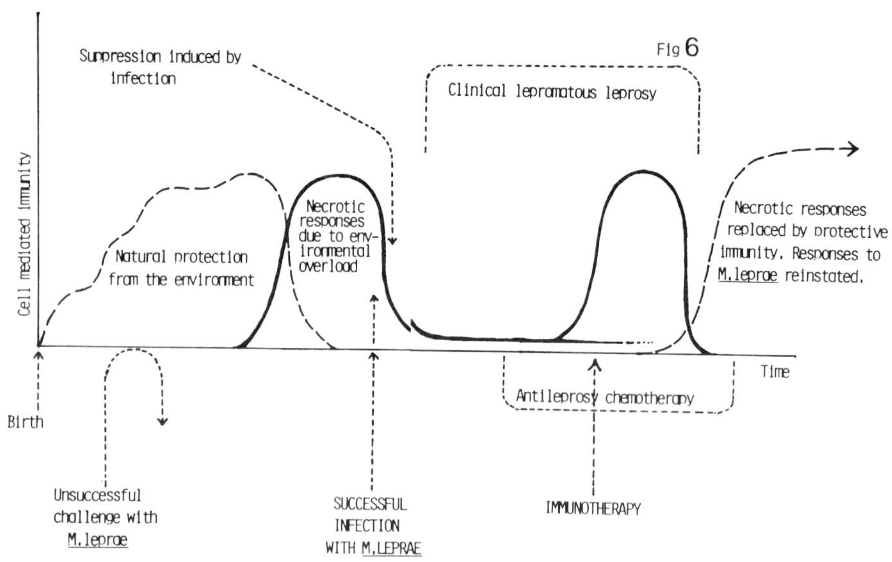

Design of the immunotherapeutic step has to depend on precise aims, and to define these we must understand the abnormality. Unfortunately, despite all the work of the last few years, we still do not know what is a normal immune state, or precisely how that of disease differs from it. Thus a potential immunotherapeutic is asked to convert one partially unknown condition to another. Sufficient is known for a start to be made, and a number of potential immunotherapeutic agents are under investigation. These include *M. leprae* itself, as given by Convit with BCG, *M. intracellulare* as given alone by Bapat and Deo, and our own efforts using *M. vaccae* with various additives as described by Torres et al. at this colloquium. If it is assumed that all three are rational approaches, to what common denominators can they be resolved?

All three posses the common (group i) mycobacterial antigens which are probably necessary for immune protection, and all three possess the adjuvant activities of mycobacteria. They differ however, in species specific (group iv) antigen content, and *M. intracellulare* also possesses slow grower associated (group ii) antigen. From these differences, two questions arise. First, are the species specific antigens of *M. leprae* required or not? Second, what constitutes a good handle for immune recognition of the immunotherapeutic?

Convit and his colleagues would say that the answer to the first question is yes, whereas Bapat's group and ourselves would say no, although perhaps for different reasons. Whereas Convit relies on BCG for the recognition of his killed *M. leprae*, Bapat and Deo work in a community highly sensitised to the species specific antigens of their reagent, and we add dilute Tuberculin to our *M. vaccae* and are considering the extra addition of a non specific inflammatory agent.

Whether all or none of these approaches are successful requires the essential test of time.

Mycobacteria of Clinical Interest. M. Casal, editor

MULTIDRUG THERAPY IN LEPROSY

J. TERENCIO DE LAS AGUAS, BERTA GERVAZIONI and ROBERTO RAVIOLI
Sanatorio de Fontilles, Alicante (Spain)

Leprosy was an incurable disease up to the introduction into clinical use of the sulphones (promin) in 1941 by Faget and Pogge, Carville, and of DDS (mother sulphone) by Cochrane in 1984, since when their use has become widespread.

After an initial optimistic period, experience soon showed that in multibacillary leprosy (LL and DL) during the first 5 years only 50% of cases become bacteriologically inactive and in the remaining cases it takes more than 5 years, even 10 years, to become negative.

Lepromatous leprosy is a systemic disease which affects several organs, with severe damage of cellular immunity and the presence of a large number of mycobacteria in the tissues, and also the persistence of dead bacilli in the patient's body: the disease is therefore prolonged for a long time, because of the body's inability to get rid of the bacilli.

Therapeutic research in leprosy, which has always followed the line of testing effective drugs in other mycobacteria (tuberculosis, lepra murina etc.) led to the use of 2 new drugs, clofazimine, utilised since 1962 by Browne and Hogerzeil, and rifampicin which was initially used experimentally and clinically by Rees and Leiker in 1970.

In recent decades, a phenomenon known as 'drug resistance' started to appear. Walgot had already observed it in 1950, but it was definitely proved in 1964 through inoculation of resistant bacilli in the mouse footpad. The phenomenon of resistance has increased in the last few years: the WHO reports 2% and 8% which represents a serious problem for the treatment and eradication of the disease.

There are different kinds of resistance: the most frequent in Spain is the secondary type, observed in patients treated for many years with sulphone (10-20 years) and who, after becoming inactive, show clinical and bacteriological relapses. Most of these patients had irregular treatment or had stopped it, or were administered low dosages. Reactions were probably the result of bacillary changes. Resistance to clofazimine (Van Diepen), rifampicin (Hastings), thiambutosine and ethionamide were also reported.

Resistance also occurs in patients who have never been treated previously: this is so-called primary resistance, but it is less frequent and is due to infection in subjects with bacilli already resistant to sulphones. There is also a third kind of natural resistance in which bacteria are insensitive to the drug.

In Spain, sulphones, clofazimine and rifampicin are the drugs of first choice. There are also other forms of chemotherapy, but these are much less effective, namely, prothionamide, ethionamide, thiacetane, delayed-sulphonamides and thiambutosine.

With regard to the importance of the therapeutic problem presented of leprosy, the reported drawbacks led us to reject monotherapy for the multibacillary forms of the disease, i.e. LL and BL, and to form our present strategy on polychemotherapy or combined therapy.

Combination of different drugs is the only way to attain the aims of efficient therapy: (1) complete healing in the shortest time, and without sequelae; (2) to reduce the patient's infectious potentiality, thus breaking the epidemiological chain; (3) to prevent drug resistance.

MULTIBACILLARY THERAPY

For several years we have based our strategy on combined therapy for multibacillary leprosy (LL, BL and BB), both in new cases and in those following monotherapy (sulphone, clofazimine, or rifampicin) but with persistent bacteriological activity, and also in cases showing relapses and secondary resistance.

Treatment must be administered regularly, using 2 or 3 drugs, at maximum doses from the beginning, with adequate bacteriological control whose lack is one of the main faults in country districts; in many cases control is poor and patients must be convinced that, at least with one of the drugs, treatment should be continued indefinitely, since even in patients whose disease has been inactive for many years bacilli can still be shown in their viscera, ganglia and nerves.

We have to avoid the exaggerated optimism of the last few decades, that is we should use the highest tolerated dosages, without relying on the minimum doses used experimentally, since high doses can kill bacteria more rapidly and blood levels are maintained for a longer time.

It is also very important, during the active period, to administer treatment daily and orally; parenteral or oral intermittent administration should be considered only after some years of chemotherapy.

It should be kept in mind that with monotherapy only 50-60% of cases become negative within 5 years.

Therapy must not be interrupted, except in special cases, not even for the Sunday test, vacations or during reactive periods.

We shall maintain the therapeutic schemes we use for our Spanish patients.

Sulphones and clofazimine: 50 or 100 mg DDS daily with 300 mg clofazimine weekly for 2 years at least, or more, until skin tests become negative. Thereafter, continue with 100 mg DDS indefinitely.

We think that this is one of the most effective forms of therapy: DDS is bactericidal, shows no toxicity, and is 500 times stronger than the minimum inhibitory concentration (MIC) in the mouse.

On the other hand, clofazimine is bactericidal at the intra-cellular, anti-inflammatory and anti-reactional levels and it tends to be stored in tissues rich in lipids and in the macrophages of the reticular endothelial system. MIC could not be obtained because of the drug's unequal distribution in the tissues.

Clofazimine is the only drug which undoubtedly causes fewer reactions than the others, and hence is advisable for reactive lepromatous cases. Compared to DDS in man, its action is slightly slower at the beginning but after a few months its effects are virtually the same as those of DDS. It can also be administered during pregnancy.

Its dosage is 50-100 mg daily. Its side effects are hyperpigmentation of the skin and conjunctiva, due to storage of a wax-like pigment caused by fatty acids saturated with bacilli. This effect is less when weekly doses of 300 mg are administered. Another side effect is persistent ichthyosis, which decreases with the above mentioned dose.

With higher doses gastrointestinal alterations may occur which may be dangerous, with enteropathy caused by the storage of crystals, and diarrhea. We have had only one case with chronic diarrhea so far.

We have been treating 45 cases monthly for 10 to 101 months with good clinical and bacteriological results as well as good tolerance:

27 patients - negative nasal mucosal smears (79.41%)

Year	No.	%
1	10	29
2	4	11.72
3	5	18.73
4	4	11.72
5	2	5.7
6	1	2.9
7	1	2.9

14 patients had negative skin smears (31.11%)

Year	No.	%
1	1	1.11
3	2	6.66
4	4	8.88
5	5	11.11
6	1	1.11
7	1	1.11

Tolerance was good and since the weekly doses of clofazimine did not exceed 400 mg weekly, the patients only presented a slight pigmentation.

Ichthyosis was present in 80% of the patients and gastrointestinal alterations in only 2 patients.

The leproreactions were mild and occurred in 74% of the patients.

Rifampicin - sulphones. Rifampicin is the most effective drug. With only one dose of 600 mg, bacilli die within 3-7 days, whereas with sulphones and clofazimine they can live for 3-4 months. This dose is 30 times stronger than MIC.

Rifampicin is effective at the nuclear level of bacilli and improvement in the morphological index is much more rapid than with other drugs; this drug is also very expensive.

When used as monotherapy, the dose is 600 mg daily, or 900 mg weekly; daily administration is not more effective than the monthly one. At present, we do not recommend it for monotherapy.

Our experience with rifampicin as monotherapy is as follows: 12 LL patients have been treated with a daily dose of 600 mg of rifampicin between 28 and 109 months. During this period, 6 cases presented negative nasal mucosal smears (66.66%) and 5 patients became skin-smear negative (41.60%). Tolerance was acceptable in 50% and the rest presented gastrointestinal intolerance. ENL reactions were present in 80% of the patients and the number of reactional episodes was also very high, some cases presenting 12, 15 or even 20 reactions.

Sulphones-rifampicin. Seventeen patients (16 LL and 1 BL) were treated with the following doses: 100 mg daily of DDS and 900 mg of rifampicin weekly for 3 to 58 months. Four patients presented negative nasal mucosal smears (23.5%) and only 1 negative skin smears (5.8%). Tolerance was worse when compared with daily rifampicin monotherapy. Some patients presented acute kidney insufficiency, 3 cases 'flu syndrome' and 2 gastrointestinal alterations. Reactions were present in 30% of the patients.

Rifampicin-isoprodian. We started this treatment in 1972 following Freekssens' outline: 600 mg of rifampicin daily and two tablets of isoprodian daily (each tablet contains 50 mg of DDS, 50 mg INH and 175 mg prothionamide). This schema has been used in several countries. The studies carried out in Malta and Bolivia are the most complete. Our experience includes 15 cases (11 LL and 1 BL) treated during a period of 21 to 116 months. Twelve patients presented negative nasal mucosal smears (85%) and 9 negative skin smears (64%). Tolerance was good and reactions were present in only 11.4% of the cases.

Rifampicin-clofazimine-sulphones. Nine patients were treated with 900 mg of RFP weekly, 300-400 mg of clofazimine weekly and 50-100 mg of DDS daily. The period of treatment was 4 to 52 months. 33.33% of the patients presented negative nasal mucosal smears and 11.11% negative skin smears. This triple drug therapy was used in patients who had become reactive due to irregular treatment, abandoned treatment or had no previous treatment. Patients who refuse clofazimine should be treated with 375 mg daily of prothionamide followed by strict control of liver function.

Our experience with all these different therapies shows that multibacillary patients should be given treatment schemes that permit the unification of criteria. We think the following scheme should be administered in multibacillary patients:

The ideal would be to use 3 drugs simultaneously in patients recently diagnosed: 600 mg daily of rifampicin or 900 mg weekly for 2 months. Clofazimine 300 mg weekly and 100 mg daily of DDS for two years or, better still, until complete negativization occurs and continue indefinitely with 50 or 100 mg daily of DDS. During the next few years drug resistance and clinical relapses can still occur despite this multidrug therapy.

PAUCIBACILLARY THERAPY

Our experience suggests that treatment for 5 years is sufficient; in TT cases we use rifampicin 600 mg daily or 900 mg weekly (for two months) and DDS 100 mg daily continued for 5 years.

In the BT forms the same combination is advisable but DDS should be continued for a longer period (about 10 years).

We think that this new multi-drug therapy is necessary, and it could be used in developed countries. However, in very endemic and under-developed countries, where there are problems of insufficient health care and economic structure, absenteeism and lack of control, this strategy may cause new troubles since the use of many drugs implies higher costs, the need for trained technical staff and the threat of side effects and intolerance.

Multibacillary leprosy, because of failure of the cellular immune system, requires combined therapy lasting for many years, and although millions of bacilli die, there is always the possibility of relapse. Hence, therapy may last indefinitely, due to the damage to cellular immunity. Complete healing can therefore be achieved only by means of effective immunotherapy which will restore the efficiency of the immune mechanisms existing before the onset of the disease.

SUMMARY

We report our personal experience, over 10 years, with different drug combinations for the treatment of leprosy, namely, rifampicin, sulphones, clofazimine, isoprodian, and their collateral reactions. The use of three drugs for the initial treatment of multibacillary patients is recommended.

REFERENCES

Alvarenga. A., Leguizamon. D., Frutos. V. v Von Rallestref. W.

O.M.S.: Comite Expertos en Lepra. quinto informe 1983.

O.M.S.: Quimioterapia Lepra para programas control 1982.

Terencio de Las Aguas. J.; Lecciones de Leprologia, 1973.

Terencio de Las Aguas, J.; Consideraciones actuales sobre la terapeutica de la lepra. Fontilles, vol. X.3. 241-249, 1975.

Terencio de Las Aguas, J.: Nuevas medicaciones en el tratamiento de la lepra. Med. Cut., 5. 366-370. 1976.

Terencio de Las Aguas, J.: Santamaria, L.: Resistencia Sulfonica de 3 casos de Lepra Lepromatosa. Fontilles, vol., X1, 2, 151-159, 1977.

Terencio de Las Aguas, J.: Gatti, C.F.: Asociacion Clofazimina- Sulifonas en el tratamiento de la lepra. Fontilles. vol. X1, 4, 371-382, 1978.

Terencio de Las Aguas, J.: Resultados del tratamiento con sulfonas a largo termino en la Lepra Lepromatosa, con especial relacion a las alteraciones renales y frecuencia de la sulfona-resistencia. Fontilles, vol. X11, 1, 31-40, 1979.

Terencio de Las Aguas, J.: Estado actual de la terapeutica de la Lepra. Fontilles, vol X1, 6, 583-591, 1978.

Terencio de Las Aguas, J.: Poliquimioterapia en la Lepra. Fontilles vol. X1V, 1, 27-35, 1983.

Terencio de Las Aguas, J.: Terapeutica combinada en la Lepra. Acta Leprologica. 2, 143-150, 1983.

Terencio de Las Aguas, J.: Treatment of Leprosy with Rifampicin and Isoprodian (L73A). Leprosy Review, 46. 165-168.

Mycobacteria of Clinical Interest. M. Casal, editor

GOAL-ORIENTED COLLABORATIVE RESEARCH IN LEPROSY

S.K. NOORDEEN

Chief Medical Officer, Leprosy Unit, World Health Organization, Geneva, Switzerland.

INTRODUCTION

Leprosy continues to remain a serious public health problem in the developing countries. The WHO global estimate of the extent of leprosy continues to remain at about 10 to 12 million patients, an estimate considered by many as conservative. The problem of leprosy is far more serious than the numbers alone indicate, particularly in terms of the social problems related to the disease.

A major problem in leprosy in recent years has been resistance of *M. leprae* to dapsone, and the prevalence is steadily increasing in many countries. The situation had been reviewed by a WHO Study Group on Chemotherapy of Leprosy for Control Programmes in 1981 and in order to prevent drug resistance, combined drug regimens based primarily on intermittent administration of rifampicin were proposed. The regimens are widely accepted in control programmes, particularly in view of their simplicity, operational ease, and increased effectiveness.

In spite of the progress with chemotherapy, the need to develop a method of primary prevention continues to be very important for leprosy control. BCG has been tested for its preventive effect against leprosy in large-scale prospective trials and found to have only a modest level of protection. Thus, BCG by itself is not a very effective tool against leprosy and the development of a sufficiently effective vaccine remains a major need.

It was in response to the above needs that two scientific working groups were set up by WHO, one on Immunology of Leprosy (IMMLEP) and the other on Chemotherapy of Leprosy (THELEP), as part of the Special Programme for Research and Training in Tropical Diseases which is co-sponsored by the United Nations Development Programme, the World Bank and the World Health Organization. The Scientific Working Groups are multi-disciplinary and carry out their tasks on a global basis, setting up their own goals and identifying their activities through strategic plans. They are responsible for planning, implementing, and evaluating progress in research related to their specific objectives which are managed through their own steering committees.

IMMLEP

The objectives of the Scientific Working Group on the immunology of leprosy are (a) development of an anti-leprosy vaccine, (b) development of immunological methods for detection of specific immune responses to *M. leprae*, and (c) an increased understanding of the immunopathological mechanism in leprosy.

Anti-leprosy Vaccine

In the 1970s it was discovered that *Mycobacterium leprae* could be grown in the nine-banded armadillo. The subsequent availability of large quantities of *M. leprae* gave impetus to the search for vaccine specific for leprosy.

An underlying premise of the approach to leprosy control through immunization is that induction of immunological reactivity to *M. leprae* antigens will lead to protection. A second premise, which is based on experimental evidence, is that *M. leprae* and other cultivable mycobacteria can produce cell-mediated immunity to antigens of *M. leprae*.

The supply of *M. leprae* is based on production in armadillos. While the possibility of using recombinant DNA technology to produce *M. leprae* antigens in *Escherichia coli* or other bacterial hosts or to make *M. leprae* grow more competently is clearly an important option for the future, the supply for the next few years will be dependent on armadillos. For the production of *M. leprae* IMMLEP currently maintains over 300 armadillos in six centres and the infected tissues are stored in a bank in London (England) to which scientists have access. A method of purification has also been developed which appears to give maximum yields of tissue-free bacteria with minimum damage to the organisms[(1)].

Extensive *in vitro* and *in vivo* testing of *M. leprae* prepared with the above purification procedure showed no loss of identifiable mycobacterial antigens. Preparation of the vaccine involved both irradiation of the infected tissue and autoclaving of the final product. The preparation, even in the absence of adjuvants, produced good delayed-type hypersensitivity in mice and guinea pigs, optimum protective immunity in mice, and some immunity in armadillos when challenged with live *M. leprae*[(2)].

Human Studies

Norway: The first Phase I study using a vaccine based on the killed *M. leprae* preparation has been completed in 31 human volunteers. The study established that the preparation did not cause any serious side-effects either systemic or regional. Local reactions in the form of ulcerations were not unacceptable and were of the same order as a smallpox vaccination.

The skin test response to soluble *M. leprae* antigen was clearly dose-related as was the ulceration. The study has shown that up to the maximum dose tested, i.e. $5x10^8$ bacilli, the side-effects were acceptable and the sensitization was found to be good even at a lower dose of $1.5x10^8$ bacilli.

Malawi: A sensitization study on healthy volunteers in Malawi in which BCG, killed *M. leprae*, and the mixture of killed *M. leprae* + BCG were compared showed that (i) the side-effects were generally acceptable to the population although a high proportion of the vaccinations caused ulcerations; and (ii) the sensitization was highest in the group receiving killed *M. leprae* + BCG when compared with killed *M. leprae* alone, and least for BCG alone.

Venezuela: The first immunoprophylactic trial, the most important field study by IMMLEP to date, began with about 69 000 contacts in three states of Venezuela. To begin with, these contacts, both household and non-household, are being tested with a soluble skin test antigen derived from *M. leprae*. The proposed vaccine trial will include all household contacts, irrespective of the skin test reaction, all non-household contacts with a skin test reaction of less than 10 mm, and a sample of non-household contacts with stronger reactions. The expected total number of study subjects for the trial is about 28 000. Half of them will receive a mixture of BCG + *M. leprae* and the other half BCG alone. Follow-up will consist of a repeat skin testing of a sample of subjects at two months, re-testing all subjects at one year, and periodic examination of study subjects for evidence of leprosy.

IMMLEP is currently initiating or planning immunoprophylactic trials in other parts of the world such as Malawi and India.

Molecular Biology

New advances in recombinant DNA technology have provided powerful tools for basic and applied research in problems like leprosy. Progress has been made by IMMLEP in the construction of clone banks of *M. leprae* DNA. DNA from *M. leprae* has been cloned using a lamda gt 11 vector known to be effective in obtaining expression of DNA from a variety of organisms. Expression has been studied and exciting progress recently reported has shown that it has been possible to isolate the genes encoding the five most immunogenic protein antigens of *M. leprae* by using monoclonal antibodies directed against *M. leprae*-specific antigens[3].

Immunodiagnosis

Monoclonal antibodies

IMMLEP has supported several laboratories for the development of monoclonal antibodies against *M. leprae*, and has set up a monoclonal antibody bank. It recently organized a Workshop in which 22 monoclonal antibodies against *M. leprae* were compared. The comparison showed that 10 out of the 22 monoclonal antibodies were specific for *M. leprae*[(4)]. The monoclonal antibodies were found to react with a broad range of different antigens including the phenolic glycolipid of *M. leprae*. The monoclonal antibodies are already proving to be extremely useful in the development of new methods for the immunodiagnosis of leprosy, in the identification of immunogenic *M. leprae*-specific gene products produced by recombinant DNA techniques and in helping to characterize those *M. leprae* antigens which are important for the stimulation of cellular immunity.

Serological test based on phenolic glycolipid of *M. leprae*

An ELISA test based on a phenolic glycolipid (PGL) produced by *Mycobacterium leprae* has now been fully biochemically and structurally characterized and synthesized[(5)]. Molecules with the same terminal trisaccaride structure have not been identified among other mycobacteria. ELISA procedures to measure antibody to PGL hold considerable promise as a means of detecting individuals with the highest bacilliary loads among patients, for the detection of subclinical infection, and perhaps as an additional parameter to follow the effectiveness of chemotherapy. With regard to the test itself, following the characterization of the chemical structure of PGL 1, several ELISA assays based on synthetic antigens are being evaluated.

THELEP

The goal of the SWG on Chemotherapy of Leprosy (THELEP) is to promote research designed to yield more effective methods of controlling leprosy by chemotherapeutic means. Lepromatous patients are the major source of infection in the community, and the aim is to discover treatment that will rapidly render such patients non-infectious and be followed by low relapse rates.

The objectives of THELEP are:

a) to assess the needs for improved chemotherapeutic methods;

b) to find better ways to use existing drugs for the control of leprosy; and

c) to promote the development of new drugs;

One of the first tasks undertaken by THELEP was to evaluate the problem of dapsone resistance through systematic studies in well defined populations. This was done through surveys on prevalence of secondary dapsone resistance and studies on primary resistance in different parts of the world.

The results of the dapsone resistance surveys, both secondary and primary, are summarized in the table. The information includes the results of some studies outside THELEP.

Improved use of existing drugs and combined chemotherapy regimens

Clinical trials

The purpose of these trials is to study, among other things, the effectiveness of different antileprosy drug regimens in eliminating "persisters" (drug-susceptible viable organisms that are found in a supposedly dormant state following chemotherapy)[6]. Trials on this have been carried out in Chingleput (India) and Bamako (Mali). Biopsies taken 3, 12 and 24 months after the start of therapy showed no clear association between clinical response and the presence of persisters as studied through inoculation into the footpads of thymectomized and irradiated (TR) mice.

Field trials on combined chemotherapy for paucibacillary leprosy have also begun in Malawi and Indonesia to evaluate the fixed duration regimen of once a month rifampicin and daily dapsone for a period of six months.

Development of New Drugs

THELEP has supported several studies on anti-microbial drugs effective in the mouse footpad model against *M. leprae*. The mouse footpad model, at present the only reliable screen for potential antileprosy compounds, has many disadvantages, including the need for a considerable quantity (10 to 20g) of test substance and the impossibility of controlling its pharmacokinetics. A number of rapid screening systems have been prepared but none are totally satisfactory. *M. leprae* isolates still have to be obtained from human or animal biopsies which contain variable proportions of viable organisms. Genetic engineering approaches to this problem are being explored.

Field trials

Trials designed to assess the effectiveness of multi-drug therapy under field conditions are now taking place in Gudiyatham Taluk and Polambakkam, in India, where two regimens are being studied among 2 300 patients. In both trials, periodic supervised drug administration is supplemented by daily self-administration. Preliminary findings show high acceptability of both regimens, with an attendance rate of over 90% in both trials. The regimens are well tolerated and side-effects minimal and rare.

PREVALENCE OF RESISTANCE TO DAPSONE

Location	Rate per 1 000 cases
Secondary Resistance	
Burkina Faso	70
Burma	200a
Burundi	37b
China - Shanghai and Municipality	86
India - Gudiyatham Taluk	95
- Trivillon Taluk	30
Mali	57b
Primary Resistance	
China - Shanghai and Municipality	500c
India - Chingleput	320d
- Gudiyatham Taluk	420e
Mali - Bamako	340d
Martinique and Guadaloupe	700b,f
Philippines - Cebu	33
Republic of Korea	220b
U.S.A. - San Francisco	18b

a. Preliminary Result

b. Study conducted outside the Programme

c. Based on 20 specimens studied

d. Data from controlled clinical trials

e. Based on 12 specimens studied

f. Based on 17 specimens studied

The search for new drugs is continuing and recent developments in this area have been quite promising. Among the promising new compounds Quinolones and Ansamycins offer considerable hope. Some of the rifamycin derivatives have longer half-lives and have shown more potency and a longer-lasting effect in the mouse footpad test than rifampicin itself. These encouraging findings are being explored, and it is hoped that some compounds may reach the clinical trial stage.

Fluorinated quinolone derivatives that are active against gram-positive organisms and penetrate readily into tissues have been found to be active against rapidly growing mycobacteria including *M. tuberculosis*, and preliminary findings in mice suggest that at least one quinolone compound is highly active against *M. leprae*. Further work is required to confirm the importance of this very promising new lead.

THE FUTURE

Future plans and expectations of IMMLEP include:

(i) Large-scale studies to be conducted in Africa and Asia on leprosy vaccines; (ii) Research on immunotherapy as a means of promoting cure of disease and prevention of relapse; (iii) Expression of *M. leprae*-specific antigens in *Escherichia coli* and other cultivable hosts for use in vaccines and diagnostic tests; (iv) Development of phenolic glycolipid derivatives for immunodiagnosis; (v) Improvement and standardization of skin-test reagents; (vi) Elucidation of the role of suppressor T-cells in immunoregulatory mechanisms underlying leprosy; (vii) Development of animal models for the study of nerve damage; and (viii) Application of serological tests to epidemiological research.

THELEP's future plans and expectations include:

(i) Development of new antileprosy drugs; (ii) Development of rapid *in vitro* methods for screening drugs, including potentially feasible techniques for cultivating *M. leprae in vitro*; (iii) Improved methods of determining *M. leprae* viability *in vitro*; (iv) Research on *M. leprae* cell-wall chemistry; and (v) Trials of immunotherapy, alone and combined with multi-drug therapy.

CONCLUSIONS

While leprosy is a major communicable disease problem in most developing countries and the research needs are quite large, the leprosy endemic countries themselves are unable to invest large resources towards this in view of other competing priorities. The developed countries have only a limited interest in leprosy research as the disease is not a problem in those countries. It is because of this that there is a great need to optimize the existing efforts and make the maximum use of available resources. The establishment of organized research in immunology and chemotherapy of leprosy by the UNDP/World Bank/WHO Special Programme for Research and Training in Tropical Diseases, and the progress made so far, have clearly demonstrated the high returns one can expect from the goal-oriented collaborative approach in research.

REFERENCES

1. World Health Organization (1980): Report of the fifth meeting of the Scientific Working Group on Immunology of Leprosy. Protocol 1/79. TDR/SWG/IMMLEP (5) 80.3

2. Shepard CC (1983) Animal vaccination studies with *M. leprae*.Int. J. Lep. 51, pp 519-523

3. Young RA, et al. (1985) Genes for the major protein antigens of the leprosy parasite *M. leprae*. Nature 316 pp 450-452

4. Engers HD, et al. (1985) Letter to the editor - results of a WHO Sponsored Workshop of Monoclonal Antibodies to *M. leprae*. Infection and Immunity 48 pp 603-605

5. Hunter SW, et al. (1982) Structure and antigenticity of the major specific glycolipid antigen of *M. leprae*. J. Biol. Chem. 257 pp 15071-15078

6. Subcommittee on clinical trials of the Chemotherapy of Leprosy (THELEP). Scientific Working Group of the UNDP/World Bank/WHO Special Programme for Research and Training in Tropical Diseases (1983). THELEP controlled clinical trials in lepromatous leprosy. Leprosy Review 54 pp 167-176

Mycobacteria of Clinical Interest. M. Casal, editor

VACCINATION IN LEPROSY

JACINTO CONVIT, NACARID ARANZAZU, MARIAN ULRICH, MARIA E. PINARDI, PEDRO L. CASTELLANOS Y MANUEL ZUÑIGA.

Instituto de Biomedicina, Apartado 4043, Caracas (Venezuela)

Effective vaccination against leprosy has long been one of the aspirations of laborers in this area of medicine, because of the difficulties encountered in control of the disease by the conventional methods of early case detection and treatment. This goal remained inaccessible for many years because no adequate source of bacilli was available and conventional methods for producing a vaccine could not be applied to leprosy. Interest in vaccination received new impetus with the successful induction of multibacillary disease in nine-banded armadillos by Kirchheimer and Storrs in 1971[1], providing a source of relatively large quantities of bacilli. The importance of preventive vaccination became increasingly apparent during the '70s, when bacterial resistance to sulfones began to emerge in many parts of the world[2]. Today, multi-drug treatment is being used which may significantly improve the therapeutic response in leprosy, but vaccination remains an important goal for several reasons. One of the peculiarities of *Mycobacterium leprae* is the capacity of sensitive bacilli to persist in the tissues for many years, in spite of the administration of appropriate chemotherapy[3]. Since the multibacillary patient does not develop cell-mediated immunity (CMI) to *M. leprae*, he remains susceptible to relapse or to re-infection after the suspension of treatment, and may be a persisting source of transmission. While there seems to be no definitive evidence, it would seem probable that transmission of the disease may occur during the incubation period of multibacillary leprosy, estimated at five to seven years. Finally, effective multi-drug therapy requires a significant economic commitment and a well developed infrastructure of health services, elements which are often deficient in the developing countries where leprosy is endemic.

In the early 1970s we began the series of investigations which culminated in the development of a new model of vaccination, originally oriented toward correcting the defect in CMI observed in leprosy. This defective response is observed in patients with lepromatous (LL) and borderline lepromatous (BL) leprosy, in persistently Mitsuda negative indeterminate leprosy (IL), and in a proportion of healthy persons in contact with leprosy patients. The development of this model was based in part on the demonstration of the specificity of the immunological defect in leprosy, in a host who responds normally toward other immunogenic stimuli. Briefly, the initial studies demonstrated that non-reactors to M. leprae in the Mitsuda test develop granulomatous reactions after the intracutaneous injection of 6.4×10^7 heat-killed M. leprae. These granulomata are composed primarily of non-differentiated macrophages, in which the acidfast bacilli injected in the test persist for months. This type of granulomatous response contrasts sharply with the immune granuloma elicited in these same individuals by the injection of BCG, or in patients with tuberculoid leprosy injected with M. leprae. The immune granuloma is formed by highly differentiated forms of macrophages including epithelioid and giant cells, with an important component of lymphoid cells; acidfast bacilli could only be demonstrated in the very early stages of the reaction[4].

Subsequently, a mixture containing 6.4×10^7 heat-killed M. leprae and 0.1 mg. of BCG was injected intradermally in non-reactors to the former micro-organism; histological evaluation demonstrated that the granuloma which developed at the site of injection possessed all of the characteristics of a typical immune granuloma as described above, and no acidfast bacilli could be demonstrated [5]. This result was interpreted in terms of a possible macrophage defect in non-reactors to M. leprae, characterized by defective recognition, digestion or processing of M. leprae and the failure to present appropriate immunogens to the lymphoid system. The non-immune type of granuloma is also observed in patients with diffuse cutaneous leishmaniasis when they are injected with inactivated leishmaniae, and may represent a characteristic response to a strong stimulus in antigen-specific CMI deficiency. When an element such as BCG which elicits a normal response is included, macrophages may be activated though usual mechanisms, resulting in the formation of an immune-type granuloma. The production of immunogens capable of eliciting a protective immune response was considered to be one of the consequences of immune granuloma formation. With this possibility in mind, we decided to test the immunotherapeutic effect of heat-killed M. leprae and BCG in Mitsuda-negative patients with leprosy. It might be emphasized that both elements of this vaccine have been used for many years in patients with leprosy and other persons; no animal model with selective immunodeficiency for M. leprae was known.

Immunotherapy was initiated in a small group of patients, including inactive LL and active IL cases, and in two persistently Mitsuda-negative individuals in 1973. The results, which included remission of indeterminate lesions and development of strong persistent Mitsuda reactivity, were published in 1979[6].

That same year a new study was begun, which included active LL and BL cases as well as persistently Mitsuda-negative IL and healthy contacts. At present, more than 700 persons have been incorporated into this study. Since 1979, M. leprae has been purified from the tissues of experimentally infected armadillos by the protocol developed by Draper[7], which includes separation in a Percoll gradient and further purification in a two-phase aqueous polymer system. Each vaccine dose, administered in three intradermal sites, contains 6. x 10^8 heat-killed M. leprae and a variable amount (0.01 to 0.2 mg.) of viable BCG, depending on the patient's response to two units of PPD. Biopsies; skin tests with soluble antigen of M. leprae (Ml-SA) and lepromin; clinical, bacteriological and neurological examination and blood extraction for serological and CMI tests were performed before and during the course of immunotherapy. A complete course of treatment consists of a series of six to ten vaccine doses administered over a period of 18 to 30 months.

Clinical examinations were made at weekly to monthly intervals, with particular attention directed not only to disease evolution, but also to possible secondary effects. Skin reactivity to Ml-SA, ELISA reactivity and lymphocyte transformation in vitro were determined at regular intervals. Tests for suppressor cells before and after vaccination were performed initially at Dr. Barry Bloom's laboratory (Albert Einstein College of Medicine, Yeshiva University, Bronx, New York) by Mehra and collaborators and subsequently in our laboratory.

Chemotherapy was continued during the course of vaccination. Prior to 1983, patients received monotherapy with sulfones; during that year, multi-drug therapy with DDS, Clofazimine and Rifampycin was initiated. Since the vast majority of the patients in this protocol has previously received chemotherapy, their reactivity before initiating immunotherapy constitutes an internal control within the study. Reversal phenomena were analysed in relation to the medical records of a group of patients who only received chemotherapy.

In general terms, clinical improvement was progressive, continuous and acceuntuated with time, even after the series of injections had been completed. The principal changes observed in variable proportions of patients included reversal reactions, to be discussed below; positivization of the 48-hour reaction to Ml-SA and the Mitsuda reaction; histological changes compatible with a shift in the sprectrum toward more resistant forms of disease, based on the Ridley-Jopling criteria; positive lymphocyte transformation tests with M. leprae and decrease in the levels of IgM antibodies to M. leprae in ELISA tests. Some of these results have been reported previously[8]; the most important aspects of a recent analysis of the data are presented in the Table.

IMMUNOLOGICAL AND CLINICAL CHANGES AFTER VACCINATION WITH M. LEPRAE-BCG

Group	Vaccine doses	Number studied	Positive after vaccination Mitsuda	Positive after vaccination M1-SA	Reversal reactions	AFB[a] negativity
LL-BL						
Active	1-7	75	10 (13.3%)	26 (34.7%)	3 (4.0%)	35 (46.7%)
	8-10	200	123 (61.5%)	109 (54.5%)	101 (50.5%)	187 (93.5%)
Inactive	1-7	60	12 (20.0%)	23 (38.3%)		
	8-10	154	125 (81.5%)	116 (73.3%)		
IL						
Active	1-3	25	25	24		
	>3	2	2	1		
Inactive	1-3	26	26	26		
	>3	1	1	1		
Contacts	1	28	27	27		
	2-3	4	2	2		

[a] Acidfast bacilli.

The reversal reactions mentioned above are quite different from those infrequently observed in LL and BL patients during chemotherapy. The former often persisted for several months and sometimes recurred. They were accompanied by notable clinical and bacteriological improvement. The histopathological examination of these lesions often revealed a borderline tuberculoid granuloma, with differing degrees of epithelioid differentiation, lymphoid cell infiltration and reduction in the bacterial population. The low frequency of neuritic phenomena associated with these reactions was of particular interest. The cubital, external sciatic-popliteal and facial nerves were occasionally affected, but only one case presents persisting facial paresis. Neuritic reactions were controlled by treatment with thalidomide or, in unresponsive cases, by corticosteroids (i.e. 8 to 12 mg. of dexamethazone for periods of 4 to 12 weeks).

Local reactions at vaccination sites were no larger than those produced by BCG alone, and left residual scars of 5 to 9 mm.

The possible protective mechanisms activated by a mixture of M. leprae and BCG have not been clearly defined, but recent evidence suggests intriguing possibilities. Lymphocytes from patients with LL do not respond to antigens of

M. leprae in transformation assays. In some patients, this lack of raeactivity can be attributed to deficient interleukin 2 synthesis[9]. Lepromatous lymphocytes with suppressor activity in a mitogen assay are induced by incubation with the Dharmendra antigen prepared from M. leprae and with specific phenolic glycolipid I antigen[10,11]. Interestingly enough, after vaccination a significant number of lepromatous patients lose this suppressor cell activity[12].

The figure below suggests a possible mechanism to explain the reactivity induced by vaccination with the M. leprae-BCG mixture in persons with selective immunodeficiency to the specific and cross-reacting antigens of M. leprae. In this figure, the normal response to BCG is represented on the right side and the defective response to M. leprae on the left. Interleukin 2, produced during the course of the normal response to BCG, may bypass the defect on the left and lead to proliferation of M. leprae-reactive lymphocytes, gamma interferon synthesis and macrophage activation.

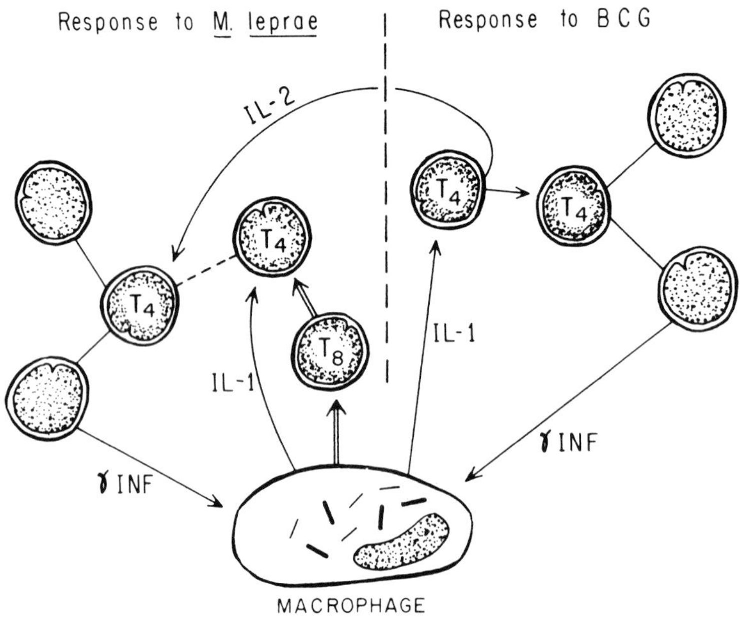

Immunotherapy of clinical leprosy may be of great importance in conferring resistance to relapse or re-infection in individual patients, but it would not be expected to significantly alter the epidemiological aspects of the disease. Perhaps one of the most important aspects of these studies is the demonstration that CMI to M. leprae can be induced in a significant number of previously unresponsive persons with the most serious and progressive forms of leprosy, and that this phemnomen is associated with protective immunity. This provides a sound immunological basis for use of the same procedure in apparently healthy individuals at risk of being infected by M. leprae.

In Venezuela, a selective approach to preventive vaccination is based on epidemiological and immunological criteria. Intra- and extradomiciliary contacts, identified in epidemiological surveys, are skin-tested with M1-SA. Non-reactors are considered to be at high risk, since they have not developed CMI to the specific or cross-reacting antigens of M. leprae in spite of contact with the disease, and they are vaccinated with a single dose of the M. leprae-BCG mixture used in immunotherapy.

Skin tests carried out in a preliminary field trial in 2659 contacts in western Venezuela revealed 78.2% giving reactions of 10 mm. or more of induration at 48 hr. The negative reactors were divided into two groups and vaccinated with BCG alone or with M. leprae-BCG. Two months after vaccination with the mixture, 98% gave positive reactions to M1-SA, and 85% were still positive at eight months. Conversion of skin test reactivity is not a definitive criterion for the induction of resistance, but is one of the parameters associated with a favorable response in lepromatous patients.

These encouraging results form the basis for a vaccine trial supported by the World Health Organization and the National Council of Scientific and Technological Research in approximately 60,000 contacts in Venezuela. Collaborative programs with Chile and Brazil are underway to evaluate immunoprophylaxis in different areas of South America.

OTHER APPLICATIONS OF THE TWO MICRO-ORGANISM VACCINATION MODEL

Hyper-reactors

An abnormal type of cell-mediated reactivity, characterized by excessive development of delayed-type hypersensitivity (DTH) and incomplete protection, is observed in some parasitic infections. This phenomenon is not frequent in leprosy; perhaps the most striking examples in Venezuela are seen in the muco-cutaneous and verrugous forms of American cutaneous leishmaniasis (ACL). These forms of ACL are characterized by chronic, progressive disease of exaggerated proportions in relation to the moderate parasitic population, with poor response to chemotherapy and frequent relapses.

A modification of the vaccine model described above, combining viable BCG with heat-inactivated promastigotes of Leishmania mexicana, has been used to treat three cases of verrugous and one of muco-cutaneous leishmaniasis. Six tc eight doses of vaccine modified the characteristics of these infections; regression and scarring of the lesions and a partial reduction of DTH as measured by the intracutaneous Montenegro reaction were observed. This type of vaccination was accompanied by a very strong local reaction and reactivation of lesions in one of the verrugous cases, an observation which indicates the need to exercise extreme caution in its use in muco-cutaneous disease affecting the nasal passages and larynx. Nevertheless, the moderate decreases in DTH and apparent increase in resistance, as indicated by scarring and inactivity of the lesions which has persisted for two years, suggest that this type of immunotherapy may help to restore a normal balance in the CMI response in selected cases.

Some cases of leprosy in the borderline-borderline tuberculoid area of the spectrum also present exaggerated manifestations of DTH. Simultaneous treatment with moderate doses of dexamethazone (8 to 12 mg./day) and immunotherapy with the M. leprae-BCG mixture has apparently corrected this immunologic imbalance, resulting in progressive and complete regression of lesions and decreased DTH.

Normal reactors

The model of vaccination using two micro-organisms has also been used in patients with localized cutaneous leishmaniasis, who suffer from limited lesions which may persist for months or years. Their immunologic response to the parasite appears to be normal. Our limited and preliminary experience in this form of ACL indicate that one or two doses of vaccine containing inactivated promastigotes and BCG produce a clear modification in the course of the infection, and might be a suitable replacement for treatment with cardio-toxic antimonials. The use of this vaccination model in the immunotherapy of cutaneous leishmaniasis is currently the subject of an extensive clinical investigation, the results of which will be published in due course.

Protective responses to Schistosoma mansoni have recently been reported in an experimental model in the mouse, using a mixture of freeze-thawed schistosomula and BCG[13]. It would be extremely premature to speculate about the mechanisms involved in immunotherapy or immunoprophylaxis with the two-microorganism vaccination model in hyper- or normal reactors, but investigations oriented toward the study of cellular subpopulations at the local level, in vitro mediators of CMI and other characteristics of the response may be expected in the near future.

Modern techniques of genetic engineering offer an extraordinary perspective in relation to the model of vaccination described in this paper. Production of the M. leprae in large quantities can only be obtained from armadillos after infection for 18 months or longer. If the genes of M. leprae responsable for inducing a protective immunological response could be introduced into BCG after cloning, a single cultivable organism combining the components present in the mixture would be available, reducing production time and costs enormously. If BCG presents particular difficulty as the bacterial host for incorporation of M. leprae genes, preliminary studies suggest that other mycobacteria might be associated with M. leprae in an active mixture. Fast-growing non-pathogenic mycobacteria such as M. phlei and M. smegmatis are being studied in combination with M. leprae, to determine if they produce similar reactions in patients with leprosy. If successful, attrempts will be made to introduce M. leprae genes into one of these fast growers. The resulting hybrid, after autoclaving, would provide a relatively inexpensive and easily produced vaccine.

REFERENCES

1. Kirchheimer WF, Storrs EE (1971) Internat J Leprosy 39:692-701
2. Pearson JMH (1981) Internat J Leprosy 49:417-420
3. Waters MFR, Rees RJW, McDougall AC, Weddell AGM (1974) Lepr Rev 45:288-298
4. Convit J, Avila JL, Goihman M, Pinardi ME (1972) Bull WHO 46:821-826
5. Convit J, Pinardi ME, Rodriguez Ochoa G, Ulrich M, Avila JL, Goihman-Yahr M (1974) Clin Exp Immunol 17:261-265
6. Convit J, Aranzazu N, Pinardi ME, Ulrich M (1979) Clin Exp Immunol 36:214-220
7. Draper P (1979) Report IMMLEP Enlarged Steering Committee Meeting. WHO, Geneva, Annex 1:4
8. Convit J, Aranzazu N, Ulrich M, Pinardi ME, Reyes O, Alvarado J (1982) Internat J Leprosy 50:415-424
9. Haregewoin A, Godal T, Mustafa AS, Belehu A, Yemaneberhan T (1983) Nature 303:342-344
10. Mehra, V, Mason LH, Fields JP, Bloom BR (1979) J Immunol 123:1813-1817
11. Mehra V, Brennan PJ, Rada E, Convit J, Bloom BR (1984) Nature 308:497-499
12. Mehra V, Convit J, Rubinstein A, Bloom BR (1982) J Immunol 129:1946-1951
13. James SL (1985) J Immunol 134:1956-1960

ACKNOWLEDGEMENTS

The studies carried out in the Instituto de Biomedicina supported in part by grants from the UNDP/WHO Special Programme for Research and Training in Tropical Diseases and the National Council for Scientific and Technological Research of Venezuela. The authors wish to thank Alfonso Sierra for preparing the drawing and Stella Millan for secretarial assistance.

Mycobacteria of Clinical Interest. M. Casal, editor

PURIFICATION OF *MYCOBACTERIUM LEPRAE* FROM INFECTED ANIMAL TISSUES

HAJIME SAITO, HARUAKI TOMIOKA, TAKASHI WATANABE and KATSUMASA SATO

Department of Microbiology and Immunology, Shimane Medical University, Izumo, Shimane 693 (Japan)

INTRODUCTION

There are at least two approaches to preparing antileprosy vaccine: one is use of mycobacteria other than *Mycobacterium leprae*, which share the antigenic determinants with *M. leprae* (1,2), and the other is the use of *M. leprae* derived from infected armadillos (3). Convit et al (4) reported that a combination of killed *M.leprae* and BCG was markedly efficacious in clinical vaccine to control borderline and lepromatous leprosy. Recently, Shepard et al (5) also reported the efficacy of the same vaccine against *M. leprae* infection in mice. Thus, a *M.leprae* vaccine with a good grade of purity and activity, and in high yield is necessary. In this study, two purification methods, Draper's method known as "IMMLEP protocol 1/79" and Mori's method, were compared, on the basis of the above criteria.

MATERIAL AND METHODS

Source of *M. leprae*

Armadillo liver (No. 825) spontaneously infected with *M. leprae* was obtained from G.P. Walsh, Armed Forces Institute of Pathology, Washington, USA. The footpads of BALB/c nude mice infected with *M. leprae* were obtained from Y. Fukunishi, Kyoto University, Kyoto, Japan.

Purification of *M. leprae*

Draper's method. The purification procedures were as described in "IMMLEP protocol 1/79" (6) with minor modifications. Liver homogenate (Fr. 1) prepared using a glass homogenizer instead of Sorvall Omnimix, was centrifuged, and the precipitate was digested with deoxyribonuclease (Fr.2). After washing with MES-buffer containing Tween 80, the bacilli were separated from the majority of tissue debris by Percoll gradient centrifugation. The resultant fraction containing bacilli (Fr. 3) was added to an aqueous 2-phase system consisting of dextran, polyethylene glycol 6000, polyethylene glycol-palmitate, potassium phosphate, and water. The purified *M. leprae* was recovered from the upper phase (Fr. 4).

Mori's method. For preparation of *M. leprae* from armadillo liver, the following procedures were used, according to Mori et al (7). Liver homogenate prepared using a Warning blender (Fr. 1) was directly subjected to Percoll gradient (from 40 to 100%, stepwise gradient) centrifugation at 100,000 x g for 60 min in a Beckman Ultracentrifuge, Swing Rotor Type 27. Leprosy bacilli distributed in the Percoll gradient were collected in 4 fractions, as shown in Fig. 1, and washed

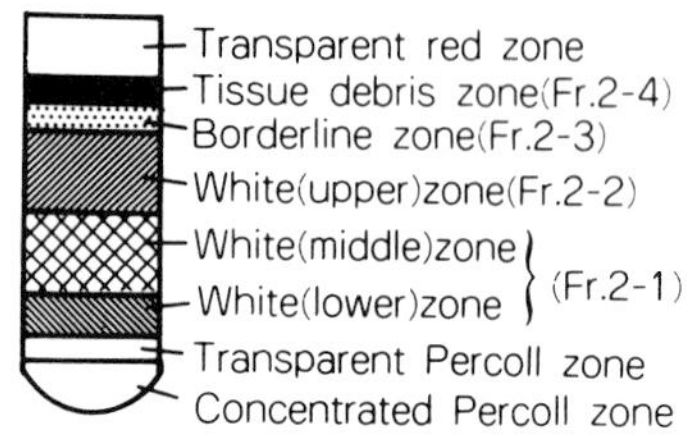

Fig. 1. Schematic drawing of ultracentrifugation pattern of *M. leprae*-infected armadillo liver homogenate on Percoll gradient according to the method of Mori et al (7).

TABLE I

PURIFICATION OF *M. LEPRAE* (ML) BY THE METHODS OF DRAPER (6)(IMMLEP PROTOCOL 1/79) AND MORI *ET AL.* (7) FROM INFECTED ARMADILLO LIVER

Method	Fraction No.	No. of Recovered ML	Yield (%)	Contaminating Debris /10^9 ML[a)]	Living Cells (%)[b)]	Cell-Associated ATP (pmole/10^9 ML) ATPase-Treatment −	+
Draper[c)]	1	1.1×10^{11}	100	8.8×10^{7}	30	7.42	0.33
	2	1.4×10^{11}	127	N D	N D[d)]	N D	N D
	3	7.7×10^{10}	70	N D	N D	N D	N D
	4	3.7×10^{10}	34	1.3×10^{6}	30	2.98	17.5
Mori *et al.*[e)]	1	5.7×10^{10}	100	7.4×10^{7}	44	7.86	0.19
	2-1	2.9×10^{10}	56	1.3×10^{6}	44	4.68	8.42
	2-2	9.0×10^{9}	17	8.7×10^{6}	44	2.46	12.2
	2-3	4.0×10^{9}	7.7	2.5×10^{7}	N D	2.28	9.39
	2-4	7.5×10^{8}	1.4	4.2×10^{8}	N D	76.0	5.63

a) Counted on the smear stained using Ziehl-Toda's method.
b) Percentage of green bacilli after FDA/EB staining.
c) Starting material was 10 g of liver.
d) Not determined.
e) Starting material was 5 g of liver.

with distilled water, by centrifugation at 20,000 x g for 60 min.

For preparation of *M. leprae* from infected footpads of nude mice, the following procedures were carried out according to Mori and Miyata (8). A homogenate of footpad tissue prepared using a mortar and pestle (Fr.1) was subjected to Ficoll gradient (from 12 to 30%, stepwise gradient) centrifugation at 100,000 x g for 80 min. Leprosy bacilli distributed in the Ficoll gradient were collected in 8

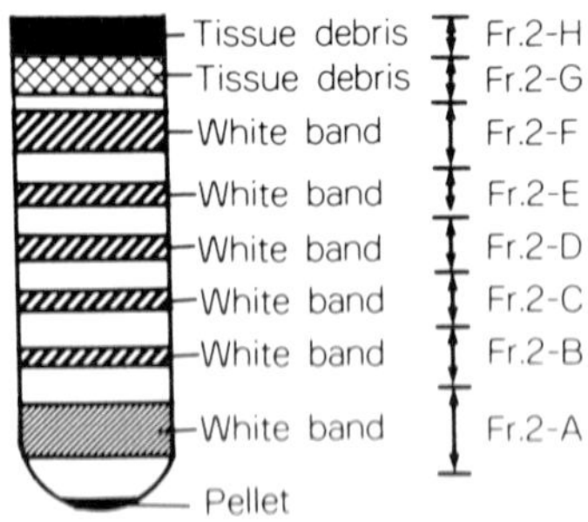

Fig. 2. Schematic drawing of ultracentrifugation pattern of *M.leprae*-infected mouse footpad-homogenate on Ficoll gradient (8).

fractions as shown in Fig. 2, and washed with distilled water, as above.

Viability of leprosy bacilli. Percentage of live bacilli in a given fraction was determined by fluorescein diacetate and ethidium bromide (FDA/EB) staining, according to Tsukiyama et al (9).

ATP content. ATP content of leprosy bacilli was estimated by the method of Dhople and Storrs with or without NaOH-trypsin-ATPase treatment (10).

RESULTS AND DISCUSSION

Table I is a summary of purification of *M. leprae* from armadillo liver, using two methods (6,7). Although the purity of the final fraction obtained by Draper's method (Fr.4) was the same, on the basis of the amount of contaminating tissue debris, compared to Fr.2-1 with Mori's method, the latter was considerably superior to the former with regard to the yield and viability of purified bacilli. The ATP content of homogenate fraction of both methods was markedly decreased after NaOH-trypsin-ATPase treatment, as described by Dhople and Storrs (10), thereby indicating abundant ATP contamination derived from liver tissue. This was also the case for Fr.2-4 (debris fraction after ultracentrifugation). In contrast, ATP content of other purified fractions increased. This indicates little contamination of debris-derived ATP in these fractions and deprivation of some luciferase inhibitors from these fractions by NaOH-trypsin-ATPase treatment.

Table II is a summary of purification of *M. leprae* from infected footpads of nude mice (8). Leprosy bacilli were distributed in a wide range of Ficoll gradient yielding at least 8 bands containing much the same amount of *M. leprae* bacilli except for bands F and G. The viability of the fractions obtained was sufficiently high (more than 63% of organisms were stained green by FDA/EB). However, the purity of these fractions on the basis of contamination with tissue debris was

TABLE II

PURIFICATION OF *M. LEPRAE* (ML) BY THE METHOD OF MORI AND MIYATA (8) FROM MOUSE FOOTPAD

Fraction No.	No. of Recovered ML	Yield (%)	Contaminating Debris per 10^9 ML	Living Cells (%)	Cell-Associated ATP (pmole/10^9 ML)
1 a)	1.1×10^{10}	100	4.7×10^{8}	64	N.D.
2-A	6.4×10^{8}	5.8	1.4×10^{9}	63	5.12
2-B	6.5×10^{8}	5.9	1.3×10^{9}	64	5.59
2-C	7.5×10^{8}	6.8	1.4×10^{9}	63	5.70
2-D	6.0×10^{8}	5.5	1.4×10^{9}	64	6.56
2-E	6.5×10^{8}	5.9	2.3×10^{9}	63	3.40
2-F	1.7×10^{9}	15.5	2.2×10^{9}	63	2.88
2-G	2.1×10^{8}	1.9	1.8×10^{10}	64	7.52
2-H	5.0×10^{8}	4.5	1.5×10^{10}	64	7.02
Pellet	2.7×10^{9}	24.5	3.0×10^{9}	63	0.18

a) Homogenate fraction was prepared from 2.3 g of *M. leprae* (Izumi strain) in-infected footpad tissue of BALB/c nude mice.

less satisfactory than purified fractions by Draper's and Mori's methods from armadillo liver. The ATP content of the fractions from mouse footpad tissue (without NaOH-trypsin-ATPase treatment) was comparable to those of *M. leprae* from armadillo liver.

Thus, the Percoll gradient centrifugation method (7) has advantages with regard to yield, safety, simplicity, and the intact state of leprosy bacilli obtained, as compared to Draper's method of purifying *M. leprae* from armadillo liver. Separate experiments showed that the speed of Percoll gradient centrifugation as well as of subsequent centrifugation is critical to achieve a high purity and yield using the method of Mori et al (7). Various approaches to improve the conditions are being studied.

We thank M. Ohara for reading this manuscript.

REFERENCES

1. Shepard CC, Van Landingham RM, Walker LL (1980) Infect Immun 29:1034-1039
2. Paul RG, Stanford JL, Carswell JW (1973) J Hyg 44:216-221
3. Godal T, Myrvang B, Stanford JL, Samuel DR (1974) Bull Inst Pasteur 72:273-310
4. Convit J, Ulrich M, Aranzau N (1980) Int J Lepr 48:62-65
5. Shepard CC, Van Landingham RM, Walker LL, Ye SZ (1983) Infect Immun 40:1096-1103
6. Draper P (1979) Protocol 1/79. Report of the Enlarged IMMLEP Steering Committee Meeting 7-8 February 1979. IMMLEP-SWG (5) 80. 3
7. Mori T, Miyata Y, Yoneda K, Ito T (1984) Int J Lepr 52:41-43
8. Mori T, Miyata Y (1984) U.S.-Japan Cooperative Medical Science Program Eighteenth Joint Conference on Leprosy, 1984, 40
9. Tsukiyama F, Katoh M, Matsuo Y (1984) Hiroshima J Med Sci 33:293-295
10. Dhople AM, Storrs EE (1982) Int J Lepr 50:83-89

Mycobacteria of Clinical Interest. M. Casal, editor

IMMUNOTHERAPY IN LEPROMATOUS LEPROSY

P. TORRES

J. TERENCIO

J. GOZALBEZ

Sanatorio de Fontilles, Alicante, Spain.

J. STANFORD

G. ROOK

Middlesex Hospital Medical School, London, U.K.

INTRODUCTION

Persistance of live bacilli despite the administration of adequate chemotherapy in lepromatous leprosy is one of the major problems of the disease, and the reason why treatment has to be given over such a long time. The major cause of these persister bacilli is the failure of the immune system to return to normal after the majority of bacilli have been killed. Immunotherapy should return the immune status to normal and allow persister bacilli to be recognised, phogocytosed and destroyed. Recognition of common mycobacterial antigens appears to be essential for protective immunity, and therefore our aim has been to restore recognition of these antigens. In the studies described, the potential immunotherapeutic agent investigated has been a suspension of ^{60}Co irradiation-killed *Mycobacterium vaccae*.

SUBJECTS INVESTIGATED

66 patients with lepromatous leprosy and 8 volunteers from the staff at Fontilles have received the immunotherapeutic agent. 48 similar patients and 6 other staff members have been investigated as controls.

Methods

1. Quadruple skin testing with Tuberculin, Leprosin A, Scrofulin and Vaccin (new tuberculins) was carried out prior to, and at intervals after immunotherapy. Reactions were read at 24, 48, and 72 hours, and in some cases after 7 days.
2. ELISA measurements of IgA, IgG and IgM antibodies to *M. Leprae*, *M. tuberculosis*, *M.avium*, *M. scrofulaceum* and *M. vaccae* have been carried out at the beginning and at intervals during the study.
3. Numerous other investigations have been carried out, but they are not reported here.
4. The potential immunotherapeutic agent, *M. vaccae*, strain R877R is harvested from Sautons medium, and suspensions prepared in borate buffered saline (pH 8.0). The suspension is then exposed to 2.5 Megarads from a ^{60}Co source. Suspensions containing 10^7, 10^8 and 10^9 bacilli per 0.1 ml have been used. The

immunotherapeutic agent has always been administered as a single intradermal injection of 0.1 ml over the left deltoid muscle. Local responses to the injection, and subsequent scar formation, are noted. Injections have never been less than 1 year apart.

Plan of the Investigation

Year 1 (1982) 30 lepromatous patients with negative bacterial indices received 10^7 *M. vaccae*.

Year 2 (1983) 31 lepromatous patients with negative bacterial indices received 10^9 *M. vaccae*.

14 patients with positive bacterial indices (+5 to +1) received 10^8 *M. vaccae*.

Year 3 (1984) 6 bacteriologically positive patients and 8 staff volunteers received 10^9 *M. vaccae*.

Each year as many immunotherapy recipients as possible have been followed up by skin testing and ELISA. Additionally monthly serum samples were obtained from the staff for the first 3 months after immunotherapy.

Results

1. Patients with zero bacterial indices.

Dose: 10^7 *M. vaccae*. 16 of the 30 patients receiving this dose had local reactions resembling mosquito bites, 2-3 days later. These resolved without ulceration or scar formation. Of the 14 without local reaction, only 1 showed some response to Leprosin A a year later, wheras of those with local reactions 3 converted to Vaccin positivity and a further 2 produced some response to Leprosin A over the same period.

We concluded from this that local reaction to the immunotherapeutic agent might be a good sign, and that the dose of bacilli was not large enough.

Dose: 10^9 *M. vaccae*. 26 of 28 patients (3 were not followed up) receiving this dose made a local response by day 7 (10.2 $\pm$ 4.0 mm) and 20 of the 26 had scars visible 1 year later (4.5 $\pm$ 2.1 mm). All 26 patients have been followed up 1 year, 2 years, or both years after immunotherapy. Of those with scars 7 of the 18 seen after 1 year produced responses to Leprosin A. Of the 18 seen in year 2, 13 had responses to Leprosin A and 6 had responses to Vaccin. Only 3 of the 18 showed no change.

All 6 without scars were followed up after 1 year, and only 1 responded to Leprosin A; 3 followed up in the second year still showed no changes.

Our conclusions were that 10^9 *M. vaccae* was an acceptable dose, and was more effective than 10^7 bacilli. The effects not only persisted, but were more marked in the second year than in the first year after the injection. Local reaction and scar formation appear to be indicative of some success.

2. Patients with positive bacterial indices.

Dose: 10^8 M. vaccae. Most of the 14 receiving this dose are out-patients and it has only been possible to follow up 7 of them so far. 2 of the 7 had small local reactions to the injection, and none produced a scar. Of 4 who were skin tested after a year, only 1 developed Vaccin positivity, and there were no responses to Leprosin A. On this fragment of data we concluded that the dose was too small.

Dose: 10^9 M. vaccae. All 6 patients receiving this dose produced a local reaction at 7 days (12.4 - 2.9 mm); their follow up is continuing.

3. Volunteer staff members.

All 8 had a local reaction after 7 days (6.3 2.5 mm). The year before their repeat skin tests and scar check is not yet completed.

ELISA results: The IgA, IgG and IgM results for the patients before initial skin testing and 1 year after immunotherapy with 10^9 M. vaccae are given in table 1 and the results for the staff over 3 months post-immunotherapy are shown in table 2.

TABLE 1

	Tuberculin	Leprosin A	Aviumin	Scrofulin	Vaccin
IgA					
Initial value	315±282	107±70	154±34	204±29	167±106
1 year later	110±86	61±50	95±50	144±74	82±64
IgG					
Initial value	1048±425	557±516	722±468	1070±412	781±408
1 year later	917±41	464±458	533±353	1113±373	790±369
IgM					
Initial value	159±164	108±111	96±108	204±203	220±293
1 year later	122±108	50±99	80±93	234±196	93±107

With 1 or 2 exceptions these results for patients show a definite trend for antibody levels in all three classes to fall over the year following immunotherapy with 10^9 M. vaccae. It is too early to be sure whether this tendency to return to more normal levels after immunotherapy can be taken as a sign of success.

TABLE 2

	Tuberculin	Leprosin A	Aviumin	Scrofulin	Vaccin
IgA					
Initial value	124±96	23±23	122±49	167±33	112±85
1 month later	192±87	46±37	198±57	185±26	155±68
2 months later	99±38	12±20	95±25	127±70	213±175
3 months later	153±95	7±2	109±43	117±58	155±80
IgG					
Initial value	777±498	0	436±294	263±164	340±153
1 month later	827±645	61±71	225±65	491±162	424±113
2 months later	958±414	75±48	252±72	765±243	436±216
3 months later	689±89	111±5	468±218	709±216	523±8
IgM					
Initial value	87±36	7±9	69±77	93±77	35±48
1 month later	138±83	18±36	76±49	130±107	62±67
2 months later	120±107	32±64	95±80	146±108	109±74
3 months later	166±88	36±55	102±92	179±135	115±109

These results for staff members show a small rise in IgA levels to all antigens in the first month after immunotherapy, followed by a fall in the second and third months. There is a more prolonged increase in both IgG and IgM levels.

CONCLUSIONS

Both skin tests with Leprosin A and antibody measurements to a series of mycobacteria show definite prolonged effects of a single dose of 10^9 killed *M. vaccae*. Whether these results really mean the success that they suggest, and whether the single dose will prove sufficient remain questions to be answered in the future.

Mycobacteria of Clinical Interest. M. Casal, editor

AN IMMUNOLOGICAL STUDY OF PAUCIBACILLARY LEPROSY

J. GOZALBEZ
J. TERENCIO
P. TORRES
Sanatorio de Fontilles, Alicante, Spain.
R. FERNANDEZ
Delegation Provincial de la Salud, Almeria, Spain.
J. STANFORD
G. ROOK
Middlesex Hospital Medical School, London, U.K.

INTRODUCTION

Paucibacillary leprosy (tuberculoid or borderline tuberculoid) is associated with an abnormal immune response to small numbers of leprosy bacilli. This abnormality may well be due to responses to antigens which do not lead to the destruction of the bacilli, or to a failure in immunological regulation. The exact nature of the abnormality has never been understood. Most students of the aetiology of leprosy have concentrated on multibacillary disease (lepromatous or borderline lepromatous), but it is clear that results obtained from such patients tend to reflect late consequences of antigen overload. Studies of paucibacillary leprosy should help to clarify the fundamental immune defect of the disease.

SUBJECTS STUDIED

These were 20 patients with paucibacillary leprosy, and 16 healthy volunteers. Seventeen of the patients were living in Sanatorio de Fontilles, or attending out-patient clinics there, and 3 attend the Public Health Clinic in Almeria. The control group were staff members or regular visitors to Fontilles.

Methods

1. Antibody measurements by ELISA.

Serum samples were obtained before skin testing from pátients and controls and tested for antibodies to *Mycobacterium leprae*, *M. tuberculosis*, *M. scrofulaceum*, *M. avium* (immunodiffusion type B) and *M. vaccae*.

The antigens used were filtered sonicate preparations, and antibodies of IgA, IgG and IgM classes were sought.

2. Blood samples were taken for HLA-DR typing and for antibody allotyping (these were carried out in Lieden and Amsterdam respectively).

3. Skin testing.

All patients and controls were tested with 4 reagents, 2 on each forearm, by the Mantoux method. The reagents used were sonicate preparations (new tuberculins).Tuberculin and Leprosin A were tested on the left arm,and Scrofulin and Vaccin were tested on the right arm. Results were read as diameters of induration after 72 hours.

On 11 of the patients data was available for skin tests with these same reagents carried out one or two years previously.

4. In vitro correlates of cell mediated mechanisms.

Several such tests have been carried out on a small number of the patients, but these are not yet ready for presentation.

Results and Discussion

The mean age of the patients investigated was 62.3 $\pm$ 15 years (range 31-85 years),and the mean age of the healthy control group was 43.1 $\pm$ 9 years(range 29-64 years).

TABLE 1

SKIN TEST Results

	Healthy Controls	Tuberculoid patients	Lepromatous(x) patients
Tuberculin % +ve	69%	65%	68%
mean +ve reaction	9.9 $\pm$ 5.5mm	11.2 $\pm$ 3.4mm	15.5 $\pm$ 5.6mm
Leprosin A % +ve	44%	50%	1%
mean +ve reaction	3.9 $\pm$ 1.2mm	10.8 $\pm$ 6.3mm	3mm
Scrofulin % +ve	56%	50%	22%
mean +ve reaction	7.4 $\pm$ 3.8mm	8.4 $\pm$ 5.7mm	8.6 $\pm$ 4.4mm
Vaccin % +ve	50%	60%	18%
mean +ve reaction	4.6 $\pm$ 1.8mm	9.1 $\pm$ 4.4mm	8.7 $\pm$ 5.1mm

(x) Data on 114 lepromatous patients included for comparison.

There are no significant differences in numbers of tuberculoid patients or controls reacting to each of the skin test reagents,however,in each case there is a tendency for the patients to produce larger reactions. This is particularly so for Leprosin A and Vaccin. The lepromatous patients(mean age 53.4 $\pm$ 15 years) show interesting differences from both the tuberculoid patients and the controls. As well as the expected lack of reaction to Leprosin A,the largest reactions to Tuberculin are amongst the lepromatous group,and numbers responding to Scrofulin and Vaccin are low.

TABLE 2
ELISA Results

	Tuberculin	Leprosin A	Aviumin	Scrofulin	Vaccin
IgG					
Tuberculoid patients	402±228	91±77	111±92	436±255	265±185
Healthy controls	112±67	3±4	79±35	94±45	61±45
IgM					
Tuberculoid patients	41±42	1.5±3.4	55±51	68±68	49±57
Healthy controls	30±27	0.7±2.2	48±40	49±40	35±25
IgA	Levels were insignificant in both groups				

Both IgG and IgM antibodies to all 5 species of mycobacteria investigated were highest among the tuberculoid patients. This was especially notable for the IgG values against Leprosin A. The wide scatter of IgG values among the patients was due to three particular individuals who had the highest antibody levels against each of the antigens. These three patients had all had their disease for at least 5 years, all were skin test positive to Tuberculin, and 2 of them were positive to Leprosin A.

COMMENT

This study has only recently started, but already there have been some interesting findings. The tuberculoid patients tend to produce larger skin test reactions to all 4 tests, not only to Leprosin A, and they have more antimycobacterial antibody than was expected. Data on HLA-DR typing and antibody allotyping is being compared with these results. It is hoped to include recently diagnosed cases as these become available, and to extend the range of investigations.

RELEVANT REFERENCES

-Shield,M.J.,Stanford,J.L.,Gallego Garbajosa,Drapper,P.,and Rees,R.J.W.(1983) The epidemiological evaluation in Burma of the skin-test reagent,LRA6;a cell -free extract from armadillo derived Mycobacterium Leprae.Part I:Leprosy patients.International Journal of Leprosy,50,436.

-Stanford,J.L. Skin testing with mycobacterial reagents in Leprosy.Tubercle,(1984),65,63-74.

-Nath,I.(1983).Immunology of human leprosy-current status.Leprosy Review (Special issue),31S-45S.

-Harboe,M.(1982).Significance of Antibody Studies in Leprosy and Experimental

Models of the Disease.International Journal of Leprosy,50,342-350.

-Melsom,R.,Harboe,M.,Myrvang,B.,Godal,T.,and Belehu,A.(1982). Immunoglobulin class specific antibodies to M.Leprae in leprosy patients, including the indeterminate group and healthy contacts as a step in the development of methods for sero-diagnosis of leprosy.Clin.exp.Immunol. 47,225-233.

-Rook,G.A.W.(1983).The immunology of Leprosy.Tubercle,64,297.

Mycobacteria of Clinical Interest. M. Casal, editor

KIDNEY TRANSPLANTATION IN LEPROMATOUS LEPROSY. A CASE REPORT

M. DEL HOYO, N. MARIN, J.M. ARRAZOLA, J.L. TERUEL, A. ROCAMORA, E. GOMEZ and A. LEDO

Department of Dermatology, Hospital Ramón y Cajal, Crta. Colmenar Viejo Km. 9,1 28034 Madrid (Spain)

INTRODUCTION

Although renal failure is a frequent cause of death in leprous patients, references to kidney transplantation in these patients are rare (1-3). The occurrence of leprosy, especially in the underdeveloped countries, probably explains this.

We describe a case of lepromatous leprosy and chronic renal failure which was successfully treated by a renal transplant.

CASE REPORT

A 40-year-old white woman was initially diagnosed as having lepromatous leprosy in 1961. After taking sulphone for 15 years, she was discharged with no medication. Chronic renal insufficiency secondary to interstitial nephritis was first recognized in February 1981. Regular hemodialysis was started at this time. During hemodialysis treatment, the dermatologic findings were negative. In September 1982 a cadaveric kidney was transplanted. Three months after transplantation she was discharged on a maintenance schedule of azathioprine (150 mg/day) and prednisone (20 mg/day); kidney function was normalized.

Fourteen months later she developed hypoesthetic, indurated, erythematous and violescent plaques over the face, trunk and limbs. Skin biopsy showed an atrophic epidermis and a nodular infiltrate in the deeper layers of the dermis, composed mainly of foamy histiocytes, lymphocytes and polymorphonuclear leukocytes. Ziehl-Neelsen stain showed numerous intracytoplasmic bacilli arranged in globi. Lepromin reaction was negative. Nasal and skin scrapings were all positive for acid-fast bacilli. A recurrence of the lepromatous leprosy was diagnosed. She continued with immunosuppressive therapy and was started on rifampicin (600 mg/day) and sulphone (100 mg/day). Three weeks later her kidney function deteriorated, rifampicin was stopped and renal function returned to previous values. After ten months on sulphone therapy total resolution of leprosy was observed. We decided to maintain her on sulphone therapy for life. At present, three years after transplantation, kidney function remains normal.

DISCUSSION

Deficiency of cellular immunity in patients with lepromatous leprosy (4) and increased tolerance to skin homografts could bring about better survival after kidney transplantation. Successful kidney transplantation in patients with lepromatous (1,2) and tuberculoid (3) leprosy have been described.

Recurrence of leprosy may be related to the use of immunosuppressive therapy. Exacerbation of cutaneous lesions, after renal transplantation, has been described in a patient with lepromatous leprosy (1).

We describe a successful kidney transplantation in a patient with lepromatous leprosy who developed a transitory recurrence of the cutaneous disease. She showed a good response to specific treatment of skin lesions.

In conclusion, renal transplantation may be useful in the treatment of patients with leprosy and chronic renal failure. Prophylactic therapy with sulphone should be initiated early and maintained indefinitely.

REFERENCES

1. Adu D, Evans DB, Millard PR, Calne RY, Shwe T, Jopling WH (1973) Renal transplantation in leprosy. Br Med J 2:280-281

2. Mocelin AJ, Ajzen H, Ancao MS, Stabile NC, Sadi A, Maluli A, Ramos OL (1979) Kidney transplantation in leprosy. A case report. Transplantation 28:260

3. Date A, Mathai R, Pandey AP, Shastry JCM (1982) Renal transplantation in leprosy. Int J Lepr 50:56-57

4. Rook GAW (1983) The immunology of leprosy. Tubercle 64:297-312

Mycobacteria of Clinical Interest. M. Casal, editor

DRUG RESISTANCE IN LEPROSY

J. TERENCIO DE LAS AGUAS
Medical Director de Fontilles (Spain)

Monotherapy with sulphones in leprosy was started in 1941 and clinical and bacteriological control corroborated its efficacy. Several decades passed before other drugs such as clofazimine (1962) and rifampicin (1970) proved to be successful.

During the last two decades, our understanding of *M. leprae* has made significant progress. Morphological studies with the electron microscope in ultra thin sections and freeze-etching techniques have revealed the specific characteristics of this bacillus such as symmetrical leaflets in the cellular wall, a trilaminar structure, spherical drops of peribacillar substances, transverse bands composed of two fine circular lines, etc.

From the biochemical point of view, the presence of a specific phenolic glycolipid and a glycanpeptide common to other mycobacteria but with alanine replacing glycine are the most important findings.

The most important tests for the identification of *M. leprae* are extraction with pyridine and loss of alcohol resistance, oxidation with DOPA, genome studies and the FLA-ABS serological test, radioimmunoassay (RIA) and the ELISA test with specific antigen (phenolic glycolipid).

Shepard (1960) started inoculating the normal mouse footpad to produce lesions that helped to indicate the efficacy of drug therapy and *M. leprae* viability even without generalization of the disease. Studies with nude mice (Prabhakaran 1975, Colston and Hilston 1976, Kohsada 1979), thymectomized neonatal Lewis mice and athymic mice have produced larger numbers of bacilli and dissemination of the disease.

In 1971 Kircheimer and Storrs transmitted leprosy to the nine-banded armadillo (*Dasypus novemcinctus*) with lesions very similar to those of human leprosy. Approximately 80% of the armadillos were receptive and 10^9 bacilli per gram were obtained from liver, spleen or lymphatic nodule. The animals with disseminated disease died within 1.5-3 years without neuropathic abnormality. It is one of the most interesting animal models for the study of *M. leprae*.

Perhaps the most important finding is that reported by Meyers et al in 1979 of a primate, the mangabey monkey (*Cercocebus atys*), with natural acquired mycobacteriosis similar to human leprosy and transmissible to other monkeys of the same species. He inoculated other mangabey monkeys with human and armadillo inoculated *M. leprae* obtaining dissemination of the disease with bacilli and lesions identical to those of human disease.

This better understanding of *M. leprae* has been very important to understand the phenomenon of drug resistance. This problem is very alarming since the number of patients with drug resistance is increasing.

The exact mechanism in resistant cases of leprosy is not well understood, although many investigators dismiss the plasmid transfer theory, among them Pattyn who explains that the frequency of mutants in leprosy in relation to different drugs is not understood (but clinical observations reveal that it is not very high, being less than 10^6) and that drug resistance in leprosy is of the penicillinic type (DDS, rifampicin, ethionamide and thiacetazone).

Multibacillary leprosy has bacteriological indices (B.I.) of 10^8 to 10^{12} per gram of tissue. This creates a serious risk for secondary resistance sometimes 10-20 years after inactivation or to irregular treatment. All this is facilitated by the long period of generation of the bacilli which is approximately 14 days.

Another important risk factor is the appearance of persistent bacilli that are not destroyed by chemotherapy, due to their low metabolic rate. These are mainly responsible for relapses, especially in patients on monotherapy.

Drug resistance in leprosy was reported by Wolcot in 1950 without experimental evidence and in 1964 using inoculation assays in mice in Malaysia Petit and Rees reported the first cases of secondary resistance in 1% of the patients treated with sulphone monotherapy. In 1981 Pearson, working in the same centre, estimated resistance at 25% which he thought was caused by low dosage of sulphones. From that time drug resistance has been studied by inoculations in mice and administration of sulphones in the diet of the animals which should inhibit *M. leprae* growth with 0.1 mg of DDS/100 g of food (0.0001%). Bacilli that are able to multiply at this concentration are considered resistant. *M. leprae* bacilli capable of multiplying with 0.1 mg/100 g but inhibited with 1 mg/100 g are considered to have low-grade resistance. Those that multiply with 1 mg/100 g but are inhibited with 10 mg/100 g present an intermediate grade of resistance. Bacilli multiplying despite administration of 10 mg/100 g are totally resistant.

Clinically, patients taking 100 mg DDS daily and presenting lesions or leproreactions and who are B.I. positive are suspected of having drug resistance.

Jacobson and Trautman studied patients who had taken promin in Carville in 1948. Thirteen are still living and 10 of them present symptoms of activation; 8 present bacilli resistant to DDS.

In recent years this phenomenon has increased considerably worldwide and the WHO has confirmed the existence of drug resistance in 25 countries.

Secondary drug resistance has also been observed against other drugs such as rifampicin (Jacobson and Hastings), clofazimine (Van Dieppen), prothionamide and thiambutosine.

Another feature of drug resistance that has appeared in leprosy is primary resistance in which patients infected with sulphone-resistant bacilli are unaffected by this drug. Londono in Columbia in 1977 and Pearson in Ethiopia were the first to report this. In 1978, Jacobson and Hastings reported this in Carville in 37% of the patients, Guinto in 1981 found 3.6% in the Philippines and also Gelber in San Francisco in 1984.

Primary and second drug resistance is increasing annually and this has made multidrug therapy necessary in multibacillary leprosy. Two or three drugs should be administered to avoid the presence of mutants, to inactivate the patient as quickly as possible; DDS should be administered indefinitely at high dosages and without interruption.

The new preventive system against resistance has changed the treatment planning and health programmes in countries where leprosy is endemic due to difficulties in using several drugs and the need for paramedical staff with higher qualifications.

It is our hope that in the years to come no further phenomena of drug resistance will appear. We think that the abnormal immune status of multibacillary patients needs to be restored to normal by immunotherapeutic treatment.

REFERENCES

Londono F(1977) Leprosy Review 48(1):51

O.M.S. (1982): Chemotherapy of leprosy for control programmes

Pattyn SR et al (1975) J. Leprosy 43:356-363

Pattyn SR et al (1984): Leprosy Rev. 55:361-367

Pearson J (1977) Leprosy Rev. 48:129-132

Petit J, Rees R (1964) Lancet 2:673-674

Petit J et al (1966) J. Leprosy 34:357-399

Terencio de las Aguas J (1973) 'Lecciones de Leprologia'

Terencio de las Aguas J (1979) Rev. Fontilles XII 1:31-40

Mycobacteria of Clinical Interest. M. Casal, editor

A MONOCLONAL ANTIBODY TO THE 36 kD GLYCOPROTEIN ANTIGEN OF M.LEPRAE

PAUL R.KLATSER, MADELEINE Y.L.de WIT, AREND H.J.KOLK & MARIA M.van RENS
Royal Tropical Institute, Laboratory of Tropical Hygiene,
Meibergdreef 39, 1105 AZ Amsterdam (The Netherlands)

INTRODUCTION

Diagnosis is the first step in the control of leprosy, both in the individual and in the community. Since antigen-sharing across mycobacterial species is extensive (Daniel & Janicki, 1978; Goren, 1982), the specificity of a diagnostic test should be high. With the use of monoclonal antibodies, through their recognition of a single epitope, the specificity of a diagnostic test would be improved (Mitchell, 1981).

Previously we reported the identification of specific M.leprae antigens which were recognized by leprosy patients' sera (Klatser, van Rens & Eggelte, 1984). We now have a monoclonal antibody (F47-9) which recognizes one of these antigens, a 36kD glycoprotein and have used this monoclonal antibody to develop an ELISA-inhibition test (Klatser, de Wit & Kolk, 1985) and to isolate the antigen (Klatser, et.al., in preparation; Ottenhoff, personal communication).

Both the ELISA-inhibition test and the purified antigen are possible candidates for the application to the diagnosis of leprosy.

MATERIALS AND METHODS

Monoclonal antibody (Moab) was prepared from M.leprae sonicate supernatant immunized Balb/c mice (Klatser, de Wit & Kolk, 1985; Kolk et.al., 1984).

The ELISA-inhibition test was performed as described before (Klatser, de Wit & Kolk, 1985). In principal, peroxidase labeled Moab (1:1250) and test sera (1:5) were added simultaneously in M.leprae extract coated polystyrene microtiter plates. The percentage inhibition was calculated using a group of sera from healthy Dutch persons as negative inhibition controls.

The 36 kD antigen was isolated using both Fast Protein Liquid Chromatography (FPLC, Parmacia, Sweden) and affinity chromatography. Purity was checked by SDS-PAGE and the SGIP technique.

RESULTS AND DISCUSSION

The 36 kD antigen was mainly present in the insoluble pellet of a M.leprae sonicate and could be solubilized by extraction with detergent. This suggests that the 36kD antigen is an integral part of the cell membrane of M.leprae. The antigen showed to be heat- and SDS-stable, but activity of the Moab was lost after trypsin treatment of the antigen, which suggest that the Moab recognizes an epitope of protein-like nature. The isolated antigen could only be stained in SDS-PAGE slab gels by a silver stain for glycoconjugates and not by coomassie. This might implicate that the 36 kD antigen is a glycoprotein.

With the ELISA-inhibition test seropositivity was found in 100% of the multibacillary leprosy patients (n=77) and in 91% of the paucibacillary patients (n=87) tested. Only 5% of the 223 control sera were positive (Klatser, de Wit & Kolk, 1985). The high seropositivity found in both pauci- and multibacillary patients suggests that the 36 kD antigen is immunodominant. This suggestion is strenghtened by the finding that T cell clones from leprosy patients preferentially recognize the purified 36 kD antigen (Ottenhof, personal communication).

There is a need for a quick, sensitive and specific test for early diagnosis of leprosy. The 36 kD antigen of M.leprae might be a suitable candidate, both for tests measuring humoral- and cellular immuneresponses to M.leprae infection.

REFERENCES

1. Daniel TM & Janicki W (1978) Mycobacterial antigens: a review of their isolation, chemistry and immunological properties. Microbiol Rev 42, 84.
2. Goren MB (1982) Immunoreactive substances of mycobacteria. Am Rev Resp Dis 125, 50.
3. Klatser PR, Rens MM van & Eggelte TA (1984) Immunochemical characterization of Mycobacterium leprae antigens by the SDS-polyacrylamide gel electrophoresis immunoperoxidase technique (SGIP) using patients' sera. Clin exp Immunol 56, 537.
4. Klatser PR, Wit MYL de & Kolk HJ (1985) An ELISA-inhibition test using monoclonal antibody for the serology of leprosy. Clin exp Immunol 62, in press.
5. Kolk AHJ, Ho ML, Klatser PR, Eggelte TA, Kuiper S, Jonge S de & Leeuwen J van (1984) Production and characterization of monoclonal antibodies to Mycobacterium tuberculosis, M.bovis (BCG) and M.leprae. Clin exp Immunol 58, 511.
6. Mitchell GF (1981) Hybridoma antibodies in immunodiagnosis of parasitic infection. Immunol Today, 2, 140.

Author index

MON PAPA NE PLEURE PAS !

À Roy,
qui m'a donné l'idée
de cette histoire.
A.P.

À mes deux filles,
de leur père
qui ne pue pas.
J.M.

D'ANDRÉE POULIN • ILLUSTRÉ PAR JEAN MORIN

Il ne reste qu'une minute.
Le compte est 2 à 2.
Un joueur fonce vers Xavier et le fait trébucher.
En tombant, Xavier pousse la rondelle
dans son propre but. Bzzzz !
La sirène annonce la fin de la partie.

Dans le vestiaire, Xavier pleure à chaudes larmes.

— Mon équipe a perdu à cause de moi.

Son papa le console :

— Ne t'en fais pas. C'est arrivé à bien d'autres joueurs.

— Je n'ai jamais compté un but de ma vie !

— Allez, allez, on ne pleure plus.

Après avoir pris son bain, Xavier sort son nouveau livre de dinosaures.

— Pas ce soir, il est trop tard, déclare papa.

Le garçon éclate en sanglots.

— Tu es fatigué, tu pleurniches pour rien, dit Vincent.

— Ce n'est pas pour rien! proteste le petit.

Vincent borde son
fils et dépose un
bisou sur son front :
— Un homme,
ça ne pleure pas !
— Je ne suis pas
un homme, je suis
un garçon !
bougonne fiston.

Xavier prend son élan et frappe la rondelle.

Boom !

Elle heurte le visage de Vincent
qui se met à saigner du nez.
Le garçon se met à pleurer.

— Pourquoi ces larmes ? demande le papa.

— Parce que je t'ai fait mal.

— Voyons, Xavier, c'est un accident.

— Mais tu saignes ! Et tu ne pleures même pas !

— Je ne suis pas un bébé, sourit le père.

Quand Xavier se pose des questions, sa voisine Marion trouve souvent une solution.

— Mon papa ne pleure pas, annonce-t-il.

— Jamais ?

— Jamais ! Il dit qu'un homme ne doit pas pleurer.

— Voyons donc ! s'exclame Marion. Les gars ont bien le droit de verser des larmes !

Soulagé, Xavier sourit.

— Moi, quand je suis triste, ça me fait du bien de sangloter, déclare Marion. Parfois, je pleure même de joie.

Xavier est vraiment décidé à faire pleurer son père. Il lui demande donc de cuisiner un chili avec **beaucoup** d'oignons. Quand Vincent a coupé tous les légumes en petits morceaux, le garçon s'étonne :

— Les oignons ne te font pas pleurer ?

Vincent éclate de rire :

— Même pas !

Xavier installe le film *Le Roi lion*.
Il l'avance rapidement à la partie la plus triste.

— Papa, viens voir !
Xavier a déjà regardé ce film avec sa mamie. Tous les deux ont pleuré quand le père du petit lion est mort. Vincent serre son garçon contre lui :
— Ne t'en fais pas, fiston, je ne vais pas mourir piétiné par les gnous.

Puisque le film n'a pas fait couler les larmes de son père,
Xavier a une autre idée.

— Papa, couche-toi sur le sofa.

Le garçon chatouille la plante des pieds de son père avec le plumeau. Il lui chatouille aussi la bedaine. Vincent frétille, gigote et rigole beaucoup. Mais Xavier ne réussit pas à le faire pleurer de rire.

Xavier cache la tuque préférée de son père puis annonce:

— Papa... quand je suis allé au musée avec ma classe, j'ai perdu ta tuque rouge.

Vincent fronce les sourcils.

— Je t'ai déjà dit qu'on n'emprunte pas les affaires des autres sans demander la permission.

— Excuse-moi.

Vincent se remet à déneiger sa nouvelle voiture.

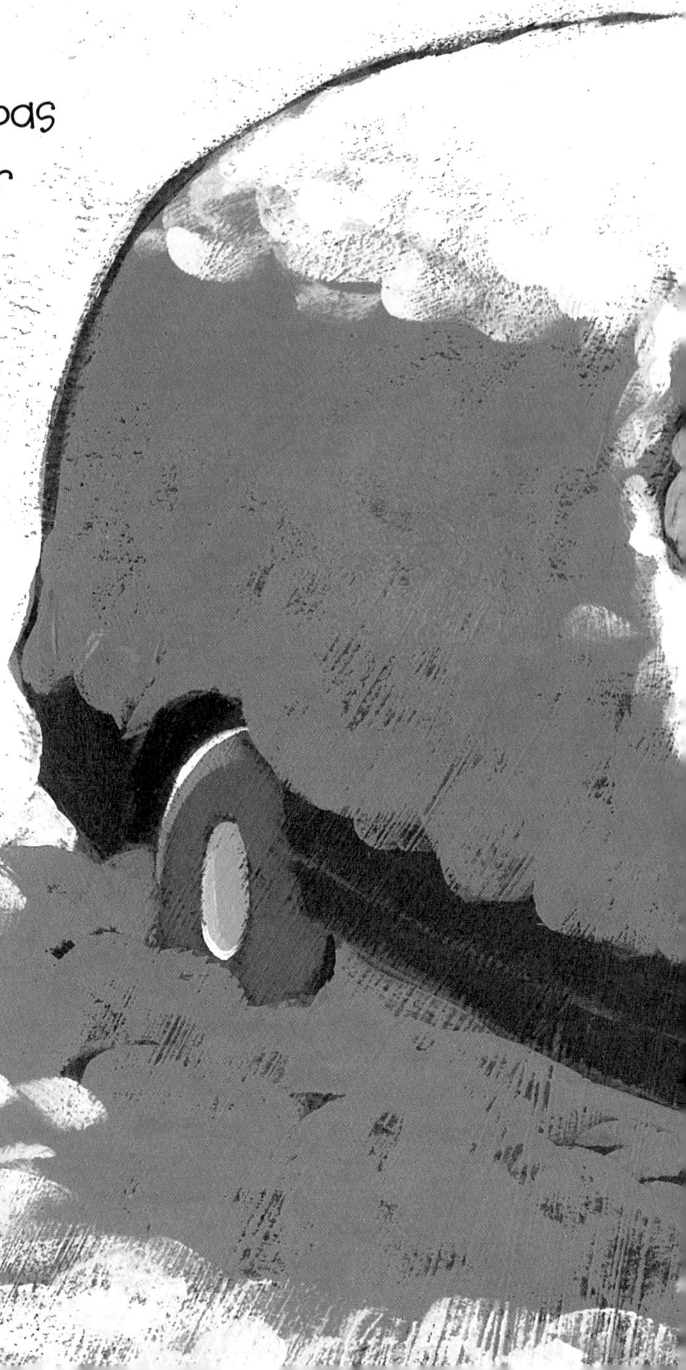

— Mais... mais... c'était ta tuque préférée.
Tu ne pleures pas? demande Xavier.

— Je ne vais pas sangloter
comme un veau
pour un simple chapeau.

Le lendemain, quand son père vient le chercher après l'école, Xavier voit que Vincent est très fâché. Et il remarque qu'une aile de leur nouvelle voiture est défoncée.
— Un conducteur distrait a oublié de faire son arrêt! rage Vincent.

Xavier glisse sa main dans celle de son père.

— As-tu pleuré papa?

Vincent ébouriffe les cheveux de son fils :

— Mais non, c'est juste de la tôle.

Découragé, Xavier raconte sa déception à Marion.

— J'ai tout essayé pour faire pleurer mon papa. Ça n'a rien donné.

— Pourquoi tu veux tellement voir ses larmes ?

— Si mon papa pleurait, moi aussi j'aurais le droit de pleurer, répond le garçon.

Xavier joue son dernier match de la saison.
Il ne reste qu'une minute à la partie.
Le compte est 2 à 2. Xavier réussit une échappée
et se retrouve seul devant le but adverse.
Il lance et compte ! Bzzzz ! fait la sirène.
Youpi ! L'équipe de Xavier a gagné !

Dans les gradins, la foule crie :
“Bravo ! Bravo !”
Xavier patine jusqu’à la sortie
où Vincent le rejoint.
— As-tu vu papa ?
Le premier but de ma vie !
Mais son papa n’arrive pas
à répondre. Il pleure de joie.

MON PAPA NE PLEURE PAS !

MON PAPA NE PUE PAS !

Éditrice : Angèle Delaunois
Édition électronique : Hélène Meunier
Éditrice adjointe : Lucile de Pesloüan

Dépôt légal : 2e trimestre 2018
Bibliothèque nationale du Québec
Bibliothèque nationale du Canada

Catalogage avant publication de Bibliothèque et Archives nationales du Québec et Bibliothèque et Archives Canada

Poulin, Andrée
[Romans. Extraits]
Mon papa ne pue pas! ; Mon papa ne pleure pas! / rédigé par Andrée Poulin ; illustré par Jean Morin.
(Tourne-pierre ; 58)
Publiés à l'origine en 2 volumes séparés : 2009 et 2011.
Public cible : Pour enfants de 5 ans et plus.
Publié en formats imprimé(s) et électronique(s).
ISBN 978-2-924769-28-7 (couverture rigide)
ISBN 978-2-924769-29-4 (PDF)
I. Morin, Jean, 1959-, illustrateur. II. Titre : Poulin, Andrée. Mon papa ne pue pas! III. Titre : Poulin, Andrée. Mon papa ne pleure pas! IV. Titre. V. Titre : Mon papa ne pleure pas! VI. Collection : Tourne-pierre ; 58.
PS8581.O837A15 2018 jC843'.54 C2018-940212-1
PS9581.O837A15 2018 C2018-940213-X

Nous remercions le Conseil des arts du Canada de l'aide accordée à notre programme de publication et la SODEC pour son appui financier en vertu du Programme d'aide aux entreprises du livre et de l'édition spécialisée et du programme de crédit d'impôt pour l'édition de livres.

Imprimé au Canada

Fiche d'activités pédagogiques disponible sur notre site www.editionsdelisatis.com à la page du livre

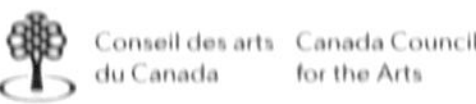

Mme Montjoie annonce l'heure du départ. Manuel s'écrie :

— Pas déjà !

— Est-ce qu'on peut revenir une autre fois ? supplie Martin.

En retournant à l'école, Magali chuchote à l'oreille de Margot :

— Ton papa sent bon.

Tout l'avant-midi, les enfants collent, bricolent et rafistolent. Chacun crée sa sculpture. Margot se promène de l'un à l'autre, aussi fière qu'une princesse dans son royaume. Elle a envie de serrer son papa dans ses bras. Elle ne se retient pas.

Aujourd'hui, vous serez des artistes du recyclage.
Fouillez dans mes trésors et fabriquez votre oeuvre d'art.

Trois jours plus tard, Mme Montjoie et ses élèves se rendent chez Margot. Lorsque les enfants entrent dans la cour, ils poussent des Oh ! et des Ah !

— Quelles superbes sculptures ! s'exclame l'enseignante.

— Bienvenue dans mon Royaume du bric-à-brac, déclare le père de Margot.

Margot ne veut pas peiner son papa, mais elle trouve que son Royaume ressemble à un dépotoir. Elle a peur que Martin et Manuel se moquent encore d'elle.
— Fais-moi confiance. Tes copains vont tellement s'amuser qu'ils ne voudront plus repartir, affirme son père.

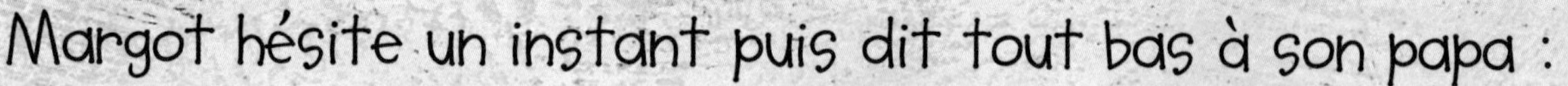

Margot hésite un instant puis dit tout bas à son papa :

— Manuel dit que tu sens le poisson pourri.

— Pourquoi faire du chichi pour une odeur ? Une petite douche et au revoir la puanteur.

Margot donne un coup de pied dans un vieux pneu.

— Si on invitait ta classe dans mon Royaume du bric-à-brac ? suggère son père.

— Euh... je ne sais pas...

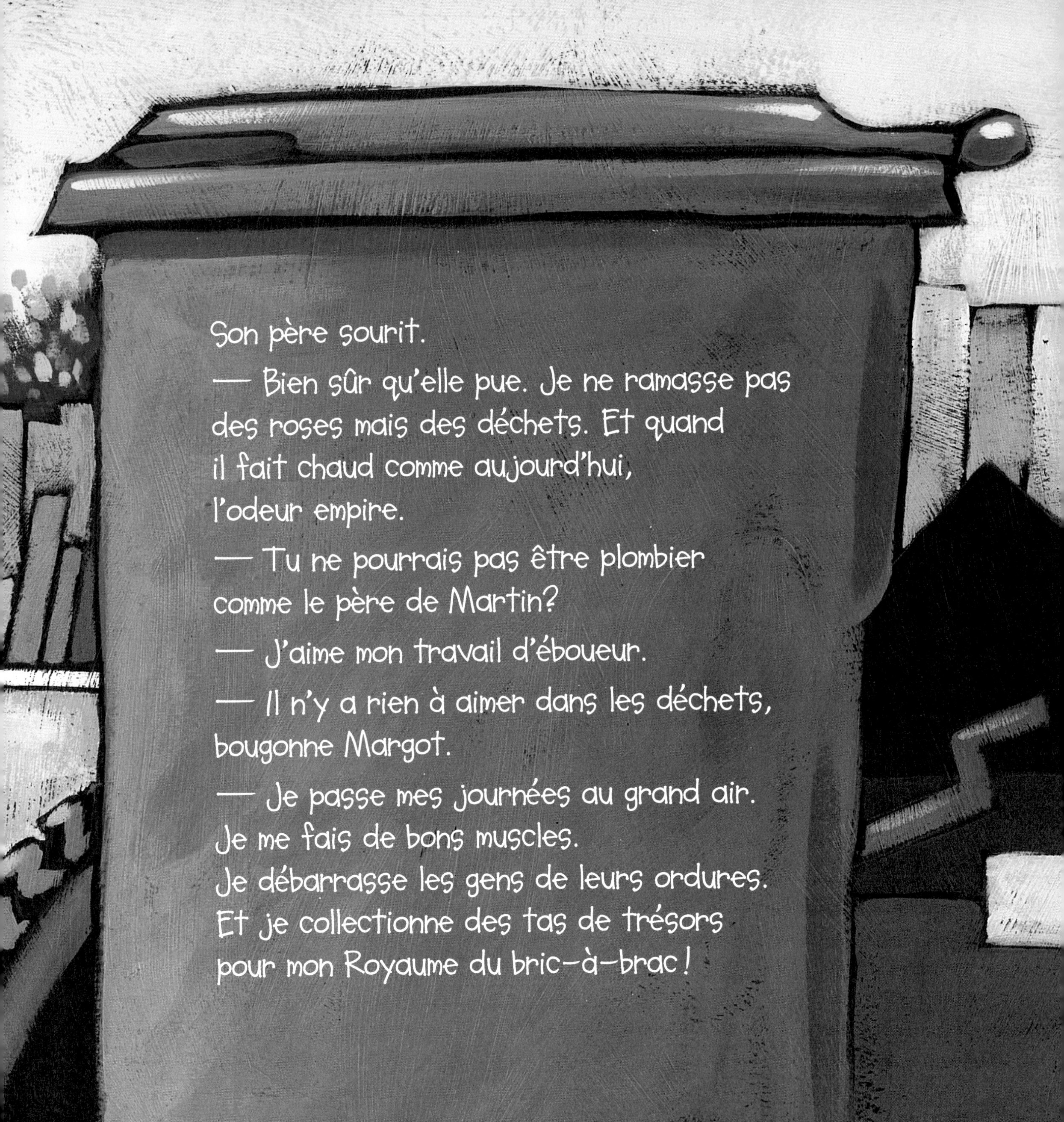

Son père sourit.

— Bien sûr qu'elle pue. Je ne ramasse pas des roses mais des déchets. Et quand il fait chaud comme aujourd'hui, l'odeur empire.

— Tu ne pourrais pas être plombier comme le père de Martin?

— J'aime mon travail d'éboueur.

— Il n'y a rien à aimer dans les déchets, bougonne Margot.

— Je passe mes journées au grand air. Je me fais de bons muscles. Je débarrasse les gens de leurs ordures. Et je collectionne des tas de trésors pour mon Royaume du bric-à-brac !

— Pourquoi tu pleures ma petite fleur ?
demande le papa.

— Ta chemise de travail pue.

Une odeur dégoûtante monte à ses narines.
La fillette tire une chemise du sac et l'approche
de son nez. Ça sent le fromage moisi.
Le poisson pourri. Quelle insupportable puanteur !
Margot éclate en sanglots.

— Peux-tu aller dans mon camion chercher mes gants de travail? demande le papa.

Margot court au camion, ouvre le sac de son père et sort les gants.

— Pourquoi tous ces mamours ?
demande-t-il étonné.
— Parce que tu sens bon !

Sitôt rentrée chez elle, Margot se précipite au Royaume du bric-à-brac. Son père y fabrique une sculpture avec des pneus de vélos.
Margot a envie de se jeter dans les bras de son papa. Elle ne se retient pas.
Elle colle son nez contre le cou de son père.
Il sent le savon à la pomme.
Rassurée, elle lui plaque un bisou bruyant sur la joue.

L'enseignante poursuit :

— Eugène Poubelle a demandé aux gens de mettre leurs ordures dans des boîtes avec des couvercles.
La ville est devenue plus propre et le choléra a disparu.

Margot murmure à l'oreille de Manuel :

— Si mon père ne ramassait pas tes déchets, tu mourrais du choléra et ce serait bon débarras.

Après la récré, Mme Montjoie raconte aux élèves l'histoire du choléra.

— Il y a très longtemps, une épidémie de choléra a frappé la ville de Paris. Cette maladie contagieuse provenait des microbes cachés dans les déchets. Des milliers de gens mouraient. Heureusement, Eugène Poubelle, un des responsables de la ville, a eu une idée fabuleuse.

Martin pouffe de rire :

— Un gars qui s'appelle Poubelle !

Magali s'approche et demande :

— Est-ce que ton papa sent les déchets ?

— Mon papa ne pue pas ! rugit Margot.

À la récréation, les garçons taquinent Margot.

— Ton père sent le poisson pourri ! crie Martin.

— Ton père sent le fromage moisi, rajoute Manuel.

Margot a envie de leur crier des bêtises, mais elle se retient.

— Ça sentirait mauvais très vite ! s'écrie Magali.

— L'éboueur est aussi important que le policier ou le pompier. Sans lui, on serait ensevelis sous les déchets, ajoute l'enseignante.

Margot a envie de se lever et de donner un bisou à Mme Montjoie, mais elle se retient.

Mme Montjoie renverse la poubelle et le bac de recyclage sur le sol.
— Depuis ce matin, on a jeté tout ça ! Imaginez si personne ne venait ramasser nos ordures à la fin de la journée.

Martin éclate de rire.

— Il joue dans la boue ?

— Il ramasse les déchets, explique Margot.

Manuel se pince le nez et s'écrie :

— Beurk ! Les poubelles, ça pue !

Aujourd'hui, dans la classe de Mme Montjoie, chaque élève présente le métier de son papa.

— Mon papa est plombier, annonce Martin.

— Mon papa est infirmier, claironne Manuel.

— Mon papa est dentiste, dit Magali.

Margot se lève et déclare fièrement :

— Mon papa est éboueur.

MON PAPA NE PUE PAS !

À mon parrain Rhéal,
artiste du recyclage.

A.P.

À mes deux filles,
de leur père qui ne pue pas.

J.M.

D'ANDRÉE POULIN • ILLUSTRÉ PAR JEAN MORIN

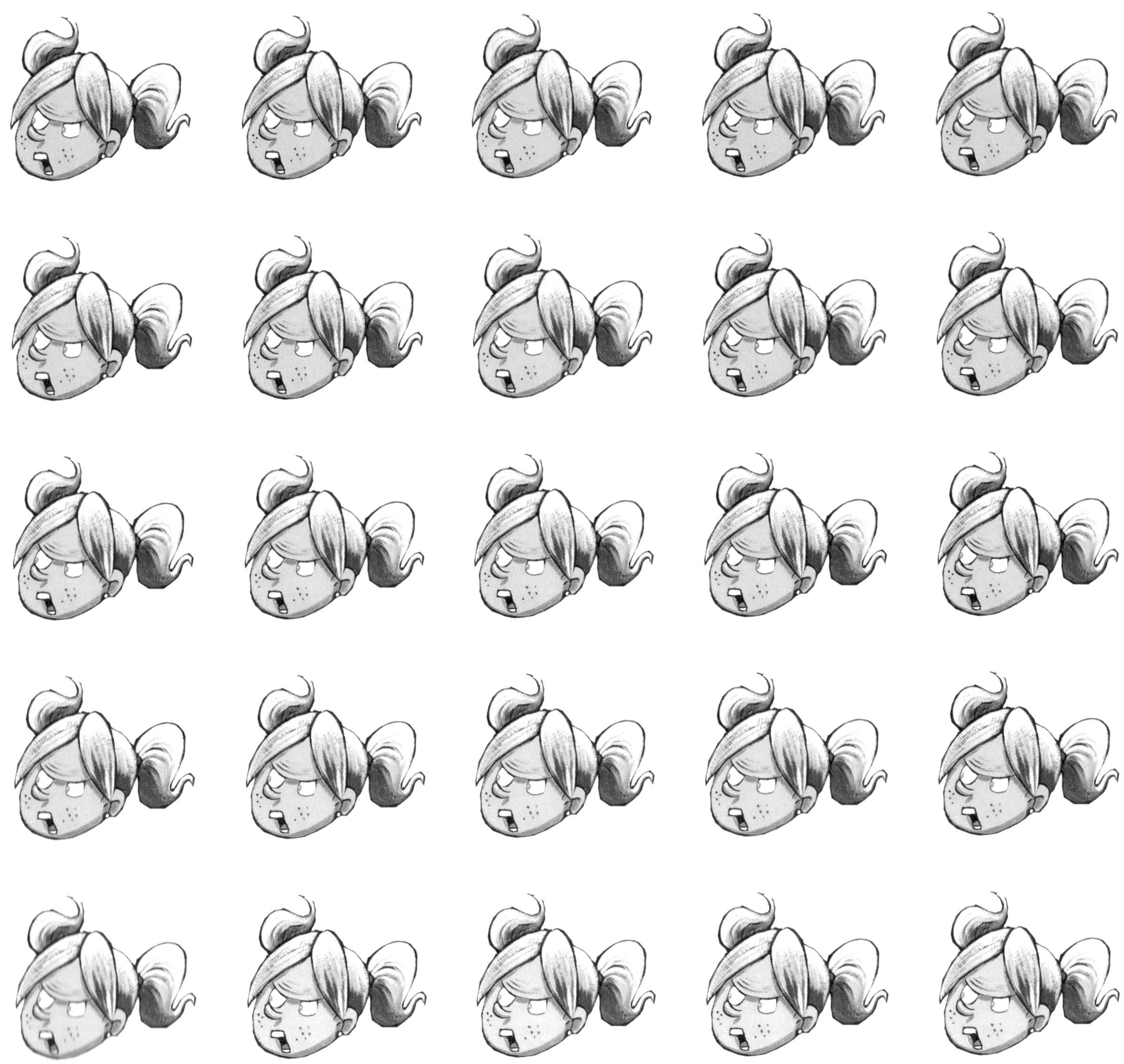